GENERALIZATION WITH
DEEP LEARNING
For Improvement on Sensing Capability

GENERALIZATION WITH
DEEP LEARNING

For Improvement on Sensing Capability

Editors

Zhenghua Chen • Min Wu • Xiaoli Li

Institute for Infocomm Research, Singapore

World Scientific

NEW JERSEY · LONDON · SINGAPORE · BEIJING · SHANGHAI · HONG KONG · TAIPEI · CHENNAI · TOKYO

Published by

World Scientific Publishing Co. Pte. Ltd.

5 Toh Tuck Link, Singapore 596224

USA office: 27 Warren Street, Suite 401-402, Hackensack, NJ 07601

UK office: 57 Shelton Street, Covent Garden, London WC2H 9HE

Library of Congress Cataloging-in-Publication Data

Names: Chen, Zhenghua, editor. | Wu, Min, 1974–   editor. | Li, Xiao-Li, 1969–   editor.
Title: Generalization with deep learning : for improvement on sensing capability /
    editors, Zhenghua Chen, Min Wu, Xiaoli Li, Institute for Infocomm Research, Singapore.
Description: Singapore ; Hackensack, NJ : World Scientific, [2021] |
    Includes bibliographical references and index.
Identifiers: LCCN 2020034548 | ISBN 9789811218835 (hardcover) |
    ISBN 9789811218842 (ebook) | ISBN 9789811218859 (ebook other)
Subjects: LCSH: Electronic surveillance--Data processing. | Remote sensing--Data processing. |
    Diagnostic imaging--Data processing. | Machine learning.
Classification: LCC TK7882.E2 G44 2021 | DDC 006.3/1--dc23
LC record available at https://lccn.loc.gov/2020034548

British Library Cataloguing-in-Publication Data
A catalogue record for this book is available from the British Library.

For any available supplementary material, please visit
https://www.worldscientific.com/worldscibooks/10.1142/11784#t=suppl

Typeset by Stallion Press
Email: enquiries@stallionpress.com

About the Editors

Zhenghua Chen received his B.Eng. degree from University of Electronic Science and Technology of China in 2011, and Ph.D. degree from Nanyang Technological University (NTU), Singapore in 2017. He has been working at NTU as a Research Fellow. Currently, he is a Scientist and Team Leader at Institute for Infocomm Research, Agency for Science, Technology and Research (A*STAR), Singapore as well as part-time Lecturer at NTU. He works as PI and Co-PI for several large projects (> S$10 million). He serves as Guest Editor for four journals, Track Chairs for two conferences, Organizer for two workshops at International Joint Conference on Artificial Intelligence (IJCAI) 2019 and 2020, one invited session at IEEE International Conference on Control, Automation, Robotics and Vision (ICARCV) 2020 and one special session at IEEE International Conference on Industrial Engineering and Applications (ICIEA) 2021, and PC member for several top AI conferences, such as IJCAI and Association for the Advancement of Artificial Intelligence (AAAI). His Google Scholar citation is 2000+ with an H-index of 18. He has been listed as World's Top 2% Scientists by researchers from Stanford University. Besides, he has won several competitive awards such as the A*STAR Career Development Award, First Runner-Up Award for Grand Challenge at IEEE VCIP 2020, Finalist Academic Paper Award of IEEE ICPHM 2020, and Chinese Government Award for Outstanding Self-financed Students Abroad. He is currently the Vice Chair of IEEE Sensors Council Chapter, Singapore. His research interests include sensory data analytics, machine learning, transfer learning and related applications e.g., smart building and smart manufacturing.

Min Wu received his B.Eng. degree in Computer Science from University of Science and Technology of China in 2006 and Ph.D. degree in Computer Science from NTU in 2011. He is currently a Senior Scientist at the Data Analytics Department, Institute for Infocomm Research, A*STAR Singapore. His research interests include sensor data analytics, graph mining, machine learning and bioinformatics.

Xiaoli Li is Head of the Machine Intellection Department and a principal scientist at the Institute for Infocomm Research, A*STAR, Singapore. He also holds an adjunct position at School of Computer Science and Engineering, NTU, Singapore. His research interests include data mining, machine learning, AI and bioinformatics. He has served as (senior) PC member/workshop chair/session chair in leading data mining and AI-related conferences, such as International Joint Conference on Artificial Intelligence (IJCAI) and International Conference on Data Mining (ICDM), to name a few. He has published more than 190 peer-reviewed papers, in top tier journals such as *IEEE Transactions TKDE, IEEE Transactions on Reliability* and *Bioinformatics*. His areas of specialization are positive unlabeled based learning and social/biological network mining. His paper "Drug-target interaction prediction via class imbalance-aware ensemble learning" won the Best Paper Award at the International Conference on Bioinformatics 2016. He has also won Best Performance Awards at the Opportunity Activity Recognition Challenge conducted by the EU Consortium and 2nd Dialogue for Reserve Engineering Assessments and Methods (DREAM 2007), respectively. With rich translational experience in working with industry, Dr. Li has led over 10 R&D projects in collaboration with industry partners across sectors, including leading aerospace companies, banks, telecom companies and insurance companies.

Preface

For years ever since the early days of AI, researchers have been struggling with the issue of machine generalization — a capability exclusive to humans and an ability to learn inductively from entity to concepts. However, AlphaGo outsmarts the human Go master and shows that it is possible for machines to learn from examples and generalize concepts. Deep learning as a game changer has set an exciting trend in the machine learning domain in recent years. It has achieved great success in many challenging research areas, such as image recognition, audio recognition and natural language processing. The key merit of deep learning is to automatically learn a very good feature representation from massive amounts of data. Nowadays, various sensors are deployed in different scenarios and thus we can easily collect large amounts of data from these sensors. Clearly, this deep learning technology can be a very good candidate for various sensing applications, including activity sensing, remote sensing and medical sensing. Various deep learning algorithms can thus be developed in many real-world scenarios to facilitate impact on practical applications.

The objective of this book is to disseminate inspiring research results and exemplary best practices of deep learning approaches to tackle the technical challenges in various sensing applications. The book will cover the fundamentals of deep learning techniques and their applications in real-world problems including activity sensing, remote sensing and medical sensing. It will demonstrate how different deep learning techniques help to improve the sensing capabilities and enable scientists and practitioners to make insightful observations and invaluable discoveries from data.

Organization of the Book

This book consists of 12 chapters, which are organized into four major parts: Introduction of Deep Learning Algorithms (Part I), Deep Learning for Activity Sensing (Part II), Deep Learning for Remote Sensing (Part III) and Deep Learning for Medical Sensing (Part IV).

Part I is composed of Chapter 1. In Chapter 1, we first briefly introduce various deep learning algorithms for improving sensing capability. We also discuss their advantages and disadvantages for different applications.

Part II is composed of Chapters 2–6 and introduces deep learning for activity sensing. In particular, Chapter 2 presents a new hierarchically aggregated convolutional neural network (CNN) architecture that can more effectively deal with long-term dynamics presented in action videos. In particular, the proposed network takes a long term video as input and exploits temporal information at multiple levels of the network hierarchy capturing more complex dynamics of the input sequence. Chapter 3 proposes to combine domain knowledge and deep learning in order to improve activity recognition models, where the domain knowledge focuses on the topological structure of on-body sensor-deployments and the mutual interactions among the sensors. Chapter 4 introduces how deep learning and unsupervised domain adaptation method further empower WiFi-based sensing. Chapter 5 focuses on how to weaken the accuracy differences among individuals in human activity recognition and improves the robustness in a single indoor environment. Chapter 6 introduces a spatial temporal graph convolutional network (GCN)-based prediction network for skeleton-based video anomaly detection.

Part III is composed of Chapters 7–9 and presents deep learning for remote sensing applications. Particularly, Chapter 7 reviews the current approaches and discusses ways forward to develop new deep learning methods for Earth sciences. Chapter 8 presents an empirical evaluation for built-up area detection from remote sensing images using deep learning methods including double-stream CNN (DSCNN), lightweight multi-branch CNN (LMB-CNN) and fully convolutional networks (FCN). Chapter 9 presents GCN variants by combining GCN with manifold assumptions to exactly acquire the spatial structure information of graph-structured data for remote sensing image recognition. In particular, the GCN variants include two-order GCN, Hypergraph p-Laplacian GCN, manifold regularized dynamic GCN.

Part IV is composed of Chapters 10–12 and focuses on deep learning for medical sensing applications. In particular, Chapter 10 proposes a deep learning-based retinal image non-uniform illumination removal called NuI-Go, which combines the powerful capabilities of CNN with the characteristics of retinal images with non-uniform illumination. Chapter 11 presents and evaluates deep learning models (e.g., SqueezeNet and GoogLeNet) to classify brain tumor MRI images. In Chapter 12, the deep learning algorithms with long short-term memory (LSTM) and bi-directional LSTM are unitized to recognize dementia-related wandering patterns based on the orientation data available in mobile devices.

In this book, we have introduced deep learning algorithms and their applications to improve the sensing capabilities, including activity sensing, remote sensing and medical sensing. Taking human activity recognition as an example, it is a crucial technology for many real-world human-centric applications, especially in the areas of eldercare, healthcare, smart home, security, pervasive and mobile computing, etc. However, it is a challenging task to achieve very accurate prediction results, due to device heterogeneity, environment changes and difficulties to recognize concurrent activities and group activities. Similarly, for remote sensing and medical sensing applications, they usually require domain knowledge and could also be affected by environment changes. We have thus covered representative latest research outcomes and practices in this book that could be very useful for addressing the challenging research problems in the above sensing applications.

We would like to take this opportunity to thank all the authors who contributed to the exciting and important research topics of developing deep learning approaches for various sensing applications. Our heartfelt thanks also go to our book reviewers who have provided very useful comments and valuable feedback, and to the publishing team at World Scientific for providing invaluable contributions and guidance throughout the whole process from inception of the initial idea to the final publication of this book.

Contents

Part I

Introduction of Deep Learning Algorithms

Chapter 1

An Introduction of Deep Learning Methods for Sensing Applications

Keyu Wu*, Wei Cui[†], Vuong Nhu Khue[‡] and Efe Camci[§]

Institute for Infocomm Research, 1 Fusionopolis Way, Singapore 138632
*wu_keyu@i2r.a-star.edu.sg
†cui_wei@i2r.a-star.edu.sg
‡vuong_nhu_khue@i2r.a-star.edu.sg
§efe_camci@i2r.a-star.edu.sg

Abstract

This chapter will briefly introduce various deep learning algorithms for improving sensing capabilities. For different sensing applications, specific deep learning architectures may be employed with proper adjustments based on the properties of sensing applications. In order to know which deep learning algorithms may be suitable for a unique sensing application, a clear understanding of different deep learning algorithms is compulsory. Currently, there are numerous deep learning algorithms. Here, we will only cover some widely used ones for sensing applications, including auto-encoder, restricted Boltzmann machine, convolutional neural network, recurrent neural network and their variants. At the end of this chapter, we will provide some discussion and a comparison of different deep learning methods.

Keywords: deep learning, auto-encoder, restricted Boltzmann machine, convolutional neural network, recurrent neural network

The four authors equally contributed to this chapter.

1. Introduction

With the development of advanced sensing technology, more and more sensory data are easily collected and have become available for various sensing applications, such as activity recognition [1], remote sensing [2], medical image processing [3], disease diagnostic and identification [4], etc. Generally, for a specific sensing application, certain suitable sensors can be selected or adopted. For instance, the sensing task of human activity recognition can be achieved by using vision sensors (i.e., cameras) [5], inertial sensors [6] and/or WiFi signals [7, 8], although different sensors may have their unique properties and limitations. For example, vision-based solutions have high accuracy for activity recognition. However, they can only cover a specific area with the requirement of sufficient illumination condition and may suffer from privacy concerns. On the other hand, wearable/smartphone inertial sensors are convenient for activity sensing [9, 10], but may be only able to identify some simple activities, such as walking, running, sitting, going upstairs/downstairs, etc. Recently, WiFi-based solutions have attracted great attention, because they are widely available and do not require users to wear any devices on their bodies [11]. Nevertheless, they often have limited performance, especially for multi-user scenarios.

To achieve real-world sensing applications, we generally can divide the existing methods into two different categories, i.e., model-based and data-driven approaches. The model-based methods require to know the physical relationship/model between sensor data and the specific task. In real life, the physical models are typically not available, especially for complicated sensing applications. Instead, data-driven methods become more and more popular, as they do not require to explicitly know the physical model of the problem. In fact, they are able to automatically learn the relationship from the data.

The most widely used data-driven models in sensing applications are machine learning methods [12]. Particularly, conventional machine learning algorithms require two steps for solving sensing application problems from given data. The first step is feature engineering which attempts to manually extract useful representations (also known as features) from the raw data. For instance, some statistical features, such as mean, variance and frequency components, can be extracted in smartphone sensor-based activity recognition [13], which are helpful in separating different physical activities of walking, running and staying static. After extracting useful features, traditional machine learning algorithms, e.g., artificial neural network (ANN), support vector machine (SVM), k-nearest neighbor (KNN), decision tree

(DT), random forest (RF), etc., can be leveraged to perform sensing or classification tasks. However, the conventional machine learning-based sensing needs to conduct manual feature engineering, which requires strong application-specific domain knowledge and will inevitably miss implicit features [14]. This limits the performance of these methods in sensing applications.

Recently, deep learning which is a new branch of machine learning has achieved great success for a variety of sensing applications [15]. Different from conventional machine learning methods which contain tedious and costly manual feature engineering, deep learning methods can automatically learn features from data and perform inference simultaneously (end-to-end model), such that the feature learning and inference can be jointly optimized, resulting in a superior performance in many application domains.

This book intends to explore different deep learning methods that enhance sensing capacity. Before that, in this very first chapter, we shall provide a brief description of various standard deep learning algorithms for sensing applications in the following sections, so that readers can have some basic understanding before zooming into more advanced deep learning approaches.

2. Deep Learning Methods

Deep learning is a new branch of machine learning. It intends to use multiple layers of neural networks to learn high-level representations of raw data and performs inference with these representations (also known as features) [15]. Many deep learning architectures have been proposed for different applications. In this chapter, as we focus on the applications of improving sensing capacities, only relevant deep learning architectures that target sensing capabilities will be introduced in detail. Specifically, we will introduce auto-encoder in Section 2.1, restricted Boltzmann machine in Section 2.2, convolutional neural network in Section 2.3 and recurrent neural network in Section 2.3.6.

2.1 *Auto-encoder*

Auto-encoder is a type of backpropagation neural network where the output and input are set to be the same. Auto-encoders aim to learn an approximation of the identity function and were first introduced by Hinton [16] to address the problem of unsupervised learning. To learn interesting properties of the data, auto-encoders are usually conditioned by placing

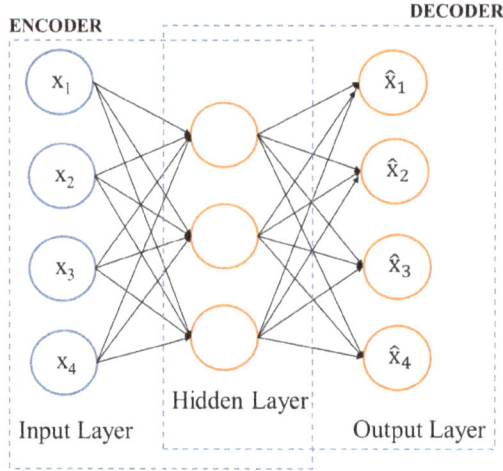

Figure 1: An example of an auto-encoder neural network.

constraints on the network such as limiting the number of hidden neurons. The reason is that learning the identity function could be trivial or meaningless without restrictions of network structure. Therefore, by imposing conditions on the network, auto-encoders are forced to prioritize which relevant aspects of the data are to be copied or learned; hence, hidden structure in the input data such as correlated input features could be discovered [17].

Figure 1 illustrates a simple feedforward auto-encoder with three layers (input, hidden and output), in which the output layer has the same number of neurons as the input layer. The network can be considered as having two parts: an encoder and a decoder. Particularly, the encoder compresses the input and retains the code to represent the input in the hidden layer.

Traditionally, auto-encoders are commonly used for dimensionality reduction or feature learning. Recently, auto-encoders and its variants which incorporate latent variable models have become more widely used for generative modeling [18]. We will introduce the two variants of auto-encoders in this chapter: sparse auto-encoder and denoising auto-encoder.

2.1.1 Sparse auto-encoder

A sparse auto-encoder is an auto-encoder in which there is an additional sparse penalty to its cost function. By regularizing the auto-encoder to be sparse, the auto-encoder can learn unique, nonreplicative statistical features of the training dataset [19]. The overall cost function of a sparse auto-encoder feedforward neural network can be written as follows:

$$J_{sparse}(W,b) = J(W,b) + \beta \sum KL(\rho \,||\, \hat{\rho}j)$$

$$= J(W,b) + \beta \left(\rho \log \frac{\rho}{\hat{\rho}} + (1-\rho) \log \frac{1-\rho}{1-\hat{\rho}_j} \right), \qquad (1)$$

where $J(W, b)$ is the cost function of the original auto-encoder in which W, b are the weights of input and the bias, respectively. β is the hyper-parameter to control the weight of the sparsity penalty term, whereas ρ is a sparsity parameter typically close to zero and $\hat{\rho}_j$ is the average activation of the hidden unit j over the training set.

$KL\,(\rho||\rho_j)$ is the Kullback–Leibler divergence which measures the difference between two Bernoulli random variables [17]. An important property of the KL function is that it attains its minimum of 0 at $\hat{\rho}_j = \rho$ and increases up to infinity when $\hat{\rho}_j$ is close to 0 or 1. Therefore, minimizing the penalty term will cause $\hat{\rho}_j$ to be close to the sparsity parameter p which in effect helps to approximately enforce the constraint $\hat{\rho}_j = \rho$.

Figure 2 shows an example of a single-layer sparse auto-encoder with bias. Straight lines represent the weights of the input data and the hidden

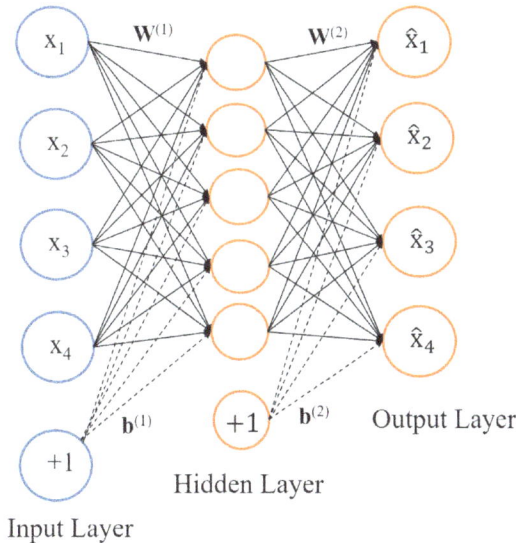

Figure 2: Spare auto-encoder with bias.

neurons ($W^{(1)}$, $W^{(2)}$), whereas dotted lines represent the weights of the biases ($b^{(1)}$, $b^{(2)}$).

2.1.2 *Denoising auto-encoder*

A denoising auto-encoder is an auto-encoder that learns to reconstruct the original, clean version of the input given a corrupted one. The corrupting process can be done by adding certain form of noise. In this chapter, we introduce a denoising auto-encoder whose input is corrupted by a stochastic mapping process [20].

Figure 3 shows a schematic representation of a denoising auto-encoder. The coding and reconstructing process of the denoising auto-encoder in Fig. 3 can be briefly summarized in the following steps:

- Step 1: Obtain a corrupted version x̃ of the initial input x.
- Step 2: Map the corrupted input x̃ to the basic auto-encoder (encoding using f_θ and decoding using $g_{\theta'}$, as shown in Fig. 3).
- Step 3: Minimize the average reconstruction error $L_H(x, z)$ to have z as close as possible to the uncorrupted input x given that z is now a deterministic function of x̃ instead of x. z is then the result of a stochastic mapping of x.

By performing stochastic gradient descent, a random corrupted version of the input is produced and the uncorrupted input is reconstructed through the gradient step. This mechanism allows denoising auto-encoders to learn a good internal representation of the input while preventing the encoder and decoder from learning a trivial identity function [18].

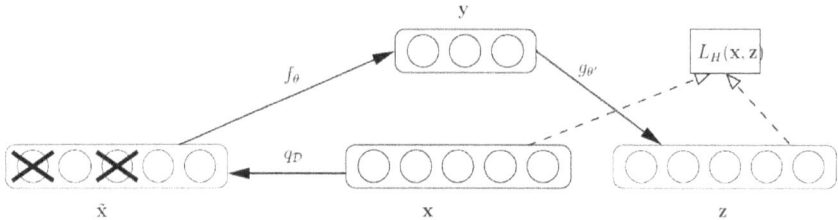

Figure 3:   Example of denoising auto-encoder. Figure is retrieved from Ref. [40].

2.2 Restricted Boltzmann machine

2.2.1 What are restricted Boltzmann machines?

A Boltzmann machine is a type of neural network which has neuron-like units that yield binary behavior (to be on or off) in a stochastic manner. It was first proposed in [21] as a parallel constraint satisfaction network that is capable of learning internal representations of a domain by only being shown examples from the domain. A Boltzmann machine essentially has two types of units, such as *visible* and *hidden* units as depicted in Fig. 4. Visible units are the ones through which the input is fed and the state is observed, whereas hidden units are the feature detectors. As we can observe clearly from Fig. 4, unlike conventional artificial neural networks, visible and hidden units are connected to each other within their own groups. Despite its success in certain case studies, this architecture will lead to an admittedly slow learning procedure and thus is less efficient [21].

A restricted Boltzmann machine (RBM) [22] is a modified version of a Boltzmann machine in which visible and hidden units are *restricted* not to form any connection within their own groups. In other words, an RBM is a two-layer artificial neural network in the form of a bipartite graph. An RBM architecture example is depicted in Fig. 4. RBMs are widely used for a variety of tasks such as filtering, dimension reduction, feature extraction and classification. In the context of deep learning, they are well known for their efficient layer-by-layer learning ability, when being stacked one after another in a deep neural network to learn hierarchical feature representations.

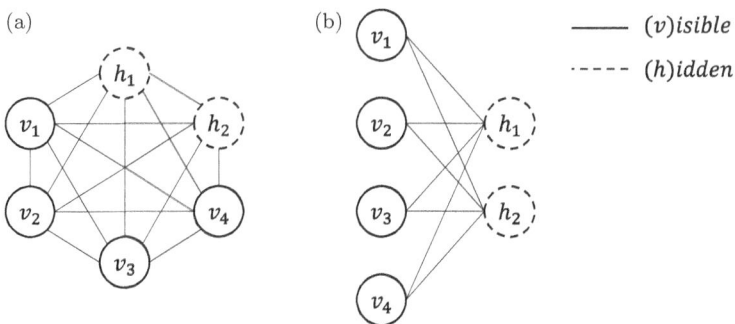

Figure 4: (a) A Boltzmann machine. (b) A restricted Boltzmann machine.

2.2.2 How do restricted Boltzmann machines work?

The main purpose of RBMs is to create a decent generative model of the input vectors. They achieve this generative modeling capability through the repetition of two main phases: forward pass and backward (reconstruction) pass.

2.2.2.1 Forward pass

The first phase refers to forward pass of the inputs that have the probability distribution of $P(v_i^{(0)} = 1)$. This phase consists of multiplying the inputs with corresponding weights, adding the respective biases, and passing them through the activation function. As a result, representative values to be passed backward in the next phase are obtained. Mathematically, this phase corresponds to:

$$P(h_j = 1) = \sigma\left(\left(\sum_{i \in visible} v_i^{(0)} w_{ij}\right) + b_j\right),$$

where $P(h_j = 1)$ are the probabilities of h_j (hidden units) to be activated, $v_i^{(0)}$ are the inputs, w_{ij} are the weights, b_j are the biases in the hidden layer, and σ is the logistic sigmoid activation function.

2.2.2.2 Backward (reconstruction) pass

The second phase refers to backward pass of the representative values that are generated in the first phase. As a result of this phase, it is attempted to reconstruct the probability distribution of the original inputs. Mathematically, this phase corresponds to:

$$P(v_i^{(1)} = 1) = \sigma\left(\left(\sum_{j \in hidden} h_j w_{ij}\right) + a_i\right),$$

where $P(v_i^{(1)} = 1)$ is the reconstructed output and a_i are the biases in the visible layer. The learning process essentially aims at bringing the probability distributions of $P(v_i^{(0)} = 1)$ and $P(v_i^{(1)} = 1)$ close to each other by learning the desirable network parameters.

2.2.2.3 Learning procedure

RBMs work based upon a notion called "energy function" as in Hopfield networks [23]. Probability distributions of visible and hidden unit vectors are determined through the energy function. Given a joint configuration of visible and hidden unit vectors (**v, h**), the energy function is defined as

$$E(\mathbf{v},\mathbf{h}) = -\sum_{i\,\in\text{visible}} a_i v_i - \sum_{j\,\in\text{hidden}} b_j h_j - \sum_{i\,\in\text{visible}}\sum_{j\,\in\text{hidden}} v_i h_j w_{ij}.$$

Accordingly, the probability assigned to each possible visible–hidden unit vector pair is

$$P(\mathbf{v},\mathbf{h}) = \frac{1}{Z} e^{-E(\mathbf{v},\mathbf{h})},$$

where Z is the partition function that acts as a normalizing constant. It is equal to the sum of $e^{(-E(\mathbf{v},\mathbf{h}))}$ for all possible (**v, h**) pairs. In the same vein, the probability assigned to a visible unit vector $P(v)$ is the sum of $e^{-E(\mathbf{v},\mathbf{h})}$ for all possible **h** vectors normalized by Z. The term $P(v)$ can be altered by tuning the network parameters. As given in Ref. [24], the derivative of log $P(v)$ with respect to the weights is

$$\frac{\partial \log P(\mathbf{v})}{\partial w_{ij}} = \left\langle v_i h_j \right\rangle_{\text{data}} - \left\langle v_i h_j \right\rangle_{\text{model}},$$

where $\left\langle v_i h_j \right\rangle$ are the corresponding expectations under the probability distribution of *data* and *model*. As explained in Ref. [24], due to the difficulty of getting an unbiased sample for $\left\langle v_i h_j \right\rangle_{\text{model}}$ which requires alternating Gibbs sampling for a long duration, it is replaced with $\left\langle v_i h_j \right\rangle_{\text{reconstruction}}$ during the weight update for a faster learning procedure. Consequently, the weight update formula becomes

$$\Delta w_{ij} = \epsilon(\left\langle v_i h_j \right\rangle_{\text{data}} - \left\langle v_i h_j \right\rangle_{\text{reconstruction}}),$$

where ϵ is the learning rate. Similarly, biases are updated using the expectations for the corresponding individual units instead of using their products. More details for the learning routine (Contrastive Divergence) can be found in Ref. [24].

2.3 *Convolutional neural network*

Convolutional neural network (CNN) is a kind of feedforward neural network that has become one of the most significant neural networks in the field of deep learning. Compared to ordinary artificial neural networks, CNNs have several advantages. For instance, in convolutional layers, each neuron is no longer connected to all neurons of the previous layer, but only to a small number of neurons. As a result, convolutional layers have sparser connections than fully connected layers, which leads to reduced parameters and faster convergence. Moreover, in a convolutional layer, the weights of each filter are moved along the input activation and thereby used at every position of the input. Through parameter sharing, less memory space is required and the number of weights is also further reduced. Meanwhile, if a filter is trained to detect certain patterns, it is capable of detecting all appearance of such patterns since it is applied to every subregion of the input. In addition, the pooling layers of CNNs are effective in absorbing shape variations and can reduce the amount of data while retaining useful information. Typically, the architecture of a CNN comprises alternate convolutional and pooling layers, followed by fully connected layers at the end. Details of these layers of CNNs are discussed in the following.

- **Convolutional layer:** Convolutional layer is the main building block of a CNN. In this layer, feature maps from previous layers are convolved with kernels and the output goes through an activation function to produce the output feature maps. The convolution operation is performed by sliding the convolution kernel over the input feature map. At every location, element-wise matrix multiplication is implemented and the obtained values are added up to the convolution result. In general, we have

$$x_j^l = f\left(\sum_i x_i^{l-1} * k_{ij}^l + b_j^l\right), \tag{2}$$

where $*$ is the convolution operator, f represents the activation function, x_i^{l-1} is the input feature map (layer $l-1$), x_j^l is the output feature map (layer l), k_{ij} and b_j^l represent the kernel and bias for the current layer, respectively. The input maps will be convolved with different kernels to generate different output maps, and the number of output feature maps depends on the number of kernels used in the convolutional mapping.
- **Pooling layer:** Pooling layer essentially performs down-sampling operation on the input feature maps. A pooling layer is usually performed after

a convolutional layer to reduce the heights and widths of the feature maps. However, in this layer, the number of feature maps is not changed. For example, if there are N input maps and a 2×2 down-sampling kernel is applied, there will be N half-size output maps. The two most common operations performed in this layer include max pooling and average pooling. In the case of max pooling, the largest value from the feature window is selected, whereas in the case of average pooling, the average value of the window is used.

- **Fully connected layer:** Fully connected layers are added after a few alternate convolutional and pooling layers to compute the score of each class from the extracted features in the preceding steps. Since fully connected layers expect 1D vectors, the outputs of the final feature maps are firstly flattened to a vector and subsequently fed into the fully connected layers. The last layer of a CNN is typically a softmax classification layer, which generates a probability distribution over predicted output classes. Finally, the class with maximal probability is predicted as the final class.

Although the basic building components are almost the same across CNN architectures, topological differences are observed in different models. Some well-known CNN models are listed in Fig. 5. In the following, several representative CNN architectures will be emphatically introduced.

2.3.1 *LeNet*

As the foundation of modern CNN, LeNet was proposed by LeCun in 1998 [25]. It was trained with the backpropagation algorithm and achieved state-of-the-art accuracy on handwritten digit recognition tasks. As shown in Fig. 6, LeNet consists of seven trainable layers containing two convolutional layers, two pooling layers, two fully connected layers and one output layer with Gaussian connection. As the pioneering CNN, LeNet combines local receptive fields, shared weights and subsampling so that it ensures scale, shift and distortion invariance to some extent. By introducing convolution with

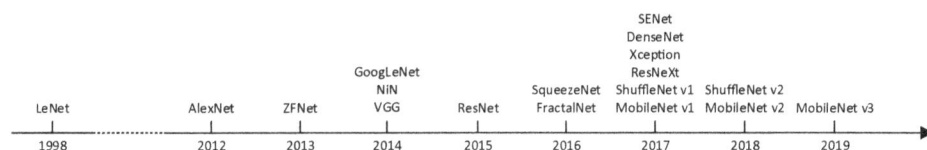

Figure 5: Representative CNN architectures.

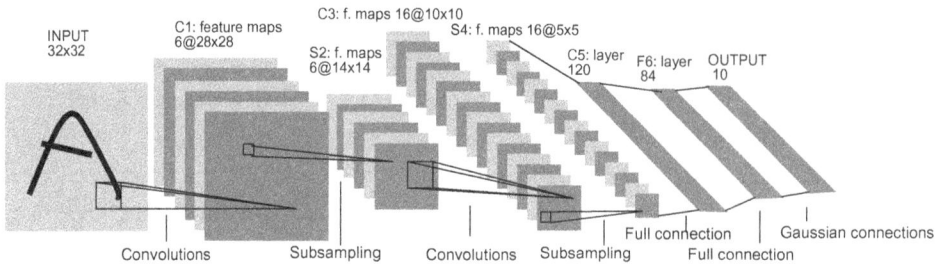

Figure 6:   Architecture of LeNet. Figure is retrieved from Ref. [25].

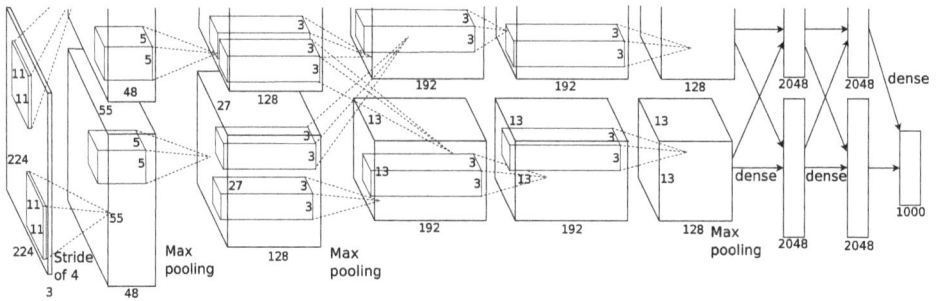

Figure 7:   Architecture of AlexNet. Figure is retrieved from Ref. [26].

learnable parameters, LeNet not only reduces the number of parameters, but also extracts features from raw pixels automatically.

2.3.2 AlexNet

Proposed in 2012, AlexNet carried forward the ideas of LeNet and was the first large-scale CNN architecture [26]. It achieved ground breaking results for image classification and recognition tasks and won the ImageNet Large Scale Visual Recognition Challenge (ILSVRC) in 2012. As shown in Fig. 7, AlexNet has eight layers, comprising five convolutional layers and three fully connected layers. In addition, the first, second and fifth convolutional layers are followed by max pooling layers. AlexNet adopts GPUs for computing acceleration. Moreover, it successfully leverages ReLU activation function, dropout, as well as local response normalization (LRN) in CNN for the first time. More specifically, ReLU activation function is used to improve the convergence rate by alleviating the gradient vanishing problem in deep networks. Dropout is proposed to avoid overfitting through ignoring some neurons randomly during training. Finally, LRN is introduced to enhance the generalization capability of the model via simulating the lateral inhibition mechanism. Its function is very similar to normalization. Therefore, LRN has become deprecated since the introduction of Batch Normalization [50]. The success of AlexNet starts the era of CNNs and triggers the deep learning boom.

2.3.3 *Visual Geometry Group*

The VGG network architecture was introduced by the Visual Geometry Group (VGG) in 2014 [27]. In their paper, a series of VGG models were presented, including VGG-11, VGG-13, VGG-16 and VGG-19. As illustrated in Fig. 8, VGG-11, VGG-13, VGG-16 and VGG-19 contain 11, 13, 16 and 19 layers, respectively, and all VGG models end the same with three fully connected layers. VGG proves that increasing the depth of a network is critical for performance improvement in CNNs. In addition, VGG removes the LRN layer since its effect in deep CNNs is not obvious. Also, it uses 3×3 convolution kernels instead of 5×5 ones because a stack of smaller kernels can achieve the same receptive field, while leading to more nonlinear variations and reduced number of parameters.

ConvNet Configuration					
A	A-LRN	B	C	D	E
11 weight layers	11 weight layers	13 weight layers	16 weight layers	16 weight layers	19 weight layers
input (224×224 RGB image)					
conv3-64	conv3-64	conv3-64	conv3-64	conv3-64	conv3-64
	LRN	**conv3-64**	conv3-64	conv3-64	conv3-64
maxpool					
conv3-128	conv3-128	conv3-128	conv3-128	conv3-128	conv3-128
		conv3-128	conv3-128	conv3-128	conv3-128
maxpool					
conv3-256	conv3-256	conv3-256	conv3-256	conv3-256	conv3-256
conv3-256	conv3-256	conv3-256	conv3-256	conv3-256	conv3-256
			conv1-256	**conv3-256**	conv3-256
					conv3-256
maxpool					
conv3-512	conv3-512	conv3-512	conv3-512	conv3-512	conv3-512
conv3-512	conv3-512	conv3-512	conv3-512	conv3-512	conv3-512
			conv1-512	**conv3-512**	conv3-512
					conv3-512
maxpool					
conv3-512	conv3-512	conv3-512	conv3-512	conv3-512	conv3-512
conv3-512	conv3-512	conv3-512	conv3-512	conv3-512	conv3-512
			conv1-512	**conv3-512**	conv3-512
					conv3-512
maxpool					
FC-4096					
FC-4096					
FC-1000					
soft-max					

Figure 8: Architecture of VGG models. Figure is retrieved from Ref. [27].

2.3.4 *GoogLeNet*

Also introduced in 2014, GoogLeNet was the first large-scale CNN created by incorporating inception modules [28]. The main objective of the GoogLeNet architecture was to reduce computation complexity and, in the meantime, achieve high accuracy. The basic structure of the inception block is depicted in Fig. 9. It constructs a "wide" network through deploying convolution kernels of different sizes to capture spatial information at different scales. The obtained feature maps are then stacked to yield a more representative one. It is worth mentioning that 1×1 convolution kernel is used in GoogLeNet for dimensionality reduction before computationally expensive layers. Moreover, it uses an average pooling layer instead of a fully connected layer before the classifier to reduce the density of connection. These tunings result in a significant decrease in the number of parameters. Although GoogLeNet is 22 layers deep even without counting the pooling layers, it has much less parameters compared to the predecessors.

2.3.5 *ResNet*

ResNet, which was proposed in 2015, revolutionized the development of CNN architectures by introducing the concept of residual learning [29]. In theory, deep networks outperform shallow ones since they can better extract features. Nevertheless, with the increase of layers, deep networks are prone

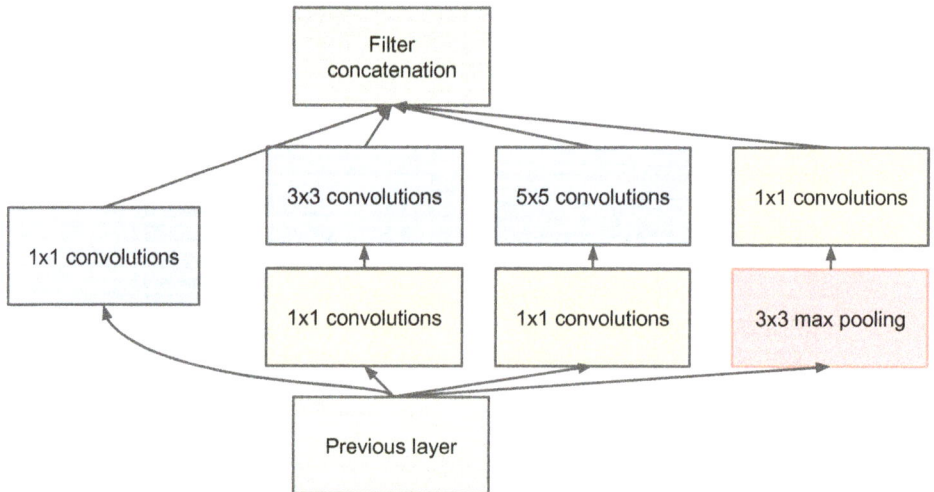

Figure 9: Inception module. Figure is retrieved from Ref. [28].

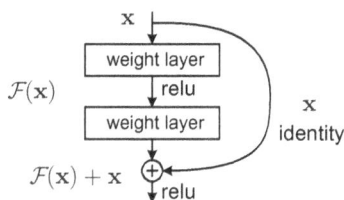

Figure 10: A residual block. Figure is retrieved from Ref. [29].

to gradient vanishing or exploding problems. In order to address these problems, ResNet exploits the residual blocks constructed by shortcut connections, as shown in Fig. 10. The mathematical definition of a residual building block can be expressed as

$$y = \quad (x, W_i) + x, \tag{3}$$

where x and y are the input and output vectors of the considered layers, respectively, and function   represents the residual mapping to be learned. The parameter-free shortcut connections enable cross-layer connectivity and speed up the convergence of deep networks. ResNet was developed with many versions, including ResNet-34, ResNet-50, ResNet-101 and ResNet-152. Although ResNet is much deeper than AlexNet and VGG, it shows less computational complexity than previous networks.

2.3.6 *DenseNet*

Similar to ResNet, DenseNet was proposed in 2017 to solve the gradient vanishing problem through cross-layer connectivity as well [30]. It consists of densely connected CNN layers and connects the outputs of each layer with all successor layers in a dense block. The conceptual diagram of a dense block is illustrated in Fig. 11. The l-th layer receives the features maps of all preceding layers as its input:

$$x_l = H_l([x_0, x_1, \ldots, x_{l-1}]), \tag{4}$$

where $[x_0, x_1, \ldots, x_{l-1}]$ represents the concatenation of previous features maps and H_l is a composite function which performs three consecutive operations, i.e., batch normalization, ReLU activation and a 3×3 convolution. The dense blocks allow DenseNet to reuse features efficiently throughout the networks and substantially reduce the number of network parameters.

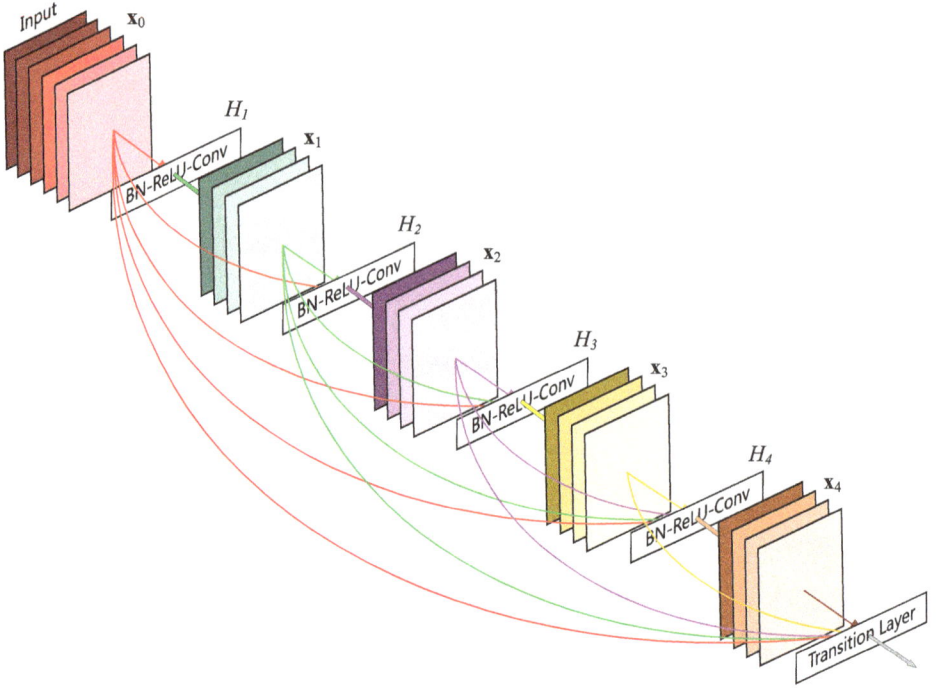

Figure 11: A dense block. Figure is retrieved from Ref. [30].

2.4 *Recurrent neural networks*

2.4.1 *Introduction*

Traditional neural network approaches such as multi-layer perceptron networks (MLP) and CNNs are incapable of dealing with the *sequential* data that have time-varying input patterns and unfixed sequential length. In 1982, John Joseph Hopfield introduced [23] a new neural network, namely recurrent neural network (RNN), to handle data with sequential structure. The RNNs are typical sequential learning models which allow previous outputs to be used as inputs while having hidden states.

The pictorial representation of a basic RNN is shown in Fig. 12, where the left part of the figure represents an unfolded network, while the right part of the figure represents a folded sequence network at time step $t - 1$, t and $t + 1$. A sequence of training data, i.e., $x_0, x_1,..., x_{t-1}, x_t, x_{t+1},...$, are given as inputs of the network, while $o_0, o_1,..., o_{t-1}, o_t, o_{t+1},...$ denote the calculated output of the network. $h_0,..., h_t$ are the hidden units, which are regarded as the "memory" of the network. As shown in Fig. 12, the current time information x_t and the previous hidden output h_{t-1} are used as the inputs to update the current hidden output h_t. Formally, it can be defined as

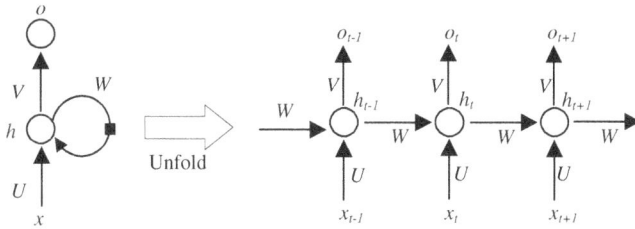

Figure 12: An abstract structure of an RNN.

$$h_t = f(Ux_t + Wh_{t-1} + b), \tag{5}$$

where b is the bias vector, U and W are weight matrices used to condition the input x_t and previous hidden state h_{t-1}, respectively. $f()$ denotes the nonlinear activation function such as tan h, ReLU or Leaky ReLU.

Ideally, the hidden state h_t could preserve the useful information of all previous time steps. The output at time step t is computed based on h_t and the corresponding weight matrix V

$$o_t = g(Vh_t), \tag{6}$$

where $g()$ is softmax function for a classification task. Unlike MLP that uses different parameters at each layer, RNN shares the parameters (W, U, V) across time.

From a theoretical point of view, RNNs can capture dependencies of any length. This characteristic made RNNs achieve superior results in the applications where the sequential data is prevalent, such as natural language processing and video analytics. However, RNNs may suffer from vanishing gradient problem during error back-propagation of training phase, which makes it hard to capture long-term dependencies in practice. To tackle this issue, some improved variants, such as long short-term memory [31] and gated recurrent neural networks (GRU) [32], have been proposed to mitigate the gradient vanishing or gradient exploding problem to facilitate capturing patterns in a longer range.

2.4.2 *Long Short-Term Memory (LSTM)*

Similar to RNNs, LSTMs also have a chain-like structure, but the repeating module has a different structure, which is shown in Fig. 13. Instead of having a single linear transformation in RNNs, there are four in LSTMs, interacting in a very special way [31]. Besides the input and hidden states, the key to

LSTMs is the adding of the cell state, the horizontal line running through the top of the diagram. The cell state is kind of like a conveyor belt. It could directly go through the entire chain to let useful information flow along it.

The ability of removing or adding information to the cell state is controlled via special structures called gates. Gates are a way to decide how much information could go through. They are actually logistic regressions, that is compositions of a linear layer followed by a sigmoid function. The sigmoid is functioned as a normalizer to make the output number of linear layer between zero and one, indicating how much of each component should be let through. A zero gate value indicates no information could go through, while a one value indicates all the information could go through.

There are three gates in an LSTM to control the information flow in the cell state. The first is called forget gate, which is to decide what information is to be thrown away from the cell state. The output of forget gate f_t is calculated by way of

$$f_t = \sigma(W_f \cdot [h_{t-1}, x_t] + b_f), \qquad (7)$$

where x_t denotes the input at time step t, h_{t-1} denotes output of hidden state at time step $t-1$. The W_f and b_f are the parameters of the linear layer corresponding to forget gate. The sigmoidal nature of f_t squashes itself into a range of $[0,1]$.

The second gate is to decide what new information is to be stored in the memory cell unit, which consists of two components, i.e., the input gate i_t modulated by a sigmoid function and input modulation $\tilde{C}_t$ modeled by a tahn function. The input gate i_t and forget gate f_t can be regarded as knobs by which LSTM learns to selectively forget its previous memory

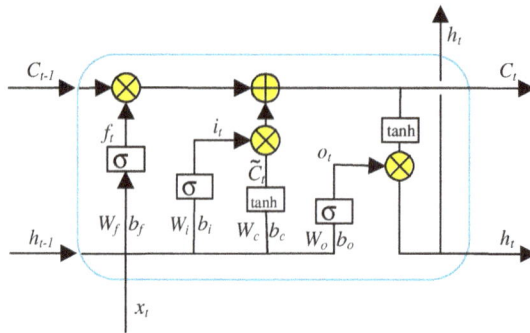

Figure 13:   llustration of LSTM.

$$i_t = \sigma(W_i \cdot [h_{t-1}, x_t] + b_i), \tag{8}$$

$$\tilde{C}_t = \tan h(W_c \cdot [h_{t-1}, x_t] + b_c). \tag{9}$$

The memory cell unit C_{t-1} is updated into C_t as follows:

$$C_t = f_t * C_{t-1} + i_t * \tilde{C}_t. \tag{10}$$

The last gate is output gate, to help generate the hidden state. Specifically, the hidden state h_t of LSTM is given in two steps. First, a gate value is computed using output gate o_t. Then, the network puts the cell unit C_t through a tan h function and multiplies it by o_t

$$o_t = \sigma(W_o \cdot [h_{t-1}, x_t] + b_o), \tag{11}$$

$$h_t = o_t * \tan h(C_t). \tag{12}$$

The adoption of the cell state along with gate operations is able to better preserve useful information in a longer range, which is critical for applications that require long-term dependencies, such as computer vision and speech recognition. Based on the initial LSTM, many variants have been developed in the past few years, such as Bi-directional long short-term memory network (Bi-LSTM), Gated Recurrent Unit (GRU) and convolutional LSTM (ConvLSTM). We will introduce these models in the following subsections.

2.4.3 *Gated recurrent unit*

GRU is a slight variation of LSTM, which was introduced by Chung *et al.* [32]. Compared with vanilla LSTM, GPU made two major changes: first, it combines the forget and input gates into a single "update gate"; second, it merges the cell state and hidden state. The resulting model is simpler than standard LSTM model. To give a clear illustration, a single cell of GRU is shown in Fig. 14.

As shown in Fig. 14, GRU consists of two gates, i.e., a reset gate r_t and an update gate z_t. The reset gate regulates how to combine the new current input with the previous memory, and the update gate decides how much of historic memory should be maintained. The reset gate and update gate are computed by

$$z_t = \sigma(W_z \cdot [h_{t-1}, x_t]), \tag{13}$$

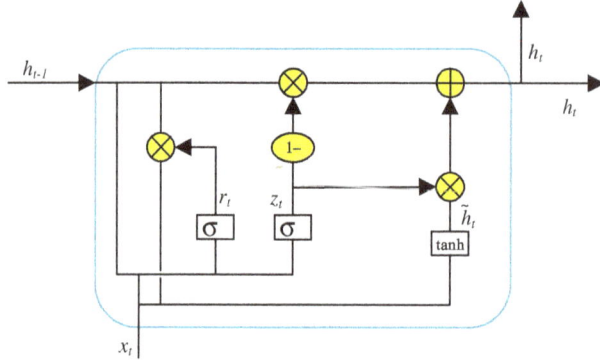

Figure 14:   Illustration of GRU.

$$r_t = \sigma(W_r \cdot [h_{t-1}, x_t]). \tag{14}$$

The hidden state h_t of GRU at time t can be given based on the previous hidden state h_{t-1} and the candidate hidden state $\tilde{h}_t$ as

$$\tilde{h}_t = \tan h(W \cdot [r_t * h_{t-1}, x_t]), \tag{15}$$

$$h_t = (1 - z_t) * h_{t-1} + z_t * \tilde{h}_t. \tag{16}$$

The GRU network is simple yet effective, and can be regarded as a light version of LSTMs in term of computation cost and complexity.

2.4.4 Bi-directional long short-term memory

The conventional LSTM network only works in one direction for processing a finite sequence, which means the current hidden state is generated only by considering the past information of the sequential data. To incorporate the information both for the past and future, Schuster *et al.* [33] proposed a bi-directional long short-term memory (BiLSTM) network to generate hidden states in both directions. Different from LSTM, BiLSTM network consists of two parallel layers propagating in two directions, i.e., a forward layer and a backward layer, which are shown in Fig. 15. In BiLSTM, the hidden state h_t of time step t is defined as the concatenation of the states of the two directions

$$h_t = \vec{h}_t \oplus \overleftarrow{h}_t, \tag{17}$$

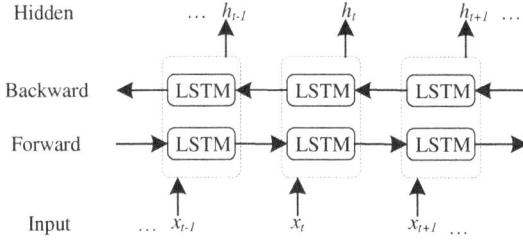

Figure 15: Illustration of Bi-LSTM.

where $\vec{h}_t$ and $\overleftarrow{h}_t$ denote the output vectors of the forward and backward layers, respectively, and $\rightarrow$ and $\leftarrow$ represent the forward and backward processes, respectively. The combination of forward and backward layers is defined as a single BiLSTM layer.

2.4.5 *Convolutional LSTM*

The LSTM model has been proven powerful for handling sequential data along one dimension, in the applications in computer vision, it requires the model to have the ability to handle 2D data sequences (e.g., video or photo).

In order to learn the long-term 2D features, Xingjian *et al.* [34] proposed a Convolutional LSTM (ConvLSTM) which has convolutional structures as shown in Fig. 16. The ConvLSTM predicts the future state of a certain cell in the grid by making full use of the spatio-temporal correlation information. This can be achieved by the convolution and recurrence operations in the input-to-state and state-to-state transitions. Formally, all the inputs $X_1,...,X_t$, the cell states $C_1,...,C_t$, the hidden states $H_1,...,H_t$ and the internal gates $i_t, f_t,$ o_t of ConvLSTM are 3D tensors, whose last two dimensions are spatial dimensions. Let $\otimes$ express the convolutional operation, and let o denote the Hadamard product. The key equations of ConvLSTM are shown as follows:

$$i_t = \sigma(W_{xi} \otimes X_t + W_{hi} \otimes H_{t-1} + W_{ci} \circ C_{t-1} + b_i), \qquad (18)$$

$$f_t = \sigma(W_{xf} \otimes X_t + W_{hf} \otimes H_{t-1} + W_{cf} \circ C_{t-1} + b_f), \qquad (19)$$

$$C_t = f_t \circ C_{t-1} + i_t \circ \tan h(W_{xc} \otimes X_t + W_{hc} \otimes H_{t-1} + b_c), \qquad (20)$$

$$o_t = \sigma(W_{xo} \otimes X_t + W_{ho} \otimes H_{t-1} + W_{co} \circ C_t + b_o), \qquad (21)$$

$$H_t = o_t \circ \tan h(C_t).\tag{22}$$

Due to the capability of capturing spatio-temporal correlations, ConvLSTM can provide decent performance for sequential datasets [35–37].

3. Discussion and Conclusion

Deep learning has been an exciting new trend in machine learning and artificial intelligence over the past decade. Conventional machine learning techniques require hand-crafted features, which are usually tedious and may limit the performance of these methods. In contrast, deep learning algorithms are very effective in automatic feature learning. They jointly optimize the feature learning and the inference in an end-to-end manner and thus lead to a superior performance. Meanwhile, researchers and practitioners are now equipped with powerful computing resources like NVIDIA GPUs. Moreover, various deep learning algorithms are publicly accessible in deep learning platforms, such as PyTorch and TensorFlow. Therefore, deep learning has a large variety of successful applications, e.g., human activity recognition [38], healthcare informatics [39], remote sensing [2], etc.

In this chapter, we briefly introduce different types of deep learning methods, including auto-encoder, RBM, CNN, RNN and their variants. Each deep learning method has its own advantages and disadvantages. For example, auto-encoders support unsupervised model training and do not require labeled data to build the models. Therefore, auto-encoders are commonly applied for dimensionality reduction, feature learning and anomaly detection [40]. RNN has a very good capability of modeling the data with temporal dependencies, e.g., time-series sensory data. Therefore, RNN has been widely applied for wearable sensory data analytics (e.g., human activity recognition [41], sleep stage classification [42], wandering pattern detection [43]) and industrial sensory data analytics (e.g., machine fault diagnosis and prognosis [44, 45]). CNN is the top choice for image classification tasks and it is also used for various sensor data applications [46, 47]. It has different network structures compared with RNN. Usually, CNN performs much faster than RNN, as it can be trained in parallel while RNN needs to be trained sequentially.

Besides the above standard deep learning algorithms, there is a clear trend to combine different deep learning architectures for some specific applications. For example, CNN auto-encoder is proposed to enhance feature learning [48] and recurrent auto-encoder is proposed for time-series outlier

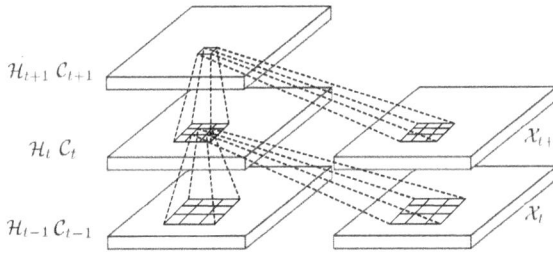

Figure 16: Inner structure of ConvLSTM.

detection [49]. A hybrid RNN and CNN approach is also able to achieve superior performance when the data are suitable for CNN and also have temporal characteristics [37]. In addition, a huge number of novel algorithms, which are based on these standard deep learning algorithms, have been proposed for different applications. We will explore some novel deep learning algorithms that have been proposed to enhance the sensing capabilities in the following chapters.

References

1. Z. Chen, C. Jiang, S. Xiang, J. Ding, M. Wu, and X. Li, "Smartphone sensor based human activity recognition using feature fusion and maximum full a posteriori," *IEEE Trans. Instrumentation and Measurement*, 2019.
2. X. X. Zhu, D. Tuia, L. Mou, G. S. Xia, L. Zhang, F. Xu, and F. Fraundorfer, "Deep learning in remote sensing: A comprehensive review and list of resources," *IEEE Geosci. Remote Sens. Magaz.*, vol. 5, no. 4, pp. 8–36, 2017.
3. H. Fu, J. Cheng, Y. Xu, C. Zhang, D. W. K. Wong, J. Liu, J., and X. Cao, "Disc-aware ensemble network for glaucoma screening from fundus image," *IEEE Trans. Medi. Imag.*, vol. 37, no. 11, pp. 2493–2501, 2018.
4. N. K. Vuong, S. Chan, and C. T. Lau, "Health sensors, techniques, and applications for managing wandering behavior of people with dementia: A review," in *Mobile Health*, Springer 2015, pp. 11–42.
5. A. Ullah, J. Ahmad, K. Muhammad, M. Sajjad, and S. W. Baik, "Action recognition in video sequences using deep bi-directional LSTM with CNN features," *IEEE Access*, vol. 6, pp. 1155–1166, 2018.
6. Z. Chen, C. Jiang, and L. Xie, "A novel ensemble ELM for human activity recognition using smartphone sensors," *IEEE Trans. Industrial Inform.*, vol. 15, no. 5, pp. 2691–2699, 2018.
7. Z. Chen, L. Zhang, C. Jiang, Z. Cao, and W. Cui, "Wifi CSI based passive human activity recognition using attention based BLSTM," *IEEE Trans. Mobile Comput.*, vol. 18, no. 11, pp. 2714–2724, 2018.

8. J. Yang, H. Zou, H. Jiang, and L. Xie, "Device-free occupant activity sensing using wifi-enabled IOT devices for smart homes," *IEEE Internet Things J.*, vol. 5, no. 5, pp. 3991–4002, 2018.

9. Q. Zhu, Z. Chen, and Y. C. Soh, "A novel semisupervised deep learning method for human activity recognition," *IEEE Trans. Indus. Inform.*, vol. 15, no. 7, pp. 3821–3830, 2018.

10. Z. Chen, L. Zhang, Z. Cao, and J. Guo, "Distilling the knowledge from hand-crafted features for human activity recognition," *IEEE Trans. Indus. Inform.*, vol. 14, no. 10, pp. 4334–4342, 2018.

11. S. Yousefi, H. Narui, S. Dayal, S. Ermon, and S. Valaee, "A survey of human activity recognition using wifi CSI," 2017, arXiv preprint arXiv:1708.07129.

12. E. Alpaydin, *Introduction to Machine Learning*. MIT press, 2020.

13. O. D. Lara and M. A. Labrador, "A survey on human activity recognition using wearable sensors," *IEEE Commun. Surve. Tutorials*, vol. 15, no. 3, pp. 1192–1209, 2012.

14. Z. Chen, Q. Zhu, Y. C. Soh, and L. Zhang, "Robust human activity recognition using smartphone sensors via CT-PCA and online SVM," *IEEE Trans. Industrial Informatics*, vol. 13, no. 6, pp. 3070–3080, 2017.

15. Y. LeCun, Y. Bengio, and G. Hinton, "Deep learning," *Nature*, vol. 521, no. 7553, pp. 436–444, 2015.

16. D. E. Rumelhart, G. E. Hinton, and R. J. Williams, *Learning Internal Representations by Error Propagation*, MIT Press, Cambridge, MA, USA, 1986, pp. 318–362.

17. A. Ng *et al.*, "Sparse autoencoder," *CS294A Lecture Notes*, vol. 72, no. 2011, pp. 1–19, 2011.

18. I. Goodfellow, Y. Bengio, and A. Cour-ville, *Deep Learning.*, MIT Press, 2016.

19. H. C. Shin, M. Orton, D. v. J. Collins, S. Doran, and M. Leach, "Organ detection using deep learning," in *Medical Image Recognition, Segmentation and Parsing*, Elsevier, 2016, pp. 123–153.

20. P. Vincent, H. Larochelle, Y. Ben-gio, and P. A. Manzagol, "Extracting and composing robust features with denoising autoencoders," in *Proc. the 25th inter. Con. on Machine Learning*, 2008, pp. 1096–1103.

21. D. H. Ackley, G. E. Hinton, and T. J. Sejnowski, "A learning algorithm for boltzmann machines," *Cognitive Science*, vol. 9, no. 1, pp. 147–169, 1985.

22. P. Smolensky, Information processing in dynamical systems: Foundations of harmony theory," Technical repat, Colorado Univ at Boulder Dept of Computer Science, 1986.

23. J. J. Hopfield, "Neural networks and physical systems with emergent collective computational abilities," *Proc. Nati. Acade. Sci.*, vol. 79, no. 8, pp. 2554–2558, 1982.

24. G. E. Hinton, "Training products of experts by minimizing contrastive divergence," *Neural comput.*, vol. 14, no. 8, pp. 1771–1800, 2002.

25. Y. LeCun, L. Bottou, Y. Bengio, and P. Haffner, "Gradient-based learning applied to document recognition," *Proc. IEEE,* vol. 86, no. 11, pp. 2278–2324, 1998.
26. A. Krizhevsky, I. Sutskever, and G. E. Hinton, "Imagenet classification with deep convolutional neural networks," *Advances in Neural Information Processing Systems,* 2012, pp. 1097–1105.
27. K. Simonyan and A. Zisserman, "Very deep convolutional networks for large-scale image recognition," 2014, arXiv preprint arXiv:1409.1556.
28. C. Szegedy, W. Liu, Y. Jia, P. Sermanet, S. Reed, D. Anguelov, D. Erhan, V. Vanhoucke, and A. Rabinovich, "Going deeper with convolutions," in *Proc. IEEE Confe. Computer Vision and Pattern Recognition,* 2015, pp. 1–9.
29. K. He, X. Zhang, S. Ren, and J. Sun, "Deep residual learning for image recognition," in *Proc. IEEE Conf. Computer Vision and Pattern Recognition,* 2016, pp. 770–778.
30. G. Huang, Z. Liu, L. Van Der Maaten, and K. Q. Weinberger, "Densely connected convolutional networks," in *Proc. IEEE Conf. Computer Vision and Pattern Recognition,* 2017, pp. 4700–4708.
31. S. Hochreiter and J. Schmidhuber, "Long short-term memory," *Neural Comput.,* vol. 9, no. 8, pp. 1735–1780, 1997.
32. J. Chung, C. Gulcehre, K. Cho, and Y. Bengio, "Empirical evaluation of gated recurrent neural networks on sequence modelling," 2014, arXiv preprint arXiv:1412.3555.
33. M. Schuster and K. K. Paliwal, "Bidirectional recurrent neural networks," *IEEE trans. Signal Proces.,* vol. 45, no. 11, pp. 2673–2681, 1997.
34. S. Xingjian, Z. Chen, H. Wang, D. Y. Yeung, W. K. Wong, and W. C. Woo, "Convolutional LSTM network: A machine learning approach for precipitation nowcasting," in *Advances in Neural Information Processing Systems,* 2015, pp. 802–810.
35. Z. Chen, R. Zhao, Q. Zhu, M. K. Masood, Y. C. Soh, and K. Mao, "Building occupancy estimation with environmental sensors via CDBLSTM," *IEEE Trans. Industrial Electron,* vol. 64, no. 12, pp. 9549–9559, 2017.
36. A. K. Bhunia, A. Konwer, A. K. Bhunia, A. Bhowmick, P. P. Roy, and U. Pal, "Script identification in natural scene image and video frames using an attention based convolutional-lstm network," *Pattern Recogn.,* vol. 85, pp. 172–184, 2019.
37. Z. Chen, M. Wu, W. Cui, C. Liu, and X. Li, "An attention based CNN-LSTM approach for sleep-wake detection with heterogeneous sensors," *IEEE J. Biomed. Health Inform.,* 2020.
38. J. Wang, Y. Chen, S. Hao, X. Peng, and L. Hu, "Deep learning for sensor-based activity recognition: A survey," *Pattern Recogn. Lett.,* vol. 119, pp. 3–11, 2019.
39. D. Ravì, C. Wong, F. Deligianni, M. Berthelot, J. Andreu-Perez, B. Lo, and G. Z. Yang, "Deep learning for health informatics," *IEEE J. Biomed. Health Inform.,* vol. 21, no. 1, pp. 4–21, 2016.

40. R. Chalapathy and S. Chawla, "Deep learning for anomaly detection: A survey," 2019, arXiv preprint arXiv:1901.03407.
41. Y. Guan and T. Plötz, "Ensembles of deep LSTM learners for activity recognition using wearables," *Proc. ACM Interactive, Mobile, Wearable Ubiquitous Technol.*, vol. 1, no. 2, pp. 1–28, 2017.
42. Z. Chen, M. Wu, K. Gao, J. Wu, J. Ding, Z. Zeng, and X. Li, "A novel ensemble deep learning approach for sleep-wake detection using heart rate variability and acceleration," *IEEE Trans. Emerging Topics Comput. Intell.*, 2020.
43. N. Bosch and S. D'Mello, "Automatic detection of mind wandering from video in the lab and in the classroom," *IEEE Trans. Affective Comput.*, 2019.
44. R. Zhao, R. Yan, Z. Chen, K. Mao, P. Wang, and R. X. Gao, "Deep learning and its applications to machine health monitoring," *Mech. Syst. Signal Process.*, vol. 115, pp. 213–237, 2019.
45. Z. Chen, M. Wu, R. Zhao, F. Guretno, R. Yan, and X. Li, "Machine remaining useful life prediction via an attention based deep learning approach," *IEEE Trans. Indust. Electron.*, 2020.
46. J. Yang, M. N. Nguyen, P. P. San, X. L. Li, and S. Krishnaswamy, "Deep convolutional neural networks on multichannel time series for human activity recognition," in *Twenty-Fourth International Joint Conference on Artificial Intelligence*, 2015.
47. S. Münzner, P. Schmidt, A. Reiss, M. Hanselmann, R. Stiefelhagen, and R. Dürichen, "CNN-based sensor fusion techniques for multimodal human activity recognition," in *Proc. 2017 ACM Int. Symp. Wearable Computers*, 2017, pp. 158–165.
48. M. Chen, X. Shi, Y. Zhang, D. Wu, and M. Guizani, "Deep features learning for medical image analysis with convolutional autoencoder neural network," *IEEE Trans. Big Data*, 2017.
49. T. Kieu, B. Yang, C. Guo, and C. S. Jensen, "Outlier detection for time series with recurrent autoencoder ensembles," in *IJCAI*, 2019, pp. 2725–2732.
50. I. Sergey and S. Christian, "Accelerating deep network training by reducing internal covariate shift," ioffe2015batch, 2015, arXiv preprint arXiv:1502.03167.

Part II

Deep Learning for
Activity Sensing

Chapter 2

Hierarchically Aggregated Deep Convolutional Neural Networks for Action Recognition

Le Zhang*, Jagannadan Varadarajan†, Yong Pei‡
and Zhenghua Chen*

*A.STAR, Singapore
†Grab, Singapore
‡Webank, P.R.C

Abstract

Deep convolutional neural networks (ConvNets) have become ubiquitous in computer vision owing to their success in visual recognition tasks on still images. This has encouraged a number of novel methods for video classification that have progressively improved over conventional handcrafted features. Existing CNN-based methods for action recognition typically train multiple streams to extract spatial and temporal characteristics and finally combine their prediction scores. However, they rely heavily on appearance and short-term temporal information. In this chapter, we present a new hierarchically aggregated ConvNet architecture that can more effectively deal with long-term dynamics present in the action videos. The proposed network takes a long-term video as input and exploits temporal information at multiple levels of the network hierarchy capturing more complex dynamics of the input sequence. This is enabled using special aggregation functions that combine information from multiple temporal segments via their feature maps. Detailed experiments show that our approach outperforms state-of-the-art methods significantly on UCF-101 and HMDB51 benchmark datasets.

Keywords: convolutional neural network, action recognition, temporal information

1. Introduction

Human action recognition has attracted great attention due to its wide applications in many fields, such as multimedia, computer vision, healthcare, smart cities, etc. Generally, it can be divided into sensor-based [1–4] and vision-based solutions [5–7]. Due to the remarkable performance of vision-based solutions, they become more and more popular in various tasks, such as visual surveillance, robotics and abnormality detection. In this chapter, we focus on vision-based human action recognition. Since the earliest methods for automated analysis of human actions, several features [5, 7–11] and learning methods [12–15] have consistently improved the recognition performance leading the research community to address more challenging scenarios [20, 50] and complex datasets [16, 17]. However, significant hurdles due to variations in appearance, view point, illumination and speed of execution come in the way of solving the problem.

Recently, deep convolutional neural network (ConvNets)-based methods have made vast inroads into computer vision due to their commendable performance in several vision tasks including image classification [18–21], segmentation [22], pose estimation [23] and face recognition [24], where (near) human performance is achieved. Inspired by this, several ConvNet models were proposed for video-based action recognition, too. Typically, these methods process videos by training two parallel networks (also called streams), where one stream (called the spatial stream) focuses on learning spatial features from RGB input images and the other temporal stream processes motion information from optical flow. The final classification scores are obtained using some weighted combination of the individual network scores.

Initially, however, ConvNet-based methods for action recognition did not lead to significant performance improvements over hand crafted features, which was in contrast to other still-image-based applications. This is attributed to (i) lack of large-scale training samples for action recognition — ImageNet challenge has 1000 samples per class whereas UCF-101 [16] has only 100 samples per class; (ii) over fitted models and (iii) large variations in the temporal dynamics of an action that is not captured well by existing ConvNet models.

Recently, methods by Refs. [25, 26] have achieved impressive results on UCF-101 and HMDB51 datasets using complex networks such as Resnets [21] and BN-Inception models [27], respectively. Nevertheless, the most popular solution proposed to deal with long-range temporal structures is to evaluate the ConvNet on multiple short-term segments, followed by a weighted combination of the scores in a late fusion strategy [25]. LRCNs were used in Ref. [28] to model long-range temporal dynamics with LSTMs using the output of the final layer of a deep ConvNet. These methods mostly focus on modeling temporal dynamics using the results of the final layer, but ignore the temporal relationships among feature maps present at each layer of deep ConvNet.

We argue that using temporal information at a single level as done in Refs. [25, 28] is not sufficient to interpret and understand the complexity of videos and that they need to be modeled at each level of the hierarchy. The *first* contribution of this chapter is, therefore, a novel hierarchical ConvNet architecture called *hierarchically aggregated convolutional neural network* (HACNN) to deal with long-term temporal dynamics. HACNN takes a long-term video consisting of multiple non-overlapping temporal segments as input and gives a single prediction score after a feed-forward pass. The network hierarchy consists of several levels of feature processing. At the lowest level, input channels from non-overlapping segments are passed through a series of convolutions, nonlinearity and pooling. Special aggregating functions are then employed to pool information from multiple temporal segments via their feature maps. When this is repeated hierarchically, we capture long-term, higher-order dynamics.

As a *second* contribution, we propose to automatically learn special *aggregation* functions that can pool information from feature maps corresponding to multiple temporal segments of the video. The aggregation functions can come from a class of commonly used pooling functions including simple and weighted average and max operation, to leverage information from multiple segments in the same video. We show that backpropagation through the aggregation function can be efficiently implemented, allowing end-to-end optimization of the recognition network.

Finally, we also incorporate the recently proposed *center loss* criteria at different levels of the hierarchy of HACNN in order to encourage low density margins and increase the overall discriminative capacity of the model.

We present detailed experimental results on two challenging benchmark datasets used widely in the literature, and show that our method improves over recently reported state-of-the-art results by 0.6% and 3.4% leading to 94.6% and 71.9% accuracy on UCF-101 and HMDB datasets, respectively, with only two input modalities consisting of RGB and optical flow information.

2. Related Work

Below we review previous works that are closely related to ours which fall in two categories:

(1) Action recognition using convolutional neural networks, and
(2) modeling long-term temporal characteristics using ConvNets.

2.1 *Action recognition using ConvNets*

One of the first attempts to perform action recognition using ConvNets was presented in Ref. [29] to recognize actions from real-life scenes, where 3D convolutions were applied on frame sequences to capture both appearance and motion properties. Following this, Karpathy *et al.* [30] released a large-scale video dataset and investigated several aspects of ConvNet architectures including early fusion, late fusion, and 3D ConvNet on classification performance. A notable development in ConvNets for action recognition was made by Simonyan *et al.* [31], where a two stream ConvNet architecture (inspired by the object-centric ventral stream and motion-centric dorsal stream of the human visual cortex) was proposed. Here, the spatial ConvNet (also called a stream) processes the RGB channels and the temporal stream processes several channels of images that encode optical flow in the horizontal and vertical directions. The outputs from the two *independent* streams are then fused to obtain the final confidence scores for an input video.

Following this, several approaches modifying the network architecture [26, 32] and (or) the input features [33, 34] were proposed. For instance, Zhang *et al.* [34] proposed to substitute optical flow vectors with motion vectors in the interest of achieving real-time speeds. In Ref. [34], strengths of handcrafted features and ConvNet-based features were combined by pooling the final convolution layer's features along dense trajectories of iDT [10]. In the C3D method [35], 3D convolutions on limited temporal support of 16 frames with $3 \times 3 \times 3$ kernels were proposed to extract generic features for video analysis tasks.

2.2 Modeling long-term dynamics using ConvNets

Modeling long-term temporal dynamics plays an important role in recognizing actions. A few attempts [28, 32, 36, 37] have been made recently in order to deal with this problem. For example, long-range temporal dependencies were learned using dynamic models such as recurrent neural networks on sequence of features in Ref. [28]. Although they proposed an end-to-end model combining ConvNets and RNNs, temporal dynamics were modeled only using high-level percepts leading to poor performance at the end. Long-term temporal convolutions involving optical flow from 100 frames were proposed in Ref. [36]. More recently, Wang *et al.* [32] proposed a network model that takes score consensus from sparse temporal segments and showed improved performance on the UCF-101 and HMDB datasets. Most of the aforementioned methods suffer from two main drawbacks: (i) temporal information is only coarsely modeled by simply evaluating the network on sparsely sampled temporal segments; (ii) consensus are built at the top layer (usually via the final scores) of the network, ignoring local relationships between the low-level features existing at different layers of deep ConvNet.

Different from these methods, we argue that hierarchically exploiting temporal interaction between the low-level feature maps is important to advance the performance of action recognition methods and propose a novel approach that takes a long-term video and pools temporal information from multiple feature maps produced by the application of the same filter on multiple segments. When this is done hierarchically, we capture long-range, higher-order dynamics present in an action video. Our method shares some similarities with the hierarchical rank pooling approach proposed in Ref. [37]. However, we differ from Ref. [37] significantly in multiple ways. Our method uses deep ConvNet architecture applied to appearance and optical flow channels and can work with any general purpose aggregation functions. We employ deeper ConvNets in each level of the hierarchy to learn high-level percepts and discriminative information for action recognition, whereas the method in Ref. [37] uses a shallow video representation method to aggregate information over time. Our hierarchically aggregated features lead to superior results as we show in Section 4.

3. Hierarchically Aggregated CNN Model

Figure 1 gives an overview of our HACNN model. The hierarchy consists of several levels, where each level is defined by a subConvNet, an aggregation

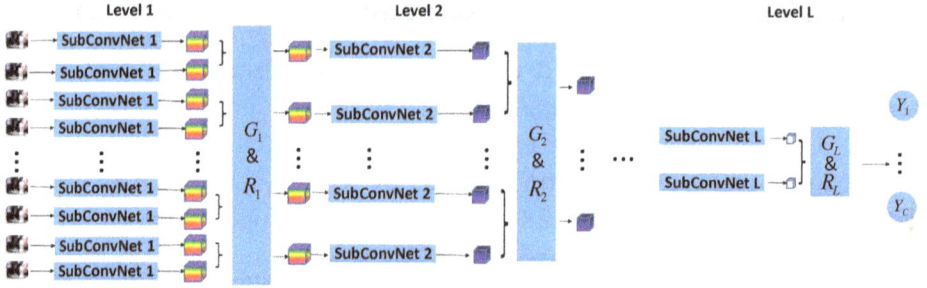

Figure 1: Illustration of HACNNs model. HACNN processes the video in a hierarchical manner. Input to HACNN consists of multiple disjoint snippets from an action video. At each level, the input video snippets or feature maps are processed via multiple parallel *subConvNets* (deep ConvNets that share the same parameters) followed by an aggregation and regularization function (G & R). This produces feature maps for the next level of hierarchy. The output of our HACNN is a C dimensional vector of scores indicating the probability of belonging to a particular action class $c \in \{1,\dots, C\}$.

function and a regularization function. At the lowest level, the model takes input from multiple temporal segments, processes them through subConvNets that share the same parameters and produces feature maps corresponding to each segment. The aggregation function pools discriminative characteristics present across multiple temporal segments and consolidates them over fewer sets of feature maps. This is repeated over each level of the hierarchy. The final layer produces probabilities for each action label.

In this section, we present the major contributions of this chapter. In Section 3.1, we present an overview of the HACNN model. Various components of HACNN, namely, the subConvNet, aggregation function and regularization function, are explained in Sections 3.2–3.4, respectively. The backpropagation steps to learn the parameters of HACNN model are given in Section 3.5.

3.1. *Background*

Given a dataset $D = \{(X_n, \Upsilon_n)\}_{n=1}^N$ of N pairs, where $X_n \in \chi^s$ represents a video of an action in the input feature space χ^s, Υ_n its corresponding action label coming from the set $\{1;\dots, C\}$ of action classes, and encoded as a C-dimensional binary vector $\Upsilon \in y \subset \{0, 1\}^C$, our goal is to learn a function $\mathcal{F} : \chi^s \to y$, representing a ConvNet. As demonstrated in Fig. 1, $\mathcal{F}$ is a nonlinear composition function that consists of multiple levels of a hierarchy indexed by $l \in \{1,\dots L\}$, where each level of the hierarchy consists of a *subConvNet*[a] function F_l, and an aggregation function G_l. More formally, given a video X consisting of multiple temporal segments, $\mathcal{F}$ is defined as

$$\mathcal{F} = H(G_L(F_L(G_{L-1}(\ldots F_2(G_1(F_1(X;W_1^s)))))))), \tag{1}$$

where F_l is the subConvNet with parameters W_l^s and H is the Softmax function. At the lowest level, the inputs to subConvNet F_1 consist of RGB or optical flow channels from T_1 temporal segments $X = \{x_1, \ldots x_{T_1}\}$, resulting in a set of feature maps $\mathbf{f}_1 = \{f_1, \ldots f_{T_1}\}$ corresponding to each input $x_t, 0 < t \leq T_1$:

$$\mathbf{f}_1 = F_1(X, W_1^s) \tag{2}$$

$$= F_1(x_1 \ldots x_{T_1}; W_1^s). \tag{3}$$

At higher levels of hierarchy, F_l is applied on feature maps resulting from the layers immediately below. SubConvNets at different levels differ based on the number of convolution and inception layers, filters and stride length. More details on the different subConvNets used are given in Section 3.2.

G_l is an aggregation function that takes the feature maps $\mathbf{f}_l$ and produces a new set of feature maps $\tilde{\mathbf{f}}_l$ that encode dynamic information present in the video in a compact form. Upon the successive application of G_l over the results of F_l, we expect to condense the information present in the original video into fewer temporal segments. More details on the choice of aggregation function are given in Section 3.3.

3.2. *SubConvNet architecture*

Our HACNN model consists of several levels in the hierarchy. Each level is endowed with a unique subConvNet function. We consider deep ConvNets for our choice of subConvNet function F at each level of hierarchy, due to their proven superior performance in computer vision problems. More specifically, a subConvNet consists of multiple layers of convolution, Relu, pooling and Batch Normalization layers and designed based on the BN-Inception deep ConvNet proposed in Ref. [27]. The subConvNet at the first layer, F_1, receives raw data (RGB or optical flow) as input and processes them through

---

[a]Although a subConvNet is by itself a deep ConvNet, we use the *sub* prefix to indicate that it is part of a larger model.

several modules of (Convolution+Batch Normalization+ReLu CBR) and pooling layer followed by eight layers of inception module. The subConvNets F_l, $1 < l < L$ in intermediate levels get discriminative features maps from lower levels and process them further with several layers of inception modules. F_L at the last level of hierarchy consists of one global pooling layer, one dropout layer with the dropout ratio set to 0.8 in all experiments, followed by 1×1 convolution to give the final prediction. Generally speaking, having more layers of hierarchy is possible to model longer and more complex dynamics in videos. However, this also leads to higher risk of over-fitting. We empirically observe that having three levels of hierarchies can lead to satisfactory results and outperforms current state-of-the-art methods. Comparative studies with other architectural choices can be found in the experimental Section 4.

More specifically, our HACNN model consists of three levels of hierarchy. Each level is endowed with a unique subConvNet function. We consider very deep ConvNets for our choice of subConvNet function F at each level of hierarchy, due to their proven superior performance in computer vision problems. More specifically, a subConvNet consists of multiple layers of convolution, Relu, pooling and Batch Normalization layers and is designed based on the BN-Inception deep ConvNet proposed in Ref. [27]. The subConvNet at the first layer, F_1, receives raw data (RGB or optical flow) as input and processes them through several modules of CBR followed by eight layers of inception module. The subConvNets F_l, $1 < l < L$ in intermediate levels get discriminative features maps from lower levels and process them further with several layers of inception modules. F_L at the last level of hierarchy consists of one dropout layer, one global pooling layer followed by 1×1 convolution to give the final prediction. Generally speaking, having more layers of hierarchy is possible to model longer and more complex dynamics in videos.

Table 1: HACNN model details.

Module	Type	#Description
		Hierarchy 1
1	CBR	(7,2,64,3)
2	Pooling	(3,2,Max,0)
3	CBR	(1,1,64,0)
4	CBR	(3,1,192,1)
5	Pooling	(3,2,Max,0)
6	Inception 1	(64,64,64,64,96,96,32,AVE)

Table 1: (*Continued*)

Module	Type	#Description
7	Inception 1	(64,64,96,64,96,96,64,AVE)
8	Inception 2	(128,160,64,96,96,MAX)
9	Inception 1	(224,64,96,96,128,128,128,AVE)
10	Inception 1	(192,96,128,96,128,128,128,AVE)
11	Inception 1	(160,128,160,128,160,160,128,AVE)
12	Inception 1	(96,128,192,160,192,192,128,AVE)
13	Inception 2	(128,192,192,256,256,MAX)
	Hierarchy 2	
1	Inception 1	(352,192,320,160,224,224,128,AVE)
2	Inception 1	(352,192,320,192,224,224,128,MAX)
	Hierarchy 3	
1	Pooling	(7,1,Ave,0)
2	Dropout	$p = 0.8$
3	1×1 Convolution	Output = Number of Classes

However, this also leads to higher risk of over-fitting. Comparative studies with other architectural choices can be found in the experimental Section 4. Detailed explanation of the network structure in each layer of hierarchy used in this study is presented in Table 1. We give detailed description on the second column for each module. For "CBR" module, four numbers in the bracket means kernel size, stride, number of output and padding, respectively. For pooling layer, four values in the bracket indicate kernel size, stride, pooling method and padding, respectively. Details of inception modules are presented in Fig. 2. For inception modules, stride for "1×1" and "3×3" convolutions are set to 0 and 1, respectively. Strides for all the convolution and pooling layers in inception module are set to 1. Kernel size for pooling layers in inception module is 3. Apart from the pooling method which is summarized in the last entry of the bracket in the second column, the main difference of different inception modules in each level of hierarchy lies in the number of outputs for convolution. We follow the rule of Refs. [25, 27] to design each inception module in HCANN. For "inception 1" module, seven numbers in the bracket from the second column stand for the number of outputs for each convolutional layer in Fig. 2 in a "left to right, bottom to top" order. In the same way, five numbers are also provided for inception 2 module.

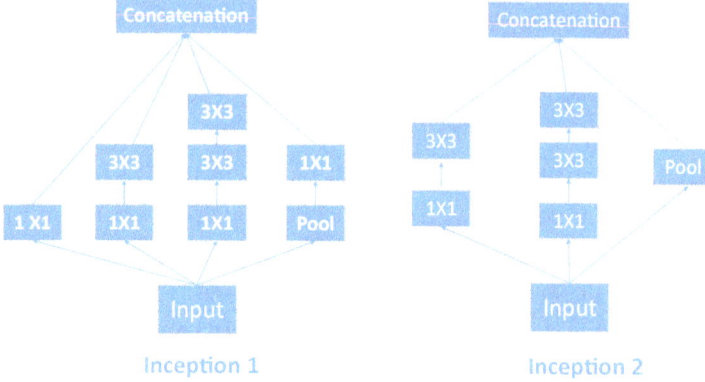

Figure 2: Demonstration of two inception modules used in this study. "1 × 1", "3 × 3" stand for convolution with kernel size of 1 and 3, respectively.

We also note that although our subConvNet model is inspired by Ref. [27], any video representation method based on architectures such as Alexnet or VGGNet [18, 20] can be utilized, in practice, for the purpose of our subConvNet function. Shallow representations such as rank pooling [16] schemes are other alternatives.

3.3. Aggregation function

Typically, a wide range of encoding methods such as BOW [38], Fisher vector [39], bilinear models [40], rank pooling [37] can be integrated in HACNN. However, they come at the cost of losing rich spatial information. Here, we propose to use an aggregation function to pool information from multiple temporal segments, while still preserving the structural information coming from spatial layout. Consider the set of feature maps $\mathbf{f}_l = \{f_{l,1}, f_{l,2}, \ldots, f_{l,T_l}\}$ of size $f \in R^{h_l \times w_l \times d_l}$ obtained from subConvNet F_l, where T_l, h_l, w_l and d_l denote the number of temporal segments, height, width and depth (number of filters) of feature maps at level l. The aggregation function G_l takes feature maps $\mathbf{f}_l$ as input and encodes them into a new set of feature maps $\tilde{\mathbf{f}}_l \in R^{h_l \times w_l \times d_l}$ with $\tilde{T}_l$ segments. By constraining $\tilde{T}_l$ to be in the range $0 < \tilde{T}_l < T_l$, we expect to learn a more compact and discriminative $\tilde{\mathbf{f}}_l$ compared to $\mathbf{f}_l$. This also reduces the overall computational complexity. More formally, the aggregation function G can be formulated as

$$\tilde{\mathbf{f}}_{l,i} = G(\mathbf{f}_l)_i = \sum_{k=i-t_l}^{i+t_l} W_k^g \circ f_{l,k}, \qquad (4)$$

where o is a binary operator denoting the Hadamard product between the input matrices. W^g is the weight vector with the same dimensions as f_l. The aggregation function pools information coming from different segments in an elementwise manner, by still preserving rich spatial information, where the amount of aggregation is determined by the receptive fields denoted by t_l. Details about the values used by t at various levels l can be found in Section 4.5. Although G in Eq. (4) is able to model the importance of each element among different feature maps, the huge number of parameters involved leads to severe over-fitting. Therefore, we propose to use a modified aggregation function given as

$$\tilde{\mathbf{f}}_{l,i} = G(\mathbf{f}_l)_i = \sum_{k=i-t_l}^{i+t_l} W_k^g f_{l,k},\qquad(5)$$

This allows the elements of W^g to be shared among all features in a particular feature map. While W^g can be learned through backpropagation, as we show in Section 3.5, we also investigate the use of simple averaging function, which treats all inputs equally, and max aggregation function: $G_{max}(\mathbf{f}_l)_i = \max\{f_{l,k}\}_{k=i-t_l}^{i+t_l}$ for G.

3.4. *Regularization function*

We also regularize the results of each aggregation function at different levels of HACNN to improve its discriminative ability. For this, we adopt the recently proposed center loss [52] function defined as:

$$R_l(\tilde{f}) = \frac{1}{T_l}\sum_{i=1}^{T_l}\left\|\tilde{f}_{li} - c_{ly}\right\|^2,\qquad(6)$$

where y is the label of the current input and c_{ly} represents the geometric mean of the input feature maps with the the the same dimensions as f_{li}. c_{ly} can be jointly learned during model backpropagation [52]. Center loss constrains the feature maps to remain close to their means within each class and encourages low-density margins. This helps in obtaining more discriminative feature maps per class. However, different from Ref. [41], which uses this regularization term on high layers of deep ConvNets, we use Eq. (6) on the results of aggregation function at each level. This is also in line with the deeply supervised Nets [42] which uses different loss functions at different layers of ConvNets to ensure sufficient gradient signal for efficient backpropagation.

3.5. *End-to-end learning*

HACNN consists of several subConvNets and differentiable aggregation functions, allowing end-to-end optimization of the network possible. Let L denote the softmax loss function. The learning algorithm aims to minimize the following loss:

$$L(F; D, W) + \sum_{l=1}^{L} \beta_l R_l(\tilde{\mathbf{f}}_l), \tag{7}$$

where β stands for a pre-defined constant and $W = \{W^s, W^g\}$. Gradients for parameters $\{W_l^s, W_l^g\}$ at the *lth* level of HACNN can be efficiently computed as:

$$\left(W_l^s\right)' = \frac{dL}{dG_l} * \frac{dG_l}{dF_l} * \frac{dF_l}{dW_l^s} + \sum_{i=L-1}^{L} \beta_i * \frac{dR_i}{dG_l} * \frac{dG_l}{dF_l} * \frac{dF_l}{dW_l^s},$$

$$\left(W_l^g\right)' = \frac{dL}{dG_l} * \frac{dG_l}{dW_l^g} + \sum_{i=L-l}^{L} \beta_i * \frac{dR_i}{dG_l} * \frac{dG_l}{dW_l^g}. \tag{8}$$

4. Experiments

In this section, we first present details about the datasets used, implementation and evaluation of the HACNN model. Next, we analyze the effectiveness of the proposed aggregation function and regularization function through detailed experimental results. Finally, we compare the results obtained from our approach, quantitatively, with various state-of-the-art methods.

4.1. *Datasets*

We conducted experiments on two commonly used action datasets, namely HMDB51 [17] and UCF-101 [16]. The UCF-101 dataset consists of 13,320 video clips from 101 classes. We follow the evaluation scheme of the THUMOS13 challenge and adopt the three training/testing splits for evaluation.

The HMDB51 dataset is a large collection of realistic videos from various sources, such as movies and web videos. The dataset is composed of 6,766 video clips from 51 action categories with large human motion. Similarly, we use the three training/test splits provided for evaluation and report the final accuracy.

4.2. *Implementation details*

We firstly adopt the commonly used two-stream approach which combines spatial and temporal networks. The spatial network receives 16 RGB images extracted from videos to learn a discriminative ConvNet to model what makes the action. The temporal network operates on 16 optical flow stacks, where each stack consists of 10 channels (five horizontal and five vertical displacements) of flow fields to represent how the action proceeds over time.[b] As mentioned in Section 3.2, the default architecture for HACNN consists of three levels of hierarchy and the receptive fields on the temporal axis of the aggregation functions are set to be 4, 2 and 2 at the 1st, 2nd, and 3rd levels, respectively. We use T_1 = 16 temporal segments from the video as input to HACNN. After the application of the first aggregation function, we obtain T_2 = 4 sets of feature maps, which are further reduced to T_3 = 2 by the second aggregation function. Finally, a single compact descriptor of the whole video is generated at the end. Detailed analysis of alternative structures is presented in Section 4.5. Finally, to predict the action class of a video, a late fusion by weighted average of the scores from the two networks is used.

As the aggregation function G at each level of the hierarchy in HACNN preserves the spatial layout of the feature maps, we can stack multiple sub-ConvNets to create a deeper network. The spatial and temporal streams presented in this manner share a very similar structure, except for the number of input channels. This similarity facilitates easy transfer of information via pre-trained model on ImageNet. To initialize the subConvNets of the spatial HACNNs, we use the BN-inception model from Ref. [27]. To initialize the temporal HACNN stream, we adopt the cross-modality training introduced in Ref. [43]. More specifically, to initialize the weights of the temporal stream, we average the weights across the RGB channels and replicate the average values through the number of channels of the temporal stream.

4.2.1. *Optimization*

Stochastic gradient descent is used to learn the parameters of the whole recognition network. We set the mini-batch size to be 4 and the momentum to 0.9. The learning rate is set to a conservatively small value, i.e., $1e^{-4}$. We further reduce the learning rate by a factor of 0.1 for every 15,000 iterations until

---

[b] For certain videos which are shorter than 80 frames, we replicate them and concatenate the frames.

reaching a maximum of 80,000 iterations for RGB streams. For the temporal stream, the maximum iterations is set to a relatively larger value, i.e., 100,000, so as to cope up with the new input modality coming from optical flow. We reduce the learning rate by a factor of 0.1 for every 20,000 iterations. For data augmentation, location and scale jittering, horizontal flipping, corner cropping are used [25, 43]. In corner cropping, we extract regions randomly from the corners and the center of the input image (either raw RGB and optical flow). For scale jittering, the input is first resized to 256×340 pixels. The width and the height of the cropped region are then randomly selected from a set of values $\{256, 224, 192, 168\}$ to insert arbitrary variations in the aspect ratio. Finally, we resize the cropped regions to 224×224 pixels and feed them into HACNN. We use the TVL1 optical flow algorithm [44] in OpenCV with CUDA to extract the optical flows for the temporal stream. We further employ a data-parallel strategy with two TitanX GPUs in a customized version of CAFFE [45] and OpenMPI. The training time of 1 epoch takes around half an hour in our platform.

4.3. *Model evaluation*

To evaluate HACNN, we consider $2 \times T_1$ RGB frames or optical flow stacks from each video. From the sampled frames or segments, we obtain crops from the four corners and the center, and use their original as well as flipped versions. In total, 10 samples are generated from each segment and in total there are 20 samples for one video. We average the scores of those 20 samples before applying a softmax operation. As both spatial and temporal HACNN are able to model long term temporal information present in an action and achieve comparable results, a simple average of the scores before softmax is used to make a final prediction.

4.4. *How to aggregate information?*

As elaborated in Section 3.3, we studied the suitability of three different aggregation functions, namely, the simple average, weighted average and max operation. The choice of the aggregation function is motivated by their ability in preserving the spatial layout information. This property enables us to adopt any existing CNN architecture easily to create a HACNN-like model. We report experimental results comparing three different aggregation functions on the first split of UCF-101 in Table 2. The baseline model in the table

Table 2: Results of different aggregation functions on first split of UCF-101 dataset.

Aggregation	Spatial (%)	Temporal (%)	Fusion (%)
Baseline	84.2	86.2	92.9
Simple Average	87.0	87.5	93.7
Weighted Average	86.3	87.7	93.8
Max + simple	87.0	88.5	94.3
Fc Aggregation	64.2	48.3	66.7

refers to the two stream BN-inception model without any aggregation, as done in Ref. [31]. We see that models with temporal aggregation improve the accuracy over the baseline method significantly, with both simple and weighted average aggregation performing similarly.

The best performance is obtained when a hybrid of max aggregation in two lower levels and a simple average at the top level is used. This is motivated by various recent works [25, 31, 46], where average class scores from multiple frames are reported to be beneficial. We will use the "max+simple" strategy in the following sections due to its superior performance.

In order to demonstrate the superiority of the choice of aggregation methods that preserve the spatial layout, we also implemented an aggregation function using a fully-connected (FC) layer, by setting the number of output nodes to $h_l \times w_l \times d_l$. This discards all spatial layout information and each element in the result is aggregated by pooling information globally from all input elements. The output activations for each sample are then re-sized to a 3D matrix of size $h_l \times wl \times d_l$ and fed into the next level of the hierarchy. Clearly, we see that this gives the worst performance in the table. This could also be due to the large number of parameters involved in the fully-connected aggregation.

4.5. *Where to aggregate information?*

Generally speaking, aggregation can take place at any layer of a typical deep ConvNet. We empirically observe that aggregating feature maps in several levels of the hierarchy is beneficial. Table 3 shows the results obtained from the first split of UCF-101 when the aggregation is done at various levels of the hierarchy.

Table 3: Results obtained from the spatial stream, temporal stream and their fusion by changing the number of levels in the hierarchy as well as where the aggregation function is applied on the hierarchy.

Aggregation	Spatial (%)	Temporal (%)	Fusion (%)
Baseline	84.2	86.2	92.9
1×1 Conv	86.0	87.0	93.6
In5+1 $\times$ 1 Conv	87.2	86.6	93.7
In4+In5+1 $\times$ 1 Conv	87.0	88.5	94.3
In3+In4+In5+1 $\times$ 1 Conv	86.9	88.5	94.0

Note: The first split of UCF-101 dataset is used for this purpose.

In Table 3, *Baseline* corresponds to results from the baseline model, with two streams of BN-inception, but without any hierarchical aggregation "1×1 Conv" in 3rd row stands for only aggregating the class scores of different video segments at the last 1×1 convolutional layer. "In5+1 $\times$ 1 Conv" stands for using the aggregation function at both "Inception5" and 1×1 Conv layers. In the 5th row, we aggregate feature maps at "Inception4","Inception5", and 1×1 Conv layers. In the last row, we further add another level to the hierarchy and use the aggregation function after "Inception3" layer.

In order to have a fair comparison, we keep the number of input video segments to be 16 for all cases. In "In5+1 $\times$ 1 Conv" and "In3+In4+In5+1 $\times$ 1 Conv", the receptive fields of each aggregation function are set to 4 and 2, respectively. In "In4+In5+1 $\times$ 1 Conv", the receptive field for each aggregation function are set to be 4, 2, 2 for each level of hierarchy, respectively. Firstly, we see that using aggregation to pool information from multiple video segments significantly outperforms the baseline method in all cases. We also observe that aggregation at each level of the hierarchy is always better than a simple score aggregation at the final level, justifying our approach to hierarchical aggregation. Finally, we note that "In4+In5+1 $\times$ 1 Conv" outperforms "In3+In4+In5+1 $\times$ 1 Conv" even though the latter has more levels in the hierarchy. In the following comparisons, we will use 'In4+In5+1 $\times$ 1 Conv" for our HACNN model.

4.6. *Hierarchical aggregation vs. recurrent structures*

Recurrent structures such as recurrent neural networks (RNNs) have been an immediate choice to model complex dynamics. Recurrent convolutional

Table 4: Comparisons with various recurrent structures on the first split of UCF-101 dataset.

Method	Spatial (%)	Temporal (%)	Fusion (%)
Baseline	84.2	86.2	92.9
Ours	87.0	88.5	94.3
LRCN [28]	70.7	69.4	80.5
LCRN-BN-Inception	84.7	82.7	92.4

Note: Results from the spatial stream, temporal stream and their fusion are compared.

models in LRCN [28] can be "doubly deep" because of joint learning of compositional representations in space and time. Long-term dependencies can be effectively modeled when non-linearities are incorporated into the network state updates. In order to compare our method with LRCN, we implemented the pioneering work of LRCN [28] by employing deeper BN-Inception model with the same input modality. Parameters for LSTMs were set based on the description in Ref. [28]. Results from the first split of UCF-101 are shown in Table 4. We also report the results originally reported in the paper [11] for a fair comparison. From this, we first observe that employing BN-Inception model for the CNN leads to much better performance compared to the original results reported in Ref. [28], which used less powerful AlexNet [47]. Thanks to the advanced data augmentation strategies adopted in feed-forward structures, we observe that the feed-forward models outperform recurrent structures for video classification. Finally, we also see that the proposed HACNN leads to much better results than the LRCN models.

4.7. *Effectiveness of regularization*

As explained in Section 3.4, we regularize the feature maps in each layer of the hierarchy to further boost its discriminative ability. The regularization parameters in "Inception4", "Inception5" and last 1×1 convolutional layers are set to $5e^{-5}$, $1e^{-6}$ and $5e^{-5}$, respectively. Using first split of UCF-101, we compare the performance of our model with and without the center-loss regularization term, and the results are presented in Table 5. Clearly, we see that the regularization leads to better results, although the improvement is small.

Table 5: Comparing the performance of HACNN method with and without center-loss regularization on the first split of UCF-101.

Regularization	Spatial (%)	Temporal (%)	Fusion (%)
No	87.0	88.0	94.1
Yes	87.0	88.5	94.3

Table 6: Comparison of our method with other state-of-the-art methods on UCF-101 and HMDB51 dataset.

Method	UCF-101 (%)	HMDB51 (%)
DT+MVSV [6]	83.5	55.9
IDT+FV [48]	85.9	57.2
IDT+HSV [38]	87.9	61.1
MoFAP [51]	88.3	61.7
Two Stream [31]	88.0	59.4
C3D (3 nets ensemble) [35]	85.2	—
Two Stream + LSTM [49]	88.6	—
FSTCN (SCI fusion) [52]	88.1	59.1
TDD+FV [33]	90.3	63.2
LRCN [28]	91.7	64.8
LTC [36]	92.7	67.2
KVMF [53]	93.1	63.3
Two Stream SR-CNN [32]	92.6	—
HRP [37]	90.7	—
ActionVlad (2 modalities) [50]	92.7	66.9
TSN (2 modalities) [25]	94.0	68.5
CNN-Fusion [54]	89.1	54.9
Conv-Fusion (2 modalities) [46]	92.5	67.3
ST-ResNet (2 modalities) [26]	93.4	66.4
HACNN (Ours) (2 modalities)	**94.6**	**71.9**

4.8. *Comparison with state-of-the-art methods*

After exploring the hierarchical aggregation approach on the two datasets and establishing the effectiveness of the approach, we summarize the results obtained by our method and compare it with several state-of-the-art methods. Table 6 presents a detailed analysis of various methods including traditional methods relying on handcrafted features such as iDT [10] and super vectors [6, 48]. We also compare with several deep learning-based methods such as the two stream architecture [31, 33], Recurrent neural networks [28], 3D convolutions [35], Trajectory-pooled deep convolutional descriptors (TDD) [33] and the more recently proposed temporal segment networks (TSN) [32] and spatio-temporal Resnets (ST-ResNet) [26]. As we can see from Table 6, deeply learned features outperformed handcrafted ones in most cases.

Notable recent works on action recognition that have improved the state-of-the-art results are: (i) ST-ResNet [26] method, which uses a very deep architecture combined with cross-stream residual connections; and (ii) TSN method [25], which uses a very deep architecture combined with a consensus function at the final layer of the network. From Table 6, we see that with the same two input modalities, HACNN outperforms ST-ResNet [26] by 1.2% on UCF-101 and 5.5% on HMDB51. Similarly, comparing with TSN method [25], we improve by 0.6% on UCF-101 and 3.4% on HMDB51. In order to confirm that the performance gain is mainly due to HACNN and not due to the additional frames used in our model, we tested the TSN with 32 frames and found no performance improvement. As explained before, we used the standard dense testing procedure to generate 10 samples for each input frame. We believe that the effect of number of frames reduces beyond a certain number. Other interesting comparisons are with the (1) two-stream LSTM [49], which attaches an LSTM to a two-stream Inception and (2) LRCN method [28], which combines CNN with an LSTM to model long-term temporal dynamics. Again, we observe that our method outperforms the former by a significant 6.0% on UCF-101 and the latter by 2.9% on UCF-101 and 7.1% on HMDB51. (3) Our method achieves much better results than more complicated aggregation approaches, such as ActionVlad [50].

This clearly demonstrates the effectiveness of the hierarchical aggregation strategy adopted in HACNN. Actions in HMDB dataset consist of complex wide ranging motions that are difficult to capture without explicit temporal modeling. The larger improvement on the HMDB dataset again strengthens our claim that HACNN is effective in modeling long-range dynamics. Note that we do not combine iDT features as done by several earlier methods [26, 46].

5. Conclusion

In this chapter, we presented a novel hierarchically aggregated deep ConvNets model to better capture long-term dynamics present in action videos. The proposed hierarchical model consists of special aggregation functions acting upon the output feature maps of a subConvNet at each level. The aggregation function has the effect of pooling most discriminative information from the neighboring segments. When repeated over multiple levels, this captures complex dynamics present in the video. We performed detailed experiments to investigate various aspects of the model such as the choice of different aggregation functions and their usage at various levels of the hierarchy. We observed that aggregation functions are useful at each level and that a spatial layout preserving function plays an important role in achieving superior performance. We also showed that the whole system is end-to-end trainable and achieves state-of-the-art performance on two challenging action recognition datasets.

References

1. Z. Chen, Q. Zhu, Y. C. Soh, and L. Zhang, "Robust human activity recognition using smartphone sensors via ct-pca and online SVM," *IEEE Trans. Indust. Inform.*, vol. 13, no. 6, pp. 3070–3080, 2017.
2. Z. Chen, L. Zhang, Z. Cao, and J. Guo, "Distilling the knowledge from hand-crafted features for human activity recognition," *IEEE Trans. Indust. Informa.*, vol. 14, no. 10, pp. 4334–4342, 2018.
3. Z. Chen, L. Zhang, C. Jiang, Z. Cao, and W. Cui, "Wifi CSI based passive human activity recognition using attention based BLSTM," *IEEE Trans. Mobile Comput.*, vol. 18, no. 11, pp. 2714–2724, 2018.
4. Z. Chen, C. Jiang, S. Xiang, J. Ding, M. Wu, and X. Li, "Smartphone sensor based human activity recognition using feature fusion and maximum full a posteriori," *IEEE Trans. Instrum. Measure.*, vol. 69. no. 7, pp. 3992–4001, 2019.
5. M. Brand and V. Kettnaker, "Discovery and segmentation of activities in video". *IEEE Trans. Pattern Analy. Mach. Intell.*, vol. 22, Aug. 2000.
6. Z. Cai, L. Wang, X. Peng, and Y. Qiao, "Multi-view super vector for action recognition," *IEEE Conf. Computer Vision and Pattern Recognition*, pp. 596–603, 2014. vol. 69, no. 7, pp. 3992–4001.
7. A. A. Efros, A. C. Berg, G. Mori, and J. Malik, "Recognizing action at a distance," *IEEE Int. Conf. Computer Vision*, Nice, France, October 2003, October 2003.
8. A. F. Bobick and J. W. Davis, "The recognition of human movement using temporal templates," *IEEE Trans. Pattern Analy. Mach. Intell.*, vol. 23, no. 3, pp. 257–267, 2001.
9. I. Laptev and T Lindeberg, "Space-time interest points," *Int. J. Comput. Vis.*, vol. 64, no. (2–3), 107–123, Sept. 2005.

10. H. Wang, A. Klaser, C. Schmid, and C.-L. Liu, "Action Recognition by Dense Trajectories," in *CVPR*, 2011.

11. M. Bruni, T. Uricchio, L. Seidenari, and A. Del Bimbo, "Do textual descriptions help action recognition?" *Proc. 2016 ACM on Multimedia Conference*, ACM, pp. 645–649, 2016.

12. J. C. Niebles, H. Wang, and L. Fei-Fei, "Unsupervised learning of human action categories using spatial-temporal words," *Int. J. Comput. Vis.*, vol. 79, no. 3, pp. 299–318, 2008.

13. M. Brand, N. Oliver, and A. Pentland, "Coupled hidden markov models for complex action recognition," *IEEE Conf. Computer Vision and Pattern Recognition*, 1997.

14. P. Turaga, R. Chellappa, V. S. Subrahmanian, and O. Udrea, "Machine recognition of human activities: A survey," *IEEE Transactions on Circuits and Systems for Video Technology*, 2008.

15. Y. Tian, Q. Ruan, G. An, and Y. Fu, "Action recognition using local consistent group sparse coding with spatio-temporal structure," *Proc. 2016 ACM on Multimedia Conference*, ACM, pp. 317–321, 2016.

16. K. Soomro, A. R. Zamir, and M. Shah, "Ucf101: A dataset of 101 human actions classes from videos in the wild." Crcv-tr-12-01, UCF, 2012.

17. E. G. T. P. H. Kuehne, H. Jhuang, and T. Serre, "HMDB: A large video database for human motion recognition," *ICCV*, 2011.

18. I. S. A. Krizhevsky and G. E. Hinton, "Imagenet classification with deep convolutional neural networks," *NIPS*, 2012.

19. C. Szegedy, W. Liu, Y. Jia, P. Sermanet, S. Reed, D. Anguelov, D. Erhan, V. Vanhoucke, and A. Rabinovich, "Going deeper with convolutions," *CVPR*, 2015.

20. K. Simonyan and A. Zisserman, "Very deep convolutional networks for large-scale image recognition," *ICLR*, 2014.

21. K. He, X. Zhang, S. Ren, and J. Sun, "Deep residual learning for image recognition," 2015, arXivpreprint arXiv:1512.03385.

22. J. Long, E. Shelhamer, and T. Darrell, "Fully convolutional networks for semantic segmentation," *IEEE Conf. Computer Vision and Pattern Recognition*, pp. 3431–3440, 2015.

23. A. J. Y. L. J. Tompson, R. Goroshin and C. Bregler, "Efficient object localization using convolutional networks," *CVPR*, 2015.

24. D. K. F. Schroff and J. Philbin, "Facenet: A uni-fied embedding for face recognition and clustering," *CVPR*, 2015.

25. L. Wang, Y. Xiong, Z. Wang, Y. Qiao, D. Lin, X. Tang, and L. Van Gool, "Temporal segment networks: Towards good practices for deep action recognition," *European Conf. Comput. Vision*, pp. 20–36. Springer, 2016.

26. C. Feichtenhofer, A. Pinz, and R. P. Wildes, "Spatiotemporal residual networks for video action recognition," *NIPS*, 2016.

27. S. Ioffe and C. Szegedy, "Batch Normalization: Accelerating deep network training by reducing internal covariate shift," 2015, arXiv preprint arXiv:1502.03167.
28. J. Donahue, L. A. Hendricks, S. Guadarrama, M. Rohrbach, S. Venugopalan, K. Saenko, and T. Darrell, "Long-term recurrent convolutional networks for visual recognition and description," *CVPR*, 2015.
29. S. Ji, W. Xu, M. Yang, and K. Yu, "3D convolutional neural networks for human action recognition," *IEEE Trans. Pattern Anal. Mach. Intell.*, vol. 35, no. 1, pp. 221–231, 2013.
30. A. Karpathy, G. Toderici, S. Shetty, T. Leung, R. Sukthankar, and L. Fei- Fei, "Large-scale video classification with convolutional neural networks," *CVPR*, 2014.
31. K. Simonyan and A. Zisserman, "Two-stream convolutional networks for action recognition in videos," *NIPS*, 2014.
32. Y. Wang, J. Song, L. Wang, L. Van Gool, and O. Hilliges, "Two-stream SR-CNNs for action recognition in videos," *BMVC*, 2016.
33. L. Wang, Y. Qiao, and X. Tang, "Action recognition with trajectory-pooled deep-convolutional descriptors," *CVPR*, pp. 4305–4314, 2015.
34. B. Zhang, L. Wang, Z. Wang, Y. Qiao, and H. Wang, "Real-time action recognition with enhanced motion vector cnns," *CVPR*, pp. 2718–2726, 2016.
35. D. Tran, L. Bourdev, R. Fergus, L. Torresani, and M. Paluri, "Learning spatiotemporal features with 3D convolutional networks," *ICCV*, 2015.
36. G. Varol, I. Laptev, and C. Schmid, "Long-term temporal convolutions for action recognition," 2016, arXiv:1604.04494.
37. B. Fernando, P Anderson, M. Hutter, and S. Gould, "Discriminative hierarchical rank pooling for activity recognition," *CVPR*, pp. 1924–1932, 2016.
38. X. Peng, L. Wang, X. Wang, and Y. Qiao, "Bag of visual words and fusion methods for action recognition: Comprehensive study and good practice," *Computer Vis. Image Understand*, vol. 150, pp. 109–125, 2016.
39. J. Sanchez, F. Perronnin, T. Mensink, and J. Verbeek, "Image classification with the fisher vector: Theory and practice," *Int. J. comput. Vis.*, vol. 105, no. 3, pp. 222–245, 2013.
40. T.-Y. Lin, A. RoyChowdhury, and S. Maji, "Bilinear cnn models for fine-grained visual recognition," *Proc. IEEE Int. Conf. Computer Vision*, pp. 1449–1457, 2015.
41. Y. Wen, K. Zhang, Z. Li, and Y. Qiao, "A discriminative feature learning approach for deep face recognition," *ECCV*, 2016.
42. C.-Y. Lee, S. Xie, P W. Gallagher, Z. Zhang, and Z. Tu, "Deeply-supervised nets," *AISTATS*, vol. 2, p. 5, 2015.
43. L. Wang, Y. Xiong, Z. Wang, and Y. Qiao, "Towards good practices for very deep two-stream convnets," 2015, arXivpreprint arXiv:1507.02159.
44. C. Zach, T. Pock, and H. Bischof, "A duality based approach for realtime tv-l 1 optical flow," *Joint Pattern Recognition Symposium*, pp. 214–223. Springer, 2007.

45. Y. Jia, E. Shelhamer, J. Donahue, S. Karayev, J. Long, R. Girshick, S. Guadarrama, and T. Darrell, "Caffe: Convolutional architecture for fast feature embedding," 2014, arXiv preprint arXiv:1408.5093.

46. C. Feichtenhofer, A. Pinz, and A. Zisserman, "Convolutional two-stream network fusion for video action recognition," *CVPR*, 2016.

47. A. Krizhevsky, I. Sutskever, and G. E. Hinton, "Imagenet classification with deep convolutional neural networks," *Advances in Neural Information Processing Systems*, pp. 1097–1105, 2012.

48. H. Wang and C. Schmid, "LEAR-INRIA submission for the THUMOS workshop," *ICCV Workshop on Action Recognition with a Large Number of Classes*, vol. 2, p. 8, 2013.

49. J. Yue-Hei Ng, M. Hausknecht, S. Vijayanarasimhan, O. Vinyals, R. Monga, and G. Toderici, "Beyond short snippets: Deep networks for video classification," *IEEE Conf. Comput. Vision and Pattern Recognition*, pp. 4694–4702, 2015.

50. R. Girdhar, D. Ramanan, A. Gupta, J. Sivic and B. Russell, "ActionVLAD: Learning Spatio-Temporal Aggregation for Action Classification," *2017 IEEE Conference on Computer Vision and Pattern Recognition (CVPR)*, Honolulu, HI, 2017, pp. 3165–3174, doi: 10.1109/CVPR.2017.337.

51. L. Wang, Y. Qiao, and X. Tang, "MoFAP: A multi-level representation for action recognition," *Int. J. Comput. Vision*, vol. 119, no. 3, pp. 254–271, 2016.

52. L. Sun, K. Jia, D.-Y. Yeung, and B. E. Shi, "Human action recognition using factorized spatio-temporal convolutional networks," *IEEE Int. Conf. Comput. Vision*, pp. 4597–4605, 2015.

53. W. Zhu, J. Hu, G. Sun, X. Cao, and Y. Qiao, "A key volume mining deep framework for action recognition," *IEEE Conf. Comput. Vision and Pattern Recognition*, pp. 1991–1999, 2016.

54. E. Park, X. Han, T. L. Berg, and A. C. Berg, "Combining multiple sources of knowledge in deep CNNs for action recognition," *WACV*, 2016.

55. A. Hernandez Ruiz, L. Porzi, S. Rota Bulo, and F. Moreno-Noguer, "3D CNNS on distance matrices for human action recognition," *Proc. 2017 ACM on Multimedia Conference*, ACM, pp. 1087–1095, 2017.

56. P. Wang, Z. Li, Y. Hou, and W. Li, "Action recognition based on joint trajectory maps using convolutional neural networks," *Proceedings of the 2016 ACM on Multimedia Conference*, pp. 102–106. ACM, 2016.

Chapter 3

Combining Domain Knowledge and Deep Learning to Improve HAR Models

Massinissa Hamidi* and Aomar Osmani[†]

LIPN-UMR CNRS 7030, Univ. Sorbonne Paris Nord
*hamidi@lipn.univ-paris13.fr
[†]ao@lipn.univ-paris13.fr

Abstract

In this chapter, we propose to combine domain knowledge and deep learning in order to improve human activity recognition (HAR) models. We explore empirically two fundamental questions: (1) which knowledge to incorporate into HAR models and (2) how to incorporate it? We focus on the knowledge about the topological structure of on-body sensor-deployments and the mutual interactions that emerge among these sensors. We report on experiments conducted on the Sussex Huawei locomotion-transportation (SHL) dataset which features a structured data generation process. Obtained results open up perspectives for the development of deep learning-based HAR models which leverage domain knowledge in order to improve both robustness and data-efflciency.

Keywords: human activity recognition, domain knowledge, deep learning, neural architecture search.

1. Introduction

The ever-increasing quantities of data sources featured by internet of things applications bring diverse and rich perspectives about monitored phenomena [1]. This is the case, for example, of applications continuously monitoring human activities from wearable sensor-deployments for diverse purposes like eHealth. While the profusion of data, both in terms of modalities and spatial perspectives, could provide many advantages like improved signal-to-noise ratio, reduced ambiguity and uncertainty, enhanced robustness and reliability, etc. [2,3], the way current machine learning processes relate information from multiple sources is not efficient [4]. The reason is that the data coming from these data sources are flattened: potentially valuable clues, related to the structure of both the sensors deployment and that of the monitored phenomena, are lost while processing these data. Even principled approaches, such as deep learning which are yet known to learn to bridge modalities and to leverage these kinds of structures [5–8], fail in practice when faced with such issues and need to be guided further to work properly [4, 9, 10].

This is the case of wearable technologies with the considered Sussex–Huawei locomotion–transportation (SHL) dataset [11] studied in this chapter. Our work focuses on recognizing mobility-related human activities from data sources materialized by on-body sensors placed at different locations of the body following a predefined and fixed topology. One way to guide these kinds of models is by providing, explicitly, additional knowledge about, e.g., the structure of the sensors deployments, the dynamics of human activities, physical models of the body movements, etc., in short, domain knowledge. These additional models have the potential to make learning models concentrate on subsets of data sources that are, for example, known, *a priori,* to be more informative, about a given human activity, than others or subsets that exhibit appropriate spatial dispositions to reliably discriminate human activities.

The main contribution of this work is to improve deep learning-based human activity recognition (HAR) models by incorporating domain knowledge to guide the learning of human activities in the context of sensor-rich environments. The first step is to derive a data generation model that captures the dynamics of the body movements pertaining to each human activity. As the dynamics of body movements are often driven by large and intricate low-level interactions involving various body parts, the derivation of the data generation model is framed as an exploration of the space of neural architectures. These architectures are composed of sensor fusion and feature extraction components controlled by sets of hyperparameters [12]. The idea has apparent complications, as we by-pass human expertise, but was encouraged by the interpretability allowed by linking the impact of each data source with the set of hyperparameters that controls its effects on the performances of the architecture. We instantiate this formulation using neural architecture search (NAS) techniques [13], namely, convolution building blocks for sensor fusion and features extraction, Bayesian optimization (BO) for surface response reconstruction [14], and an efficient implementation of the functional analysis of variance (fANOVA) to diagnose the resulting high-dimensional function of dependent variables [15]. The derived data generation model is used to constrain the training of simpler HAR models.

We report on experiments conducted on the SHL dataset featuring a sensor-rich environment in real-life settings. We empirically show that the derived data generation model is consistent with empirical results in the literature. Furthermore, the effectiveness of the obtained domain models is demonstrated via a setting that exploits subsets of the most important and interacting data sources, extracted from the data generation model, in order to constrain the learning process of a neural network. The proposed setting showed substantial improvements over the baseline where all data sources of the sensor-rich environment are used. In particular, we get an improvement of the recognition performances, measured by the f1-score, of up to 17.84% over the baseline and this using solely 12 data sources among those available.

This chapter is organized as follows. First, we introduce in Section 2 the SHL dataset which features a sensor-rich environment. Then, we present in Section 3 the general formulation for deriving the data generation model followed by the proposed instantiation using NAS techniques in Section 4. Section 5 details how the derived data generation model is used to constrain the learning of human activities followed by some discussions in Section 6.

2. Sensor-Rich Environments

Using multiple motion sensor modalities is probably one way to go in order to get more robust models and to deal with ambiguity when deriving high-level context [16]. Multi-sensor fusion or the use of multiple positions can disambiguate or reinforce the situational context that is inferred as opposed to relying solely on one sensor. The use of multiple sensors is, indeed, made possible by the characteristics of the sensors which are becoming more and more lightweight, small-sized, cheap and oriented towards small energy

Table 1: Some representative related datasets of human activities.

Dataset/Study	Multi-modality	Multi-location	Modality	Location
GeoLife [19]	✗	✗	*GPS*	—
USC [20]	✓	✗	*acc, gyr, mag, GPS, WiFi, ECG, GSR,* etc.	*hip*
HTC-TMD [17]	✓	✗	*acc, gyr, mag, GPS, WiFi*	—
US-TMD [21]	✓	✗	*acc, gyr, sound*	—
SHL [18]	✓	✓	*acc, gyr, mag, lacc, ori, gra,* etc.	*hand, hips, bag, torso*

Notes: Comparison based on the availability of multiple modalities in multiple locations simultaneously. Motion-based modalities which are referred to with the following abbreviations: *acc* (accelerometer), *gyr* (gyroscope), *mag* (magnetometer), *lac* (linear accelerometer), *gra* (gravity), *ori* (orientation) and *pre* (pressure). In addition, *GPS* refers to global positioning system, *GSR* to galvanic skin response and *ECG* to electrocardiogram.

footprint. Paradoxically, rather little effort has been put into this aspect and current approaches in mobility-related activity recognition continue to rely solely on a small subset of sensor modalities. The current situation is partly because these studies take into account the computational and energy footprint of the approaches they develop. For example, in the challenge organized around the HTC transportation modes dataset [17], in addition to the recognition performances, ranking took into account the energy footprint of the proposed solutions. To that matter, the SHL dataset [18] circumvent such limitation by providing data from built-in sensors of 4 smartphones, each positioned in a different predefined body location. Table 1 summarizes the main characteristics of some representative related datasets of mobility-related human activities.

2.1 *Sussex–Huawei locomotion–transportation dataset*

The SHL dataset [18][a] is a highly versatile and precisely annotated dataset aiming to overcome the lack of such datasets dedicated to mobility-related activity recognition. The total amount of collected data corresponds to 2812 h of labeled data and 17,562 km of traveled distance.

In contrast to related representative datasets like [19], which includes solely global positioning system information, the SHL dataset contains

---

[a]The preview of the SHL dataset can be downloaded from http://www.shl-dataset.org/download/.

multi-modal locomotion data recorded in real-life settings. There are in total 16 modalities including accelerometer, gyroscope, cellular networks, WiFi networks, audio, etc. making it suitable for a wide range of applications and in particular the task we are interested in which is the recognition of mobility-related human activities (transportation modes). Indeed, there are eight primary categories of transportation that we are interested in: *Still, Walk, Run, Bike, Car, Bus, Train, Subway (Tube).*

The locomotion data were recorded by three participants, referred to in the dataset as *User 1, User 2* and *User 3*, involved full-time. Data collection was performed by each participant using four smartphones simultaneously placed in different body locations where people are commonly carrying phones, *Hand, Torso, Hips,* and *Bag* (see Fig. 1). Featuring a broad range of perspectives for the locomotion modes constitutes another strength that is credited to the SHL dataset in contrast to related datasets like [17]. These four positions define the topology of the underlying infrastructure that we study in this chapter. Among the 16 modalities of the original dataset, we select the body-motion modalities to be included in our experiments, namely: accelerometer, gyroscope, magnetometer, linear acceleration, orientation, gravity, and in addition, ambient pressure.

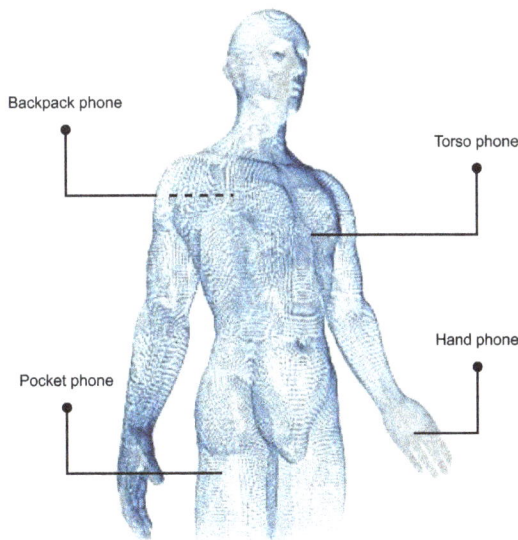

Figure 1: Schematic representation of the sensor-rich environment's topology induced by the locations of the on-body sensors (data sources).

3. Data Generation Model

In this section, we present the details around the derivation of the most important and interacting data sources. We start by introducing the notation, and formulate the problem of exhibiting a model for the data generation process in sensor-rich environments. We provide some challenges that led to the definition of our approach.

In this chapter, we consider settings where a collection S of M sensors (also called data generators or data sources), denoted $\{s_1,\ldots, s_M\}$, are carried by the user during daily activities and capture the body movements. Each sensor s_i generates a stream (or sequence) $\mathbf{x}^i = (x_1^i, x_2^i, \ldots)$ of observations of a given body-motion modality. Each observation is composed of a given number of channels, e.g., x, y and z axes of an accelerometer.

We recall that, in this chapter, what we consider as a modality is a form of perception that conveys a particular perspective about a given phenomenon. In this matter, we exploit mainly body-motion modalities that are often used in human activity recognition applications. For example, acceleration, gyroscopic and magnetometric observations are different modalities each describing, in a particular way, the motions of the body. A given data source is, then, characterized by two main attributes: the first is the *modality* being produced by the sensor and the second one is the *position* where the data source is located on the body. A data source is then uniquely defined with these two attributes.

3.1 *Modeling the data generation process*

We are interested in deriving a model for the data generation process in order to exploit this knowledge to build more robust and data-efficient HAR learning systems. The main assumption is that the recognition of human activity from wearables is a profoundly structured issue. Indeed, numerous works such as [22–24], and more recently [25], where authors consider multimodal and multilocation data sources, provide evidence of the large influence of subsets of data sources and specific interactions that emerge towards the recognition of particular human activities. Primarily, what characterizes these dynamics is the fact that they define precisely the activity in question. From this observation follows the concept of subsets of data sources, the main characteristic of which is that they allow a robust recognition of human activities. In this work, we model the data generation process by deriving the subsets of data sources that exhibit important interactions and substantial influence on the performances of the learning models. For this, we formalize the notions of *importance* of and *interaction* between data sources.

> *Importance.* The importance of a data source is defined as a quantity that corresponds to the level of its influence on the performances of a learning model. We denote the importance of a given data source s_i by $\mu_i \in [0,1]$.

> *Interaction.* An interaction, on the other hand, involves two or more data sources and is defined as their degree of dependence regarding their influence on the performance of a learning model. Given a set of interacting data sources, I, their degree of interaction is denoted by μ_I, and specifically, for two interacting data sources, s_i and s_j, it is denoted by $\mu\langle s_i, s_j \rangle$, or simply $\mu_{i,j}$.

In addition, we define two parameters, τ_{imp} and τ_{int}, that correspond to the importance and interaction thresholds, respectively. These two parameters determine limits above which a given data source, with a given importance, or a set of interacting data sources, can be included in a subset of data sources. It follows that the subsets of interacting data sources, denoted χ, is defined as

$$\chi := \{s_i \in \mathcal{S} | \mu_{s_i} \geq \tau_{\text{imp}}\} \cup \{I \subsetneq \mathcal{S} | \mu_I \geq \tau_{\text{int}}\}. \tag{1}$$

3.2 *Exploration of the architecture space*

As mentioned above, the goal is to find subsets of data sources and specific interactions that emerge towards the recognition of particular human activities. In order to obtain such subsets, we have to be able to leverage information contained *within* and *across* data sources using efficient features extraction and sensor fusion schemes.

This requires modeling the dynamics, that involve the subsets of data sources, directly from the body movements of each human activity. This is a rather complex task which, beyond the huge amount of heavy experiments that have to be conducted in real settings, requires human expertise which is known to be scarce and limited in terms of the insights it may convey. Thus, rather than relying on a direct modeling of these dynamics, we recast this problem into an analysis of the behavior of the architectures which are responsible for modeling these dynamics through feature extraction and sensor fusion schemes. In other words, the problem is reframed as an exploration of the architecture space and for which adequate tools are available to solve it.

(Neural) architecture. An architecture is defined as a set of components, e.g., analysis unit, feature fusion unit, decision fusion unit, etc. [12], responsible for extracting valuable insights, in the form of features, from the observations and efficiently fusing different data sources carrying different modalities and various spatial perspectives. Architectures are parameterized by a set $\mathcal{H}$ of N hyperparameters, $\{h_1, ..., h_N\}$, controlling the effects of the various components, and eventually impacting the architecture performance, denoted v.

We focus, particularly, on the insights that stem from tuning and adapting these architectures, through their hyperparameters and those controlling specifically the influence of the data sources. At each layer of a given architecture, setting the right combination of hyperparameters is critical. In particular, choosing the right instantiation for the features learning and sensor fusion stages can lead to an architecture capable of building, from the various data sources, an original set of features which is suitable for recognizing a given activity.

Assumption. Let $\mathcal{H}_S$ ($\mathcal{H}$ be the set of hyperparameters controlling the impact of a given data source s. The impact of the data source s is represented by the global impact of $\mathcal{H}_S$ on the recognition performances. Framed differently, the importance of the data source s as part of the dynamics of body movements, pertaining to a given human activity, is represented by the global effect of $\mathcal{H}_S$ on the recognition performances.

4. NAS-based Derivation of the Data Generation Model

In order to derive a model for the data generation process underlying the sensorrich environment, we describe here an original instantiation of the previously stated re-formulation. The re-formulation we proposed in the last section leads us to exhibit three main challenges.

First, the various sensing modalities featured by sensor-rich environments carry information with various perspectives. However, building appropriate features extraction and sensor fusion schemes is not trivial. So, how can we leverage information contained within and across data sources when we know limits of human expertise in terms of features extraction and sensor fusion schemes?

Neural networks hold important properties which are advantageous to multimodal recognition tasks. They are able to construct, or learn, hierarchies of abstract features and relate efficiently modalities between them. In order to replicate such capabilities within our architectures, we use convolutional neural networks.

Second, modeling of the data generation processes being framed as an exploration of an induced architecture space, how can one explore such, potentially, very large spaces? Stated differently, the exploration of the whole space being infeasible, how can we get a fairly precise picture of it given a constrained exploration budget?

> Recent advances in neural architecture search demonstrated noticeable successes in many fields leading in certain cases to state-of-the-art architectures using Bayesian optimization, in particular [14]. We propose, in this instantiation, to tune hyperparameters via Bayesian optimization. A given configuration of the hyperparameters is modeled as a sample from a Gaussian process, referred to as the response surface. Given a constrained budget, Bayesian optimization offers a good trade-off between exploration and exploitation.

And third, how can we quantify the individual and global impact of the data sources, based on the exploration results, in order to form subsets that satisfy the hypothesis we formulated previously?

> Functional analysis of variance (fANOVA) models have emerged as a prominent class of models useful to capture nonlinear relations in the data [26]. The principle behind fANOVA models is to approximate the underlying multivariate functional relation by an additive expansion. We use this principle, in conjunction with the assumption that we stated earlier, in order to quantify the individual and global impact of the data sources.

The main idea is, then, to use architectures based on neural networks in order to overcome the aforementioned limits of human expertise and to come-up with genuine features extraction and sensor fusion schemes. Exploration of the induced space is then recast as a problem that can be treated with NAS techniques. Figure 2 shows a schematic representation of the proposed instantiation.

These are the three aspects we will be discussing in the following sections. We describe the convolutional modes and the resulting neural architectures space in Section 4.1. We discuss how we explore the neural architectures space in Section 4.3 and how we come up with the global impact of data sources using the functional analysis of variance in Section 4.4.

4.1 *Convolutional-based architecture components*

Beyond the frequent application of convolutional neural networks for the recognition of human activities, which show, by the way, good performances,

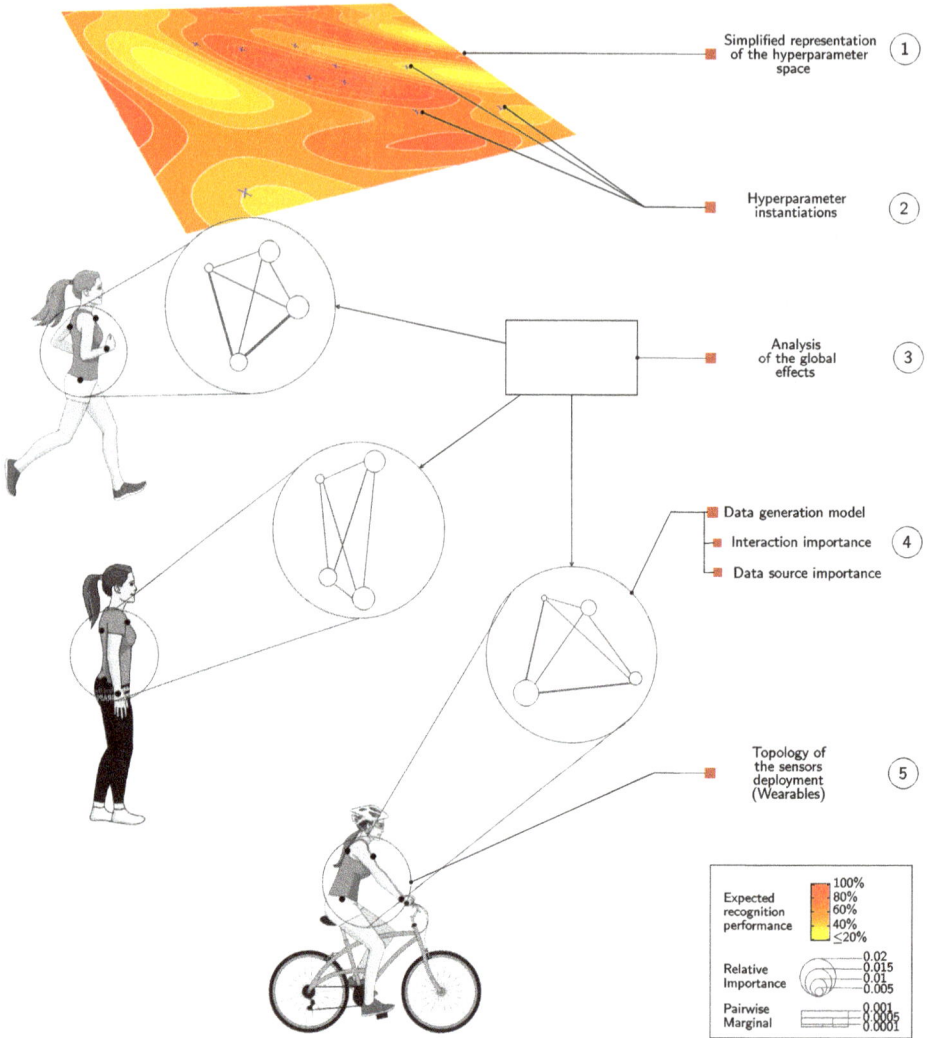

Figure 2:   Schematic representation of the proposed instantiation. Deriving the data generation model that captures the dynamics of body movements pertaining to each human activity is framed as an exploration of the neural architecture space. The goal of using neural architecture search techniques, in our case, is not to find a golden architecture, but rather to explore the space of architectures, to get insights regarding the hyperparameters that are the most influential w.r.t. each human activity. Precisely, we obtain subsets of data sources that are found to be relevant both in terms of importance and interaction to discriminate a given human activity. Here we depict an example of the relative importance of each position and the level of interactions between modalities for the recognition of running, staying still and bicycling.

e.g., [2, 8, 25, 27, 28], these kinds of networks are being adopted, primarily, for their ability to efficiently aggregate heterogeneous data from different sources. In Ref. [27] for example, authors proposed various convolutional architectures featuring an explicit mechanism for partial and full-weight sharing, by placing separate convolution kernels on each modality and in the upper layers responsible for aggregating features maps.

In this work, we construct architectures by stacking Conv/ReLU/MaxPool blocks. These blocks are followed by Fully Connected/ReLU layers. In order to allow for the emergence of cross-modal relationships at both low and high-levels of abstractions, we define three *convolutional modes* of the input sequences with each set of filters of the first layer (input layer): (1) whole modalities grouped and convolved, which we refer to as *grouped modalities*, (2) each modality convolved apart, which is designated by *split modalities*, and (3) each channel convolved apart, referred to as *split channels*.

These various convolutional modes can be considered as different levels of sensor fusion. From this perspective, *split channels* would correspond to a late fusion and, on the contrary, *grouped modalities* would correspond to an early fusion scheme.

4.2 *Architecture training*

In order to train a given architecture, we frame recognizing human activities as a sequence classification problem, where the goal is to learn a function $f \in \mathcal{F} : X \rightarrow Y$ mapping inputs to outputs. In order to simplify the given sequences of observations $x^i|_{i \in \{1...M\}}$, (defined above), we thus introduce a segmentation step, which will be responsible for decomposing the stream of observations, x^i, into sequences of fixed size, and optionally overlapping. Each sequence of observations, $x_j \in X$, is assigned with a label, $y_j \in Y = \{1,..., A\}$, corresponding to the human activity being performed. As in the traditional classification setting, performance of the neural architecture is quantified with a loss function $l : X \times Y \rightarrow \mathbb{R}$, and a mapping is found via

$$f* = \underset{f \in \mathcal{F}}{\arg\min} \sum_{j=1}^{n} l(f(x_j), y_j), \qquad (2)$$

which can be optimized using a gradient descent algorithm over a predefined class of functions $\mathcal{F}$. In our case, $\mathcal{F}$ will be convolutional neural networks parameterized by their weights and the loss function will be $l(f(x_j), y_j) = \mathbb{1}\{f(x_j) \neq y_j\}$. For a fixed architecture, i.e., a particular instantiation of the hyperparameters, the optimization process will tune the weights of the network and, by the same occasion, the subsequent uni-modal and multi-modal features that are extracted from the input signals.

But beyond the optimization of the parameters of an architecture, one of the most important aspects, which defines to an extremely great extent its potential performances, is the set of hyperparameters that determines and forges the shape of the architecture. In the case of convolutional neural networks, the set of hyperparameters includes the filters of the different layers, both in terms of size and number, the strides, and so on.

At each layer of a given architecture, setting the right combination of hyperparameters is critical. In particular, setting the right instantiation for the features learning and sensor fusion stages can lead to an architecture capable of building an original set of features from the various data sources which are suitable for recognizing a given activity. This is particularly appealing in the sense that by varying (particular sets of) hyperparameters (independently or not), we can discover some interesting insights regarding the interactions and phenomena that emerge in these high dimensional spaces. Moreover, recently, the field of neural networks has seen a resurgence of approaches focusing on this aspect, i.e., architecture engineering, hyperparameters tuning, etc. Where the performance of a model is largely determined by the architecture. These approaches are grouped under the term of NAS.

The combination of the large-scale optimization of the hyperparameters, provided by NAS techniques, and the previously introduced convolutional modes defines what we call the *neural architecture space*. In the following, we present how we explore this space using BO which offers a good trade-off between exploration and exploitation.

4.3 *Exploration of the neural architecture space*

In this work, the performance of an architecture is denoted v_k and is obtained using a performance metric such as the f1-score or accuracy. We also exploit the partial performances obtained during the training of the architectures for each of the human activities considered. These are denoted v_k^a, where k refers to the configuration and a to the activity.

Models development and BO process are based on off-the-shelf implementations, all of which are free software. In particular, we use TensorFlow [29] for building the neural architectures, and the scikit-optimize library [30] specialized in optimizing cost functions (Table 2 provides the list of optimized hyperparameters). Inputs are segmented into sequences of 6,000 samples which correspond to a duration of 1 min. given a sampling rate of 100

Table 2: Summary of the different hyperparameters assessed during Bayesian optimization process along with their respective bounds.

Hyperparam. (sym.)	Low	High	Prior
Kernel size **1**ˢᵗ layer (ks_m^1)	9	15	—
Kernel size **2**ⁿᵈ layer (ks_m^2)	9	15	—
Kernel size **3**ʳᵈ layer (ks_m^3)	9	12	—
Number of filters (nf_m)	16	28	—
Stride (s_m)	0.5	0.6	log
Learning rate (lr)	0.001	0.1	log
Dropout probability (p_d)	0.1	0.5	log
Number of units dense layer (n_u)	64	2048	—

Note: m corresponds to the data source.

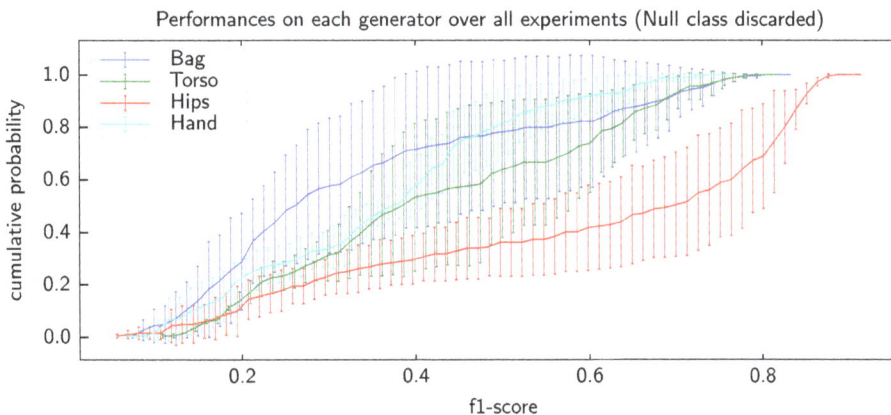

Figure 3: Cumulative distribution of the performances of the models trained on each body-position. Bayesian optimization results averaged across users and positions.

Hz. We use batch normalization on top of each convolutional layer. This procedure makes the neural networks more stable by normalizing the inputs of the different layers [31]. Each hyperparameter instantiation is trained for a maximum of 52 epochs and a minimum of 12 epochs. After 13 epochs, if there is no improvement in the recognition performances over the best score for eight subsequent epochs, we stop training.

In this part, we want to delve further into the exploration of the neural architectures space and, specifically, the response surface of the BO process. We assess this aspect with respect to each position of the sensors-deployment.

Recognition performances of the different architectures are presented in Fig. 3 in the form of cumulative distributions. These compare in particular the robustness of each position towards the overall recognition task and are, in addition, intended to provide a sense of the underlying response surface induced by the various configurations. Results are grouped based on the respective position of each data source.

Results exhibit, in particular, a noticeable variability concerning the hyperparameter space induced by the data sources located in the bag and on the torso. Indeed, the number of architectures yielding, for example, a given recognition performance varies to a critical large extent (±0.8 standard deviation points) and underlines how complex the subspaces generated by these configurations are. In other words, the generated subspaces can be viewed as spanning the entire recognition performances range and encompassing many plateaus of variable sizes scattered all over it, each of which yield equivalent performing architectures. This contrasts with architectures trained with data generators located in the hand which rather show clearly a more stable behavior with a variability that is confined to ±0.2 standard deviation points all over the recognition performances range.

On the other hand, data sources located in the hips give the best architectures and achieve more than 90% fl-score. This corroborates findings in Refs. [22,23] on the suitability of the hip location for recognizing different kinds of human activities. Surprisingly, although achieving the best recognition performances, these exhibit a large spread in terms of performances which range from 0.05 to 0.92, suggesting that the induced response surface is characterized by a certain complexity contrasting with the relatively constrained architectures' configurations that we set up.

Beyond the observed variability and large spread of recognition performances, the fundamental aspect that these results outline is the ability of the joint exploitation of convolutional modes and neural architecture search techniques to yield such complex phenomena. Indeed, it would be difficult to derive models describing the dynamics underlying the body movements based on handcrafted configurations and domain expertise solely.

4.4 *Global effects of data sources*

Here, we describe how we go from individual importance of hyperparameters, measured by the recognition performance of the corresponding architecture, to the notion of importance of a data source and the potential interactions that emerge among them. Functional ANOVA models have emerged as a prominent class of models useful to capture nonlinear relations in the data [26].

The principle behind functional ANOVA models is to approximate the underlying multivariate functional relation by an additive expansion. Given the set of data sources S with their corresponding subsets, $\mathcal{H}_s$, of the hyperparameters $\mathcal{H}$, functional ANOVA decomposes the obtained performance data for the neural architecture space $\hat{y} : \Theta = \theta_1 \times \ldots \times \theta_n \rightarrow \mathbb{R}$, with θ_i denoting the domain of the hyperparameter h_i, into how much each data source and each combination of data sources contribute to the variance of the neural architecture space [15].

Here, we evaluate the ability of the proposed approach to derive an effective model of the data generation process that underlies the considered sensor-rich environment. We conducted a large-scale analysis of interactions through the proposed approach as well as an extended literature review around the recognition of human activities from various modalities and spatial perspectives. We make use of an efficient implementation of the fANOVA framework proposed by Ref. [15] which is based on random forests.

Figure 4 shows how data sources, grouped by their respective positions, contribute to the overall recognition of some activities (running, biking, and staying still). Figure 5 summarizes results of the hyperparameters importance estimation conducted using the fANOVA framework. The estimated first- and second-order effects of the hyperparameters controlling the importance of each considered modality are illustrated, respectively.

Overall, the contributions of data sources for recognizing bus, train and subway-related activities are equivalent. More variability appears in the case of the bus activity. Data sources located on the hips, for their part, yield overall the smallest variability. This variability is to some extent more important in the case of bus and run activities, but stays in fairly acceptable terms. In the case of car-related activities, relying on the data sources located on the hips seems to be sufficient, this position yielding the best models overall (90%–95% fl-score). The same observation can also be made regarding bike and walk activities where hips data sources seem to discriminate them accurately. This may be explained by the tight link that exists between these activities and the hips position: biking, walking and conducting a car involve specific repetitive patterns that are their hallmark [21].

From the modalities' perspective, data sources carrying gravity, gyroscope and magnetometer account for a large part of the variability that is observed on the recognition performances. Surprisingly, another set of modalities emerges from the derived model rather than the accelerometric data which is considered to be one of the most important modalities in representative related work [24, 32]. Globally, the respective individual marginal

M. Hamidi & A. Osmani

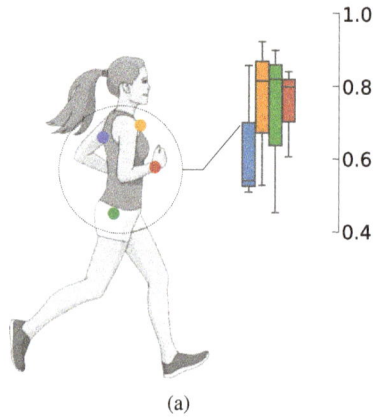

(a)

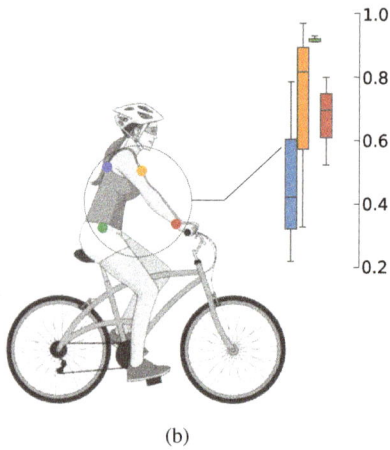

(b)

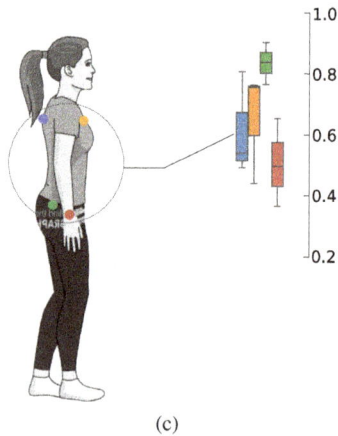

(c)

Figure 4: Contribution of the data sources to the overall recognition performances of each locomotion mode. Data sources are grouped by their respective positions. The analysis combines the entire configurations obtained via the Bayesian optimization runs.

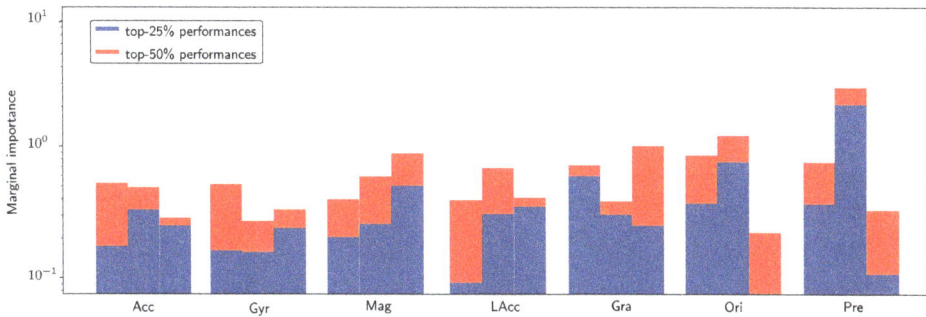

Figure 5: Individual marginal importance of the kernel size hyperparameters controlling the impact of each modality. Comparison between top-25% and top-50% performing architectures.

importance of the accelerometer-related data lies approximately around 0.004 and does not exceed 0.006, while those of gravity, gyroscope, as well as magnetometer reach 0.01 and almost 0.02. This observation is further confirmed when we analyze the pairwise marginals of the hyperparameters controlling the set of three modalities mentioned above. In the same vein, when comparing the impact of hyperparameters on the top-25% and top-50% performing architectures (see Fig. 5), we notice again that these same modalities have a more pronounced impact and undergo, as well, a substantial evolution. Note also that, in contrast to the global observations made earlier, the more we focus on the top-performing architectures, the impact gets spread over all hyperparameters. In particular, we observe the emergence of the hyperparameters controlling the effect of pressure.

If we focus on the computed marginals (interactions) between pairs of hyperparameters (kernel size) grouped by their respective modalities, we can similarly observe that gyroscope, gravity and magnetometer are among the most interacting data sources. For example, here are the most important pairwise marginals computed via the fANOVA framework: $((ks^2_{gyr}, ks^2_{gra}), 9.2778)$, $((ks^1_{mag}, ks^2_{ori}),$ $7.0166)$, $((ks^2_{gur}, ks^2_{ori}), 5.5122)$, $((ks^1_{acc}, ks^1_{mag}), 4.0382)$, $((ks^1_{pre}, ks^3_{gyr}), 2.3154)$, $((ks^3_{gyr}, ks^1_{mag}), 2.2472)$, $((ks^1_{mag}, ks^1_{ori}), 2.1216)$, $((ks^3_{pre}, ks^2_{gyr}), 1.76305)$, etc.

5. Incorporation of the Data Generation Model into Learning Models

In the previous experiments, we evaluated empirically the ability of the proposed approach to make explicit the subsets of interacting data sources that emerge in a sensors-deployment, and this, w.r.t. the recognition of each particular human activity. We saw, in particular, the emergence of some interactions that are found to be consistent with empirical results in the literature. In addition, as we set up the neural architectures to construct abstract features and novel sensor fusion schemes, we see the emergence of some less common forms of interactions.

We can leverage all these interactions in order to constrain the training of (simpler) architectures. During this constrained training phase, these architectures are encouraged to concentrate on the provided subsets of data sources to learn the corresponding human activities. This could substantially improve the robustness of these architectures as only curated subsets of inputs are provided making it easy to learn patterns. Thus, in this second experiment, we want to evaluate the effectiveness of the obtained interactions model via a simple setting where the training of our neural architectures is constrained with these interactions. Specifically, our main assumption is that the subsets of interacting data sources, which are obtained using our approach, are able to define precisely a given human activity.

We construct neural networks, similar to the architectures used to derive the data generation model, with exactly 3 Conv/ReLU/MaxPool stacked blocks. These blocks are followed by Fully Connected/ReLU layers. The weights of the layers corresponding to all inputs are optimized during training without distinction, the constraining being specified via data augmentation. Indeed, in this setting, for each subset of interacting data sources, we perform data augmentation by assigning values drawn from a normal distribution, to the unimportant data sources. The goal is to make the neural network insensitive to the noisy inputs. We provide training examples to the neural network according to the given pairs and triplets of interacting data sources that we extract from the derived model. Training of the neural networks was performed using subsets of data sources parameterized with different values of τ_{int} and τ_{imp}.

Figure 6 shows the evolution of the neural network depending on the parameters τ_{int} and τ_{imp}. In addition, this figure illustrates the average number of data sources that are included in the subsets, depending on these two thresholds. In particular, when τ_{imp} and τ_{int} are set, for example to 0, all data sources are included. We find that the neural networks trained with smaller subsets of data sources perform better than the baseline and most of the settings relying on a higher number of data sources. Noticeably, we get a recognition performance of 88.7%, measured by the fl-score, over the baseline using a subset of the data sources of size 12. Thus, an improvement of 17.84% in terms of recognition performances and a reduction of one-half

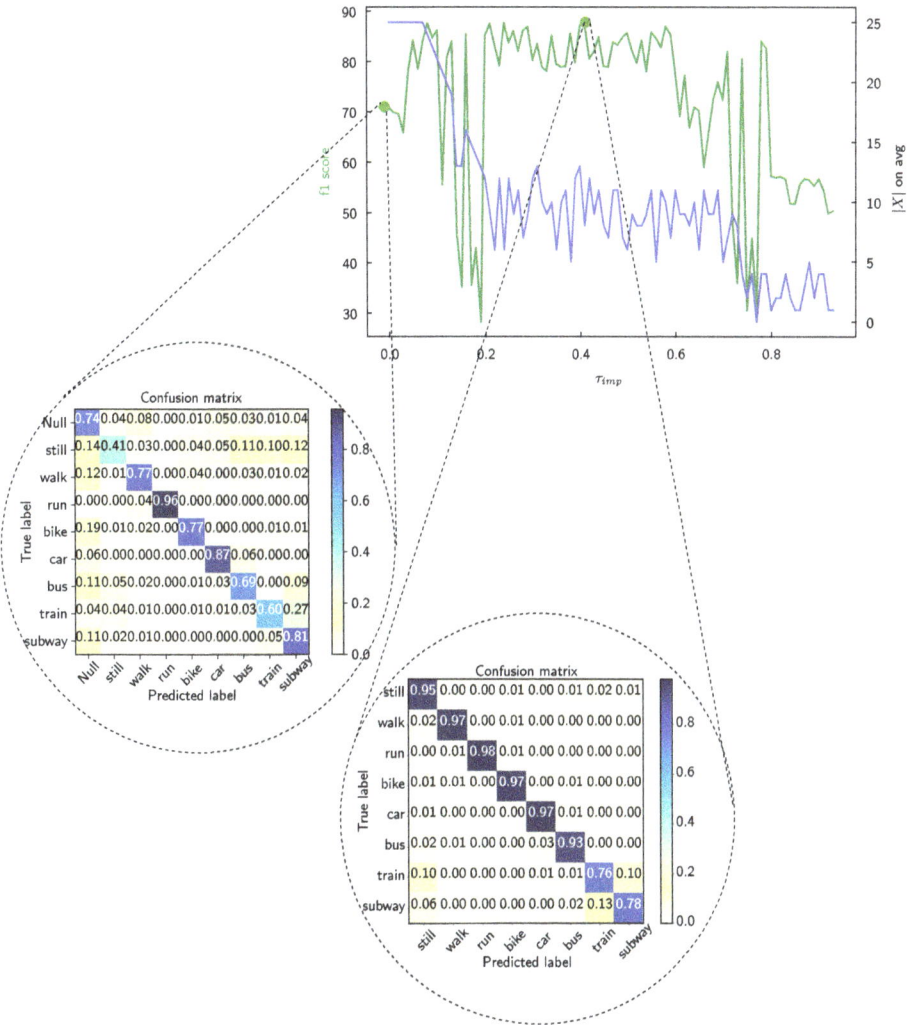

Figure 6: Recognition performance measured by the fl-score (in green) obtained in the constrained learning setting and average cardinality of the subsets of data sources (in blue), as a function of τ_{imp} and τ_{int}. Confusion matrices corresponding to (a) the model trained on all data sources fused and (b) the best performing model trained on a constrained number of data sources. The model trained on all data sources fused together resulted in 70.86 % recognition performance measured by the fl-score. Both models use the following set of hyperparameters: $l_r = 0.1$, $nf = 28$, $ks^i_{acc} \in \{15,9,9\}$, $ks^i_{gyr} \in \{9,15,9\}$, $ks^i_{mag} \in \{13,15,9\}$, $ks^i_{lac} \in \{10,14,12\}$, $ks^i_{gra} \in \{15,9,10\}$, $ks^i_{ori} \in \{9,9,12\}$, $ks^i_{pre} \in \{15,15,9\}$, $s = 0.6$, and $n_u = 2048$. The best performing model yields a recognition performance of 88.7%, and an improvement of 17.84% over the baseline, using a subset containing solely 12 data sources, hence reducing by one-half the required quantities of data.

concerning the required quantities of data. Surprisingly, we do not see a lot of bad subsets of interacting data sources for $0.2 \leq \tau_{imp} \leq 0.6$, where the number of data sources per subset is confined between 5 and 12.

For comparison, we also trained a neural network on the whole data sources. This network achieved a performance of 70.86 % measured by the fl-score. We used for this neural network the set of hyperparameters that were obtained for the best configuration during the exploration of the neural architecture space. From the rightmost confusion matrix which is embedded in Fig. 6, we notice a more pronounced confusion for still class which is often confused with public transportation-related activities (bus, train and subway with 11%, 10% and 12% confusion, respectively). Analysis of the evolution of the confusion of this class over the BO runs shows an fl-score of approximately 26.28% ± 0.13, with a maximum of 43% recorded at the 59th run. A similar observation can be made for train class which is often confused with subway and reflects a clear link between the two activities.

6. Discussion

We proposed, in this chapter, to leverage domain models in order to improve the way neural architectures relate data sources together in a sensor-rich environment. First, we presented a novel approach for deriving subsets of important and interacting data sources, namely, the data generation process that is intended to capture the dynamics of body movements pertaining to each human activity. As deriving such models using human expertise is hard, we framed this task as an exploration of the space of neural architectures. These architectures are composed of sensor fusion and feature extraction components controlled by sets of hyperparameters, which in turn control the impact of the corresponding data sources processed by these components. Derivation of the subsets of important and interacting data sources is based upon this assumption. We confronted the obtained data generation model to empirical results from the literature. In addition to some specific insights, such as the emergence of genuine set of modalities that are found to impact substantially the dynamics of human activities, the obtained model is consistent with those from the literature. These specific insights can be explained by the type of architecture components that we used in the proposed instantiation and which are based on convolutional neural networks. This part opens up perspectives to modeling real-world phenomena like the dynamics of human activities using a large-scale exploration of the space of architectures. Such models could constitute general wisdom, similar to human expertise, that can be leveraged as domain knowledge.

Second, the effectiveness of the obtained domain models is demonstrated via a setting that exploits the subsets of interacting data sources in order to constrain the learning process of a neural network. Indeed, the proposed setting showed substantial improvements over the baseline where all data sources of the sensor-rich environment are used. In particular, we get an improvement of the recognition performances, measured by the f1-score, of up to 17.84% over the baseline and this using solely 12 data sources among those available. The experimental results on a fully-featured dataset of human activities prove the effectiveness of the proposed approach and open up perspectives regarding the incorporation of richer domain models in order to relate data sources more efficiently and develop data-efficient learning systems. Another perspective is related to the way these kinds of domain models are incorporated into the actual learning processes. One can derive from these domain models an additional term, $R : \mathcal{F} \to \mathbb{R}$, that can be plugged to the original optimization objective (Eq. (2)), which, then, becomes

$$f^* = \underset{f \in \mathcal{F}}{\arg\min} \ \mathcal{L}(f) + \lambda R(f), \tag{3}$$

where $\lambda \in [0; 1]$ is a weight parameter used to control the impact of the regularization term. This term can be used precisely at the interfaces between data sources and the components that process them. The goal is to impose the learning model to reproduce, in its internals, the insights that are provided as domain knowledge.

References

1. J. K. Aggarwal and Q. Cai, "Human motion analysis: A review," *Computer vision and image understanding*, vol. 73, no. 3, pp. 428–440, 1999.
2. V. Radu, C. Tong, S. Bhattacharya, N. D. Lane, C. Mascolo, M. K. Marina, and F. Kawsar, "Multimodal deep learning for activity and context recognition," *Proc. ACM Interact. Mobile, Wearable Ubiquitous Technol.*, vol. 1, no. 4, p. 157, 2018.
3. R. Kurle, S. Günnemann, and P. van der Smagt, "Multi-source neural variational inference," in *Proc. the AAAI Conf. Artificial Intelligence*, vol. 33, pp. 4114–4121, 2019.
4. M. Zeng, H. Gao, T. Yu, O. J. Mengshoel, H. Langseth, I. Lane, and X. Liu, "Understanding and improving recurrent networks for human activity recognition by continuous attention," in *Proc. 2018 ACM Int. Sympo. on Wearable Computers*, ACM, pp. 56–63, 2018.

5. K. He, X. Zhang, S. Ren, and J. Sun, "Deep residual learning for image recognition," in *Proc. IEEE Conf. Computer Vision and Pattern Recognition*, pp. 770–778, 2016.

6. C. Szegedy, W. Liu, Y. Jia, P. Sermanet, S. Reed, D. Anguelov, D. Erhan, V. Vanhoucke, and A. Rabinovich, "Going deeper with convolutions," in *Proc. IEEE Conf. Computer Vision and Pattern Recognition*, pp. 1–9, 2015.

7. T. Plötz, N. Y. Hammerla, and P. Olivier, "Feature learning for activity recognition in ubiquitous computing," in *Int. Joint Conf. Artificial Intelligence*, p. 1729, 2011.

8. F. J. Ordóñez, and D. Roggen, "Deep convolutional and LSTM recurrent neural networks for multimodal wearable activity recognition," *Sensors*, vol. 16, no. 1, p. 115, 2016.

9. M. T. Luong, H. Pham, and C. D. Manning, "Effective approaches to attention-based neural machine translation," arXiv preprint arXiv:1508.04025, 2015.

10. Z. Chen, L. Zhang, Z. Cao and J. Guo, "Distilling the knowledge from hand-crafted features for human activity recognition," *IEEE Trans. Indust. Inform.*, vol. 14, no. 10, pp. 4334–4342, 2018.

11. H. Gjoreski, M. Ciliberto, F. J. O. Morales, D. Roggen, S. Mekki, and S. Valentin, "A versatile annotated dataset for multimodal locomotion analytics with mobile devices," in *Proc. 15th ACM Conf. Embedded Network Sensor Systems*, ACM, 2017, p. 61.

12. P. K. Atrey, M. A. Hossain, A. El Saddik, and M. S. Kankanhalli, "Multimodal fusion for multimedia analysis: a survey," *Multimedia Syst.*, vol. 16, no. (6), pp. 345–379, 2010.

13. T. Elsken, J. H. Metzen, and F. Hutter, "Neural architecture search: A survey," *J. Mach. Learni. Res.*, vol. 20, no. 55, pp. 1–21, 2019.

14. J. Snoek, H. Larochelle, and R. P. Adams, "Practical bayesian optimization of machine learning algorithms," *Adv. Neural Inform. Process. Sys*, vol. 2, pp. 2951–2959, 2012.

15. H. Hoos and K. Leyton-Brown, "An efficient approach for assessing hyperparameter importance," in *Int. Conf. Machine Learning*, pp. 754–762, 2014.

16. L. Bao and S. S. Intille, "Activity recognition from user-annotated acceleration data," in *Int. Conf. Pervasive Computing*, Springer, 2004, pp. 1–17.

17. M. C. Yu, T. Yu, S. C. Wang, C. J. Lin, and E. Y. Chang, "Big data small footprint: The design of a low-power classifier for detecting transportation modes," *Proc. VLDB Endowment*, vol. 7, no. 13, pp.1429–1440, 2014.

18. H. Gjoreski, M. Ciliberto, L. Wang, F. J. Ordonez Morales, S. Mekki, S. Valentin, and D. Roggen, "The University of Sussex-Huawei locomotion and transportation dataset for multimodal analytics with mobile devices," *IEEE Access*, 2018.

19. Y. Zheng, X. Xie, and W. Y. Ma, "Geolife: A collaborative social networking service among user, location and trajectory," *IEEE Data Eng. Bull.*, vol. 33, no. 2, pp. 32–39, 2010.

20. M. Zhang, and A. A. Sawchuk, "USC-HAD: a daily activity dataset for ubiqui-tous activity recognition using wearable sensors," in *Proc. 2012 ACM Conf. Ubiquitous Computing*, pp. 1036–1043, 2012.

21. C. Carpineti, V. Lomonaco, L. Bedogni, M. Di Felice, and L. Bononi, "Custom dual transportation mode detection by smartphone devices exploiting sensor diversity," in *2018 IEEE Int. Conf. Pervasive Computing and Communications Workshops (PerCom Workshops)*, IEEE, pp. 367–372, 2018.

22. F. Foerster, M. Smeja, and J. Fahrenberg, "Detection of posture and motion by accelerometry: A validation study in ambulatory monitoring," *Comput. Human Behav.*, vol. 15, no. 5, pp. 571–583, 1999.

23. J. Mantyjarvi, J. Himberg, and T. Seppanen, "Recognizing human motion with multiple acceleration sensors," in *2001 IEEE International Conference on Systems, Man, and Cybernetics*, vol. 2, IEEE, 2001, pp. 747–752.

24. S. Reddy, M. Mun, J. Burke, D. Estrin, M. Hansen, and M. Srivastava, "Using mobile phones to determine transportation modes," *ACM Transas. Sensor Networks (TOSN)* vol. 6, no.2, p.13, 2010.

25. A. Bevilacqua, K. MacDonald, A. Rangarej, V. Widjaya, B. Caulfield, and T. Kechadi, "Human activity recognition with convolutional neural networks," in *Joint Eur. Conf. Machine Learning and Knowledge Discovery in Databases*, Springer, pp. 541–552, 2018.

26. G. Hooker, "Generalized functional anova diagnostics for high-dimensional functions of dependent variables," *J. Comput. Graph. Statist.*, vol. 16, no. 3, pp. 709–732, 2007.

27. S. Ha and S. Choi, "Convolutional neural networks for human activity recogni-tion using multiple accelerometer and gyroscope sensors," in *2016 Int. Joint Conf. Neural Networks (IJCNN)*, IEEE, pp. 381–388, 2016.

28. A. Osmani and M. Hamidi, "Bayesian optimization of neural architectures for human activity recognition," in Kawaguchi N., N. Nishio, D. Roggen, S. Inoue, S. Pirttikangas, K. Van Laerhoven (eds.), *Human Activity Sensing Corpus and Applications*, Springer Nature, 2019.

29. M. Abadi, P. Barham, J. Chen, Z. Chen, A. Davis, J. Dean, M. Devin, S. Ghemawat, G. Irving, M. Isard, *et al.* "Tensorflow: A system for large-scale machine learning," in *12th {USENIX} Symp. Operating Systems Design and Implementation ({OSDI} 16)*, pp. 265–283, 2016.

30. F. Pedregosa, *et al.*, "Scikit-learn: Machine learning in python," *J. Mach. Learn. Res.*, vol. 12(Oct), pp. 2825–2830, 2011.

31. S. Ioffe and C. Szegedy, "Batch normalization: Accelerating deep network train-ing by reducing internal covariate shift," in *ICML*, vol. 37, pp. 448–456, 2015.

32. S. Wang, C. Chen, and J. Ma, "Accelerometer based transportation mode recognition on mobile phones," in *2010 Asia-Pacific Conf. Wearable Computing Systems (APWCS)*, IEEE, pp. 44–46, 2010.

Chapter 4

Deep Learning and Unsupervised Domain Adaptation for WiFi-based Sensing

Jianfei Yang*,‡, Han Zou†,§, Lihua Xie*,¶ and Costas J. Spanos†,∥

*School of Electrical and Electronic Engineering,
Nanyang Technological University, Singapore
†Department of Electrical Engineering and Computer Sciences,
University of California, Berkeley, USA
‡yang0478@e.ntu.edu.sg
§enthalpyzou@gmail.com
¶elhxie@ntu.edu.sg
∥spanos@berkeley.edu

Abstract

With the wide deployment of WiFi infrastructures, wireless signals fill the air in most commercial regions and homes. Human activities usually affect the propagation of WiFi signals, which can be recorded by channel state information (CSI) data. A CSI sample is a 2D matrix that records the temporal and spatial dynamics in the current environment. Recent works exploit how to leverage machine learning and pattern recognition for WiFi-based sensing technology. In this chapter, we introduce how deep learning and unsupervised domain adaptation method further empower WiFi-based sensing. Specifically, convolutional neural network (CNN) extracts discriminative features automatically via supervised learning. Unsupervised domain adaptation (UDA) transfers knowledge from a label-rich source domain to an unlabeled target domain. For example, we investigate how to leverage CNN and UDA for adaptive WiFi-based gesture recognition. As CSI data is sensitive to temporal and spatial dynamics, the deep model trained in the original environment (source domain) cannot achieve accurate recognition in a new environment (target domain), which is referred to as distribution shift or domain shift. An adversarial domain adaptation approach is introduced to tackle this problem. Apart from classic CNN architecture and adversarial domain adaptation, much recent deep learning progress can be connected with WiFi-based sensing. We introduce these potential directions including transfer learning, few-shot learning, cross-modality supervision, multimodal learning and model compression, which will lead to more robust and comprehensive smart sensing technology.

Keywords: channel state information, domain adaptation, deep learning, gesture recognition

1. Introduction

In the era of smart sensing, people usually think of various sensors such as Inertial Measurement Unit (IMU), infrared device and radar [1]. IMU is equipped in smart phones, which record triaxial acceleration data that can be used for human activity recognition. Infrared devices are commonly deployed in smart buildings for occupancy detection and lighting control. Radar that is accurate but expensive plays an essential role in robotics and autonomous driving. In this chapter, a new sensing technique is introduced for human activity or gesture recognition in a cost-effective and device-free manner — WiFi-based sensing technology.

WiFi-based sensing leverages WiFi signal variations to recognize human activities or more fine-grained occupancy information [2]. When WiFi signals propagate from a transmitter to a receiver, due to multi-path effect, some paths of the signals can be affected by human activities. Such signal variations are reflected by channel state information (CSI) that refers to known channel properties of a communication link. CSI data is extracted by particular tools of network cards such as Intel 5300 NIC [3] and Atheros Tools [4]. Patterns of activities hidden in CSI data are able to be extracted by traditional statistical learning or modern deep learning techniques. Therefore, WiFi-based sensing technology is enabled by accurate CSI extraction tools and progressive machine learning approaches.

Employing wireless signals for smart sensing has many advantages, especially in smart buildings. Firstly, WiFi access points are very common infrastructures in indoor environments. As more and more intelligent appliances are equipped with WiFi modules for communications, many WiFi communication links are formed, which provides plenty of CSI data for human activity recognition. Leveraging existing infrastructures is cost-effective. Secondly, compared to IMU, wireless signals render sensing to be device-free, which is convenient for users. Thirdly, compared to infrared sensors and radiofrequency (RF) signals, CSI data contains high-resolution signal data that depicts the spatial and temporal states by an image-like 2D matrix. Hence, WiFi-based sensing can be utilized for fine-grained human activity recognition, gesture recognition and even breath detection. As shown in Fig. 1, compared to other intrusive and non-intrusive sensing approaches, WiFi-based sensing method is more cost-effective but has high granularity of sensing.

As mentioned above, the development of machine learning improves the capacity of classification model and enables more fine-grained WiFi-based sensing applications. Deep neural networks are able to automatically extract robust representations from CSI data and obtain an accurate classifier by backpropagation [5]. However, to deal with wireless signals for practical scenarios, there still exist some limitations for typical deep models. For example, deep networks require large scales of annotated training data, the collection of which is time-consuming and labor-intensive. How to tackle the problem of lack of sufficient training data requires to be investigated. Another example

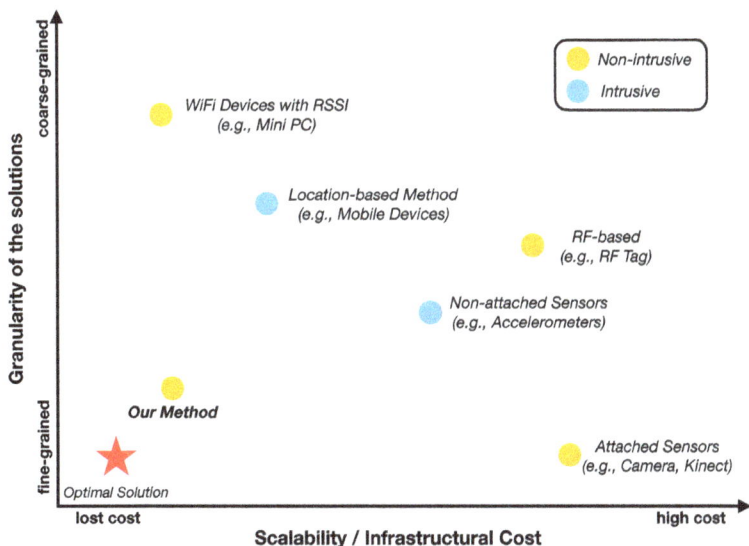

Figure 1: Comparison with prevailing sensing techniques.

is that deep models have strong fitting ability, but are often overfitting existing data, leading to unsatisfactory generalization ability. How to enhance the adaptability of deep models also remains a challenge.

In this chapter, we introduce a CSI-based gesture recognition method in a real-world scenario when the trained-once deep model is deployed in a new environment [6]. As wireless signals are very sensitive to environmental changes, deep models trained in one room usually fail to work in a new room due to data distribution difference between two circumstances. In practical applications, assuming that we have collected plenty of data in the factory and have obtained a well-trained model, when we apply this model in a new environment such as a user's house, how is the model adapted to the new circumstance using little effort? How can we improve the system portability and robustness over spatial and temporal dynamics? The method introduced in the chapter relies on domain adaptation to solve this problem. Domain adaptation refers to transferring knowledge from a label-rich source domain to a label-scarce or even unlabeled target domain [7]. In our scenario, the recognition model is trained in one room (i.e., source domain) and is adapted to a new room (i.e., target domain).

Domain adaptation plays an important role in computer vision and neural language processing. As datasets for training could be possibly domain-specific, such as a dataset that consists of images from only one country, domain adaptation adapts the trained-once model to new environments (e.g., other countries) in an unsupervised manner. It usually minimizes the distribution discrepancy of data from two domains, or leverages advanced adversarial training to generate a domain-invariant feature space for classifier [8]. In our scenario, we introduce a novel adversarial domain adaptation approach for adapting CSI-based sensing model only using unlabeled data in the new environment. Through adversarial training, deep neural networks that serve as the feature extractor can adaptively extract robust CSI patterns from the target domain, leading to significant improvement [6].

In subsequent sections, we begin with a preliminary introduction that details the properties of CSI data, how CSI data perform with human activity and what WiFi systems can be used for sensing. Then we discuss the basic deep neural networks for CSI-based gesture recognition and the adversarial domain adaptation for model transfer. Extensive experiments are conducted to demonstrate the superiority of the method. After discussing some potentialites of how recent deep learning methods can be integrated with WiFi-based sensing, we summarize the whole chapter in the last subsection.

2. Preliminaries

In wireless communications, received signal strength indicator (RSSI) and Channel State Information (CSI) are two kinds of useful information for environment perception at the physical layer. RSSI is a scalar value that exhibits the signal strength of wireless signal, and therefore RSSI is only utilized for indoor localization, occupancy detection or other coarse activity recognition. Whereas CSI is more fine-grained data that depicts the quality of communications. Wang *et al.* build a theoretical model between CSI phase information and object moving speed [9] and then the theory of Fresnel zone is adopted for human respiration detection [10]. These physical models are explainable and inspire us to develop recognition methods based on them. However, in the real world, the environment is complicated and many irrelevant signals interfere with the sensing signals, which hinders the availability of these physical models. Different from them, to model CSI patterns under complex circumstances, we introduce the data-driven model based on deep learning. Therefore, we firstly investigate the properties of our data — what CSI is and how CSI is extracted. To use this deep learning model, we need to show how CSI patterns are connected with human activities or gestures.

2.1 *Channel state information*

In wireless communication, channel state information reflects the channel properties of a communication link. The propagation of wireless signals are affected by the physical environment, leading to variations of CSI due to reflections, diffractions and scattering of signals [11]. Furthermore, modern WiFi modules apply orthogonal frequency division multiplexing (OFDM) at the physical layer and follow IEEE 802.11n/ac standard that permits multiple transmit and receive antennas for multiple input, multiple output (MIMO) communication. In this manner, CSI consists of more information that describes the conditions of communication links of each transmitter–receiver (TX–RX) pair. Therefore, CSI reveals fine-grained characteristics of wireless signals combining effect of time delay, amplitude attenuation and phase shift of multiple paths on each communication subcarrier. Compared to RSSI, which is only a superimposition of multi-path signals, CSI contains more communication information and owns higher resolution. In communications, the WiFi signals can be modeled as channel impulse response (CIR) $h(\tau)$ in frequency domain:

$$h(\tau) = \sum_{l=1}^{L} \alpha_l e^{j\phi_l} \delta(\tau - \tau_l), \tag{1}$$

where τ_l is the time delay, L indicates the total number of multi-path, $\delta(\tau)$ denotes the Dirac delta function, and α_l and ϕ_l represent the amplitude and phase of the lth multi-path component, respectively. Nevertheless, due to limited WiFi bandwidth, only some multi-path components are differentiable. In this case, to obtain complete CSI data, the OFDM receiver offers a sampled version of the signal spectrum at subcarrier level, which comprises amplitude attenuation and phase shift via a complex number. These estimation can be represented by

$$H_i = \|H_i\| e^{j\angle H_i} \tag{2}$$

where $\|H_j\|$ and $\angle H_j$ are the amplitude and phase of ith subcarrier, respectively. Thus, the total estimated CSI data describes the communications of multiple TX–RX pairs at subcarrier level. The size of CSI streams is $N_{TX} \times N_{RX} \times N_{sub}$ where N_{TX}, N_{RX} and N_{sub} denote the number of transmitter antennas, receiver antennas and subcarriers, respectively.

Theoretically, CSI phases are supposed to be more robust with less intrinsic noises than CSI amplitudes. However, in the current CSI data, the opposite situation occurs due to hardware imperfections and environmental variations [12]. The carrier frequency of a device can drift by up to 100 kHz for 5 GHz band, which hinders the availability of CSI phase. In practical applications, accurate CSI phase can be obtained by manual phase calibration and complicated denoising techniques, but these difficulties bring much inconvenience to users. Though CSI phase is often drifted for various WiFi devices, phase difference is exploited to be robust enough, which will be shown empirically in Fig. 5. Hence, we leverage CSI phase difference information as our WiFi sensing data.

2.2 System architecture

Extracting CSI from network card requires particular tools. For commercial off-the-shelf (COTS) WiFi access points, there are two main streams of tools using different hardware:

- **Intel 5300 NIC card:** Linux 802.11n CSI Tool is based on Intel WiFi Wireless Link 5300 802.11n MIMO radios. The tool consists of a custom modified firmware and open source drivers for Linux. In 20 MHz or 40

MHZ bandwidth, it reports the channel matrices for 30 subcarrier groups. For each channel matrix entry, a complex number is assigned with signed 8-bit resolution for both the real and imaginary parts. This tool has been widely adopted for CSI-based sensing and localization. The common devices equipped with this network card include mini-PC and laptops.

- **Atheros CSI Tool:** It is also an open source 802.11n measurement tool that enables extraction of CSI, the received packet payload and the RSS of each antenna from the physical layer. The tool is built on top of ath9k, which is a Linux kernel driver that supports Atheros 802.11n PCI/PCI-E chips. It theoretically supports all types of Atheros 802.11n WiFi chipsets such as AR9590 and QCA9558 so CSI data can be extracted from WiFi routers with Atheros network card. Different from Intel 5300 NIC tool, Atheros CSI Tool extracts complete CSI data at subcarriers level, i.e., 56 subcarriers for 20MHz and 114 subcarriers for 40 MHz. More subcarriers means better resolution of CSI. The tool has been investigated to conduct occupancy detection [13], crowd counting [14], activity recognition [15, 16], gesture recognition [6, 17] and person identification [18].

These tools help extract CSI data from network card, and some works on CSI-based sensing platforms are built on top of these kernels [2]. In this chapter, the CSI data for evaluation are collected by Atheros CSI Tool [4] on the sensing platform [2]. Fine-grained CSI data are the prerequisite for subtle gesture recognition.

2.3 *How human gesture affects wireless signals*

When wireless signals propagate from a transmitter to a receiver in an indoor environment, as shown in Fig. 2, the propagation usually has line-of-sight (LOS) path and multiple none-line-of-sight (NLOS) paths. Any human activities such as walking and waving hands in the sensing range can affect these propagation paths, i.e., communication links, which is reflected and recorded by CSI data. Naturally, active motions bring more influences to CSI data and relatively static motions only lead to slight variations. Intuitively, we visualize CSI data of different levels of motions, and try to find out some regular patterns.

2.3.1 *Occupancy influence*

The simplest case is to observe CSI variations when a room is occupied or not. As shown in Fig. 3, we show the CSI amplitude of seven timestamps

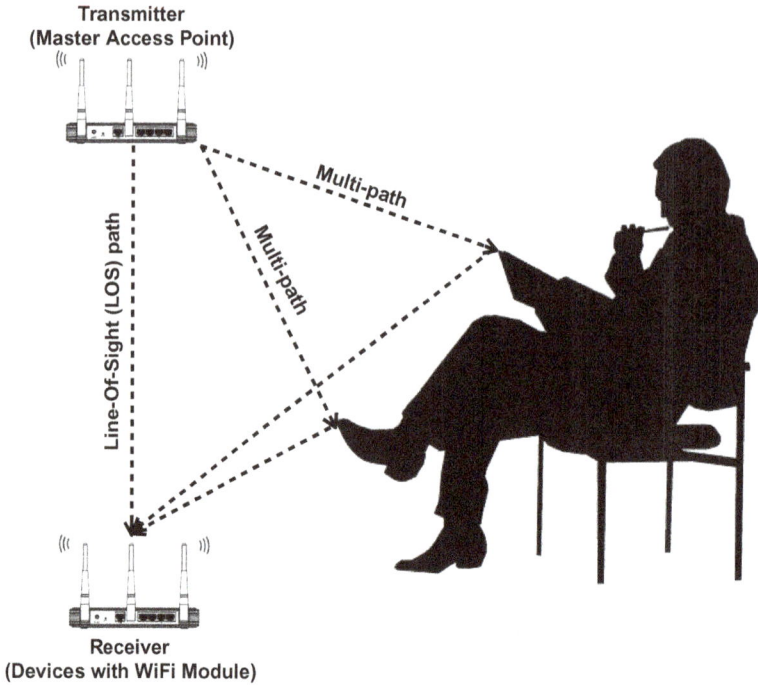

Figure 2:   Multi-path effect in the propagation of WiFi signals.

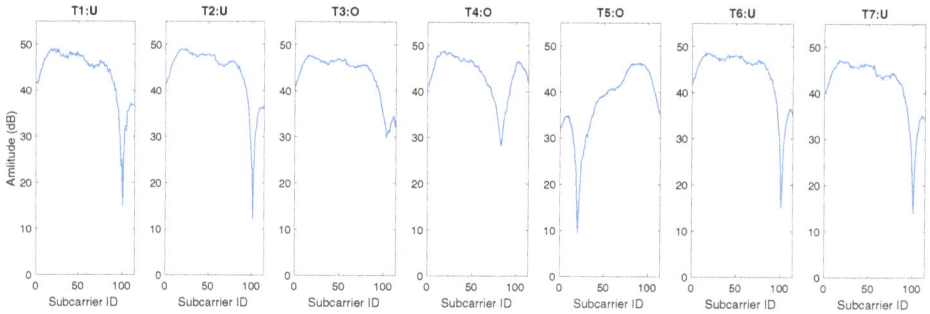

Figure 3:   CSI amplitude of seven timestamps when the rooms state transfers from unoccupied (U) to occupied (O), and back to unoccupied (U).

using one TX–RX pair with 114 subcarriers. It is noteworthy that when a person came into a room in T3–T5, the CSI amplitudes of these subcarriers changed a lot. Furthermore, when the person left the room in T6, the CSI wave shape recovered to T1. These observations indicate that using the similarity of the CSI wave shape can help justify whether a room is occupied [13]. Moreover, we can infer that the CSI amplitudes of multiple subcarriers describe the spatial information. Intuitively, CSI records the state of communication links under the current circumstances, and any obstacles such as human bodies that alter the propagation paths lead to the changes of CSI.

2.3.2 *Human activity*

A more difficult application is how human activities affect CSI. We investigate very common activities in daily life such as walking and running using three TX–RX pairs (with 1 antenna on the transmitter and 3 antennas on the receiver). In Fig. 4, we draw the 3D graph of CSI matrix that includes three axes: time, CSI amplitude and subcarrier index. In Fig. 4(a), a person walked through the TX–RX pairs, and we observe that this moment is captured by CSI in around 1s timestamp. In Fig. 4(b), a person ran around and we can see more and larger variations than those of walking activity. Different activities were also observed and where these activities happen are reflected by different antenna. As multi-path effects in real-world scenarios are too complicated to analyze by physical model, we seek for help from deep neural networks and data-driven solutions that extract discriminative patterns in an automatic way.

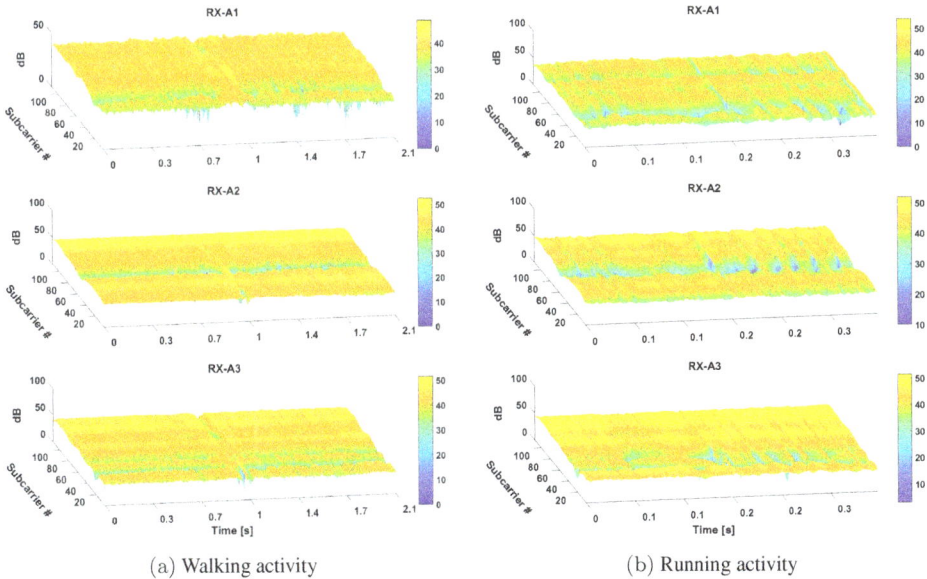

(a) Walking activity   (b) Running activity

Figure 4: The 3D visualizations of temporal CSI stream of walking and running.

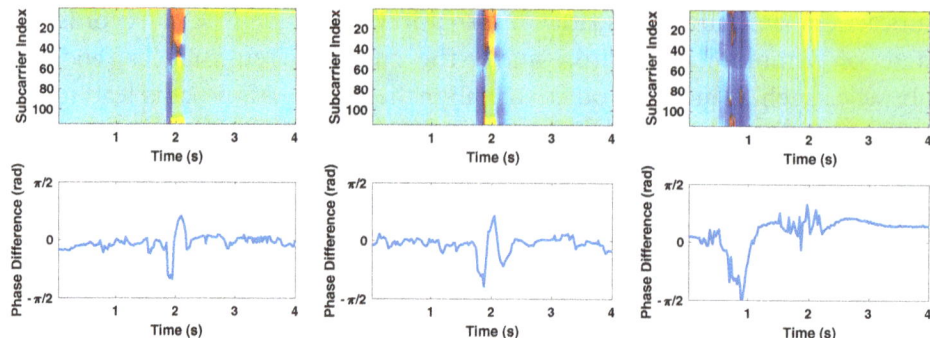

Figure 5: CSI phase difference. The left two columns are *downward* and the last one is *rightward*.

2.3.3 *Human gesture*

Human gestures are more subtle and thereby the fluctuations of CSI made by gestures should be quite small. To capture these kinds of subtle patterns, fine-grained CSI phase difference and deep learning strut their stuff. We compare the *downward* and *rightward* gesture performed by one hand. In Fig. 5, the top row shows the phase difference of two antennas across 114 subcarriers, and the darker color indicates larger value. The bottom row shows the specific value of CSI phase difference of one subcarrier. We can see that CSI phase difference shows robust shapelets for different gestures. These observations demonstrate the rationality of using CSI phase difference and further imply that a powerful representation learning model should be established for these sensitive CSI data.

3. Deep Neural Networks for CSI-based Sensing

For modern deep neural networks, supervised representation learning mainly relies on CNN for spatial features and long short-term memory (LSTM) for temporal features. For CSI data, each sample is written as a matrix $x \in \mathbb{R}^{N_s \times N_t}$ where N_s is the number of subcarriers and N_t is the time duration. Therefore, for each subcarrier, temporal patterns hide in the row of CSI matrix, while for each timestamp, spatial features hide in the column of CSI matrix that refers to the relationship among all subcarriers. In this fashion, one CSI sample is similar to a video sequence in which each timestamp of CSI data describes the current space. Therefore, using 2D CNN has already considered both temporal and spatial features since a 2D convolutional kernel acts on both

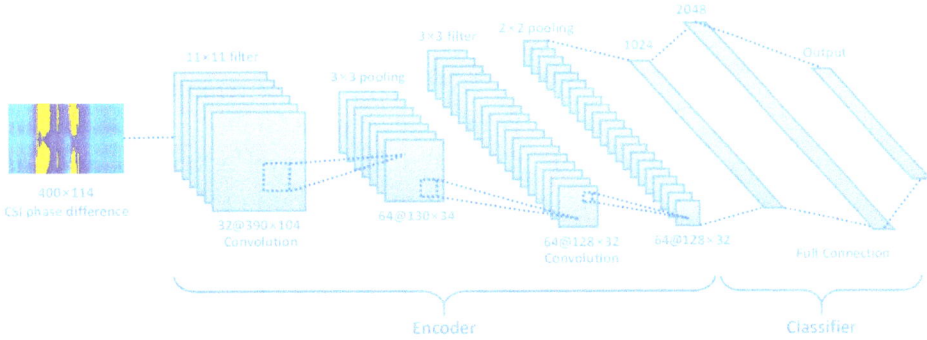

Figure 6: Network architecture of the encoder and the classifier.

temporal and spatial dimensions. Moreover, as the model is designed to adapt across domains, CNN normally performs well and is easier to train.

CNN has been successfully used in computer vision and neural language processing [19]. It enables high-level tasks such as semantic segmentation and vision-based action recognition. Inspired by CNN for image classification, we design the CNN model for CSI data. Figure 6 illustrates the CNN architecture which consists of an encoder (i.e., feature extractor) and a classifier. The proposed CNN architecture includes two convolutional layers and three fully connected layers. It is noteworthy that the first convolutional layer has very large size of kernels (11×11), which aim to extract robust temporal features across more timestamps. After each convolution operation, a pooling layer is used to reduce the dimensions and prepares for the next convolutional layer that is able to learn higher-level features. The activation function is rectified linear unit (ReLU) that follows each pooling layer.

The proposed CNN model is trained in a supervised manner as a great deal of labeled CSI data can be collected in the lab or factory. The model has strong capacity to fit the training data. Nevertheless, when the model is tested in a new environment (such as a user's house) or the layout of the original training environment changes, the model fails due to the spatial shift of CSI data.

4. Adversarial Training for Domain Adaptation

As discussed before, when deep neural networks or traditional methods such as SVM are employed for CSI data, their performance is vulnerable to temporal and spatial dynamics. The reason is that these methods fail to adaptively learn the classifier. In this work, we introduce a device-free **WiFi**-enabled **ad**aptive **g**esture recognition system (**WiADG**) that is aimed at robust and adaptive human gesture recognition consistently under temporal and spatial dynamics by means of adversarial domain adaptation [6].

To overcome the problem of spatial shift, a novel adversarial training approach is introduced, which eliminates the distribution difference of CSI data caused by environmental changes. The target of domain adaptation in our approach is to learn an adaptive encoder in the new environment (target domain) using labeled data in the original environment (source domain) and unlabeled data in the new environment [8]. To this end, adversarial training plays an essential role in unsupervised domain adaptation. It forces the encoder (i.e., CNN) to generate CSI representations whose distributions are similar across two domains. The whole algorithm is summarized in Fig. 7, and we introduce the three steps in detail.

Step 1: Assume we collect N_s CSI frames X_s with labels Y_s in the source domain (referred to as the original environment). The first step of WiADG is to obtain a robust and accurate model in the source domain. The source model is composed of a source encoder E_s for feature extraction and a source classifier C_s, which can be trained in a supervised way via backpropagation:

$$\min_{E_s, C_s} \mathcal{L}_C(X_s, Y_s) = -\mathbb{E}_{(x_s, y_s) \sim (X_s, Y_s)} \sum_{l=1}^{N} \left[\mathbb{I}_{[l=y_s]} \log C_s(E_s(x_s)) \right]. \quad (3)$$

The architectures of source encoder and classifier have been detailed in Fig. 6. The optimization is to minimize the cross-entropy loss between the output of the network and the ground truth using a one-hot vector and we utilize Adam optimizer to efficiently conduct gradient descent, updating the parameters in CNN. In this fashion, the classifier C_s performs accurately in the original environment.

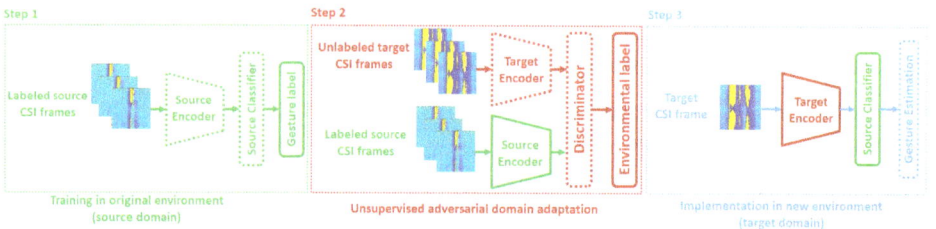

Figure 7: The training and testing procedure of WiADG.

Step 2: *The crucial step of WiADG is the adaptation design that aims to enable the model to recognize gestures in a totally new environment (target domain) without the labor of annotations in that domain. Since our CSI platform can collect data in a non-intrusive manner with high sampling rate, unlabeled CSI frames in the target domain can be easily obtained while user is performing gestures. Denote these unlabeled CSI data as X_t. Step 2 aims to train a new target encoder E_t (with the same architecture as the source encoder E_s) that minimizes the distance between the source and target mapping distributions $E_s(X_s)$ and $E_t(X_t)$. As a result, the source classifier C_s is able to be directly applied if the target encoder maps the target CSI data to the source feature space. To this end, we rely on adversarial training based on a domain discriminator D that learns to identify whether the features come from the source domain or the target domain [8]. The idea is similar to the vanilla GAN that aims to generate real-like images that cannot be distinguished from the real images by a discriminator [20]. In our scenario, since we know the domain label for the CSI frames (source or target), the discriminator and the target encoder play an adversarial game that can be formulated as*

$$\min_D \mathcal{L}_D(X_s, X_t, E_s, E_t) = -\mathbb{E}_{x_s \sim X_s}[\log D(E_s(X_s))]$$
$$- \mathbb{E}_{x_t \sim X_t}[\log(1 - D(E_t(X_t)))]. \tag{4}$$

The domain discriminator D plays a binary classifier role that learns to distinguish the features generated by the source encoder E_s and the target encoder E_t using the domain labels. To generate features that can fool the domain discriminator D, the target encoder is trained in an opposite way, which is formulated as

$$\min_{E_t} \mathcal{L}_{E_t}(X_s, X_t, D) = -\mathbb{E}_{x_t \sim X_t}[\log D(E_t(X_t))]. \tag{5}$$

In the adversarial domain adaptation, we initialize the target encoder E_t using the parameters of E_s obtained in Step 1, as these parameters provide the capacity of feature extraction and make the adversarial training more smooth and stable. In WiADG, we simply design 3 fully connected layers for the discriminator D: 1024 hidden units — 2048 hidden units — binary output layer. ReLU is employed as the activation function. Through Step 2, the parameters in E_t and D are jointly updated. D only plays an auxiliary role while E_t is utilized for feature extraction in the new environment.

Step 3: *As the target encoder E_t can map the target CSI frames to a shared feature space with the source encoder E_s, we can directly adopt the pre-trained source classifier C_s in the new environment. The combination of these two components is the target gesture recognition model that can accurately identify the gestures in the new environment (target domain).*

In summary, as described in Fig. 7, the Step 1 trains a source encoder E_s and a classifier C_s in a supervised way. Then we fix E_s and conduct an adversarial game between the new target encoder E_t and the domain discriminator D to force E_t to generate source-like representations for the target CSI frames. In the phase of testing, we simply use the target encoder and the original classifier in the new environment.

5. Experiments

5.1 *Setup*

We implemented WiADG using 2 TP-LINK N750 access points, one of which serves as TX and the other as RX. Then the method is evaluated in real indoor environments including an office and a conference room. Using the Atheros CSI tool [4] on the sensing platform [2], we obtain 400×114 CSI frame for each gesture sample, in which 400 is the time duration and 114 is the phase difference of two antennas in the RX. The sampling rate is 100 packets/s and we employ linear interpolation to ensure the stationary interval of CSI values to overcome packet loss. The method is implemented by Pytorch and runs on a Think-pad laptop with Intel i7-4810MQ 2.80GHz CPU and 16GB RAM.

5.2 *Data collection*

To simulate the CSI sensing in the real-world scenario rather than ideal environment, we choose 2 typical indoor environments. As shown in Fig. 8, the system was deployed in a 7 m × 5 m conference room and a 4.5 m × 5.6 m office zone. The distance between TX and RX routers was 1 meters and they were put on a table as shown in Fig. 9. Two volunteers were involved and six common gestures were collected including *right, left, rolling right, rolling left, push* and *pull* using one hand. For each gesture at each testbed, we collected 100 samples to train the classifier on one day, and other 100 samples were obtained on different days with temporal dynamics. The total number of CSI gesture frames is over 2500 which were used to evaluate the effectiveness of WiADG.

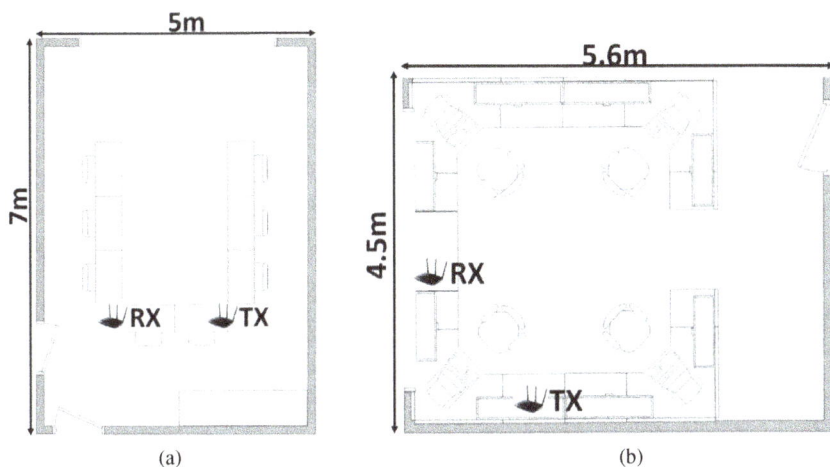

Figure 8: The layouts of the conference room (a) and the office (b).

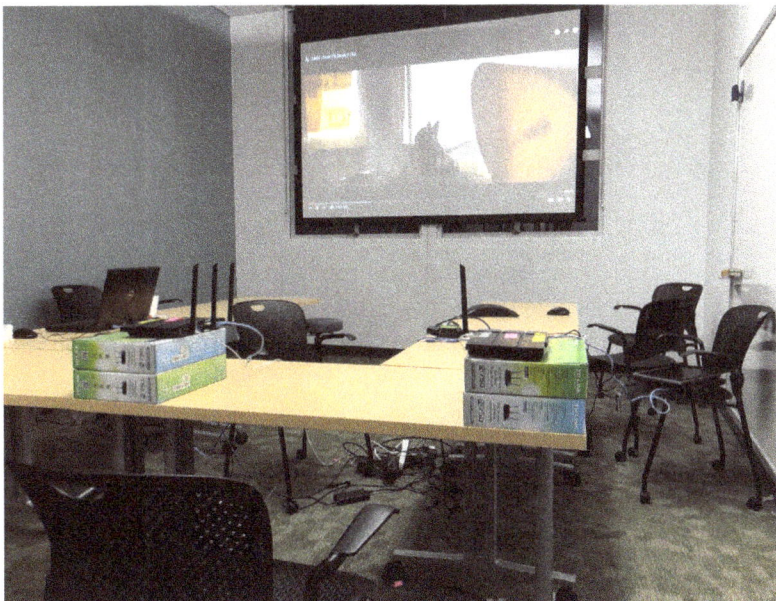

Figure 9: Experimental setup in the conference room.

5.3 *Evaluation in the original environment*

First we need to validate the superiority of deep neural networks when they are compared to traditional statistical learning approaches including WiG [21] and WiAG [22]. WiG utilized wavelet denoising and an SVM classifier for gesture recognition. WiAG is based on principle component analysis

(PCA), discrete wavelet transform and KNN. To compare with these approaches, we report the true positive rate (TPR) in two environments. Fig. 10 shows the results of these approaches, and we can see that WiADG significantly outperforms these approaches for all cases. It improves the overall TPR over WiAG and WiG by 7.7% and 9.5% in the conference room, and 8.6% and 10.9% in the office zone, respectively. These results validate that deep neural networks have better capacity to extract CSI patterns for gesture recognition in the same environment.

We can further explore the confusion matrix for the accuracy of WiADG. The cross-validation accuracy for WiADG is 98.3% and 98% in the conference room and the office zone, respectively. As shown in Fig. 11, the performance

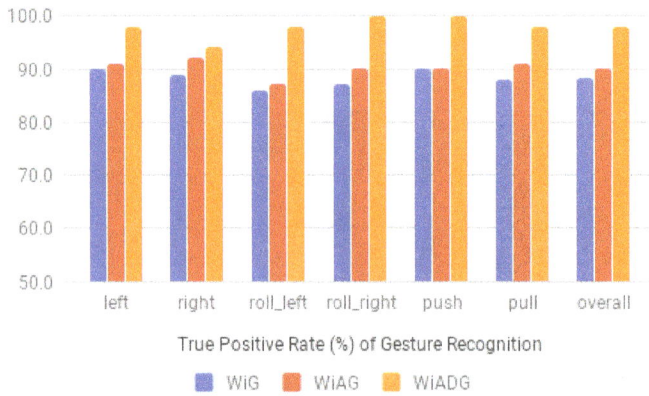

Figure 10: Comparison of TPR among three WiFi-based gesture recognition methods in the two testbeds.

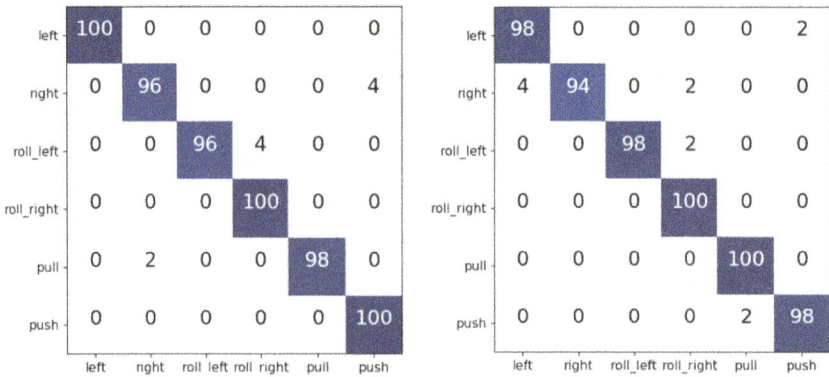

Figure 11: Confusion matrices of gesture recognition accuracy using WiADG in the original environments.

in the conference room is slightly better than that of office zone, which might be due to the complicated layout and obstacles in the office zone generating more propagation paths for the wireless signals. In total, the accuracy of each gesture is not less than 94%, which demonstrates the high stability of WiADG under different circumstances.

5.4 *Evaluation under environmental dynamics*

We further investigate the performance of WiADG in a more realistic scenario where the training environment is different from the testing environment. As we collected data from two environments, two evaluation protocols are formed: conference room (source domain) to office zone (target domain) and the opposite mode. Table 1 compares the TPR of WiADG and other two methods. It is easily observed that the performances of WiG, WiAG and WiADG (source only) degrade severely, which shows that spatial and temporal dynamics affect both traditional machine learning and deep learning approaches. Though WiADG still has better performance than WiG and WiAG, around 50% accuracy is still too low for practical applications.

Then we conducted adversarial domain adaptation method as described in Section 4 using unlabeled data in the target domain. We learned a target encoder E_t by forcing it to fool the domain discriminator D. As the discriminator learns to distinguish the features from the source encoder E_s and the target encoder E_t, the adversarial training successfully enhances the distribution similarity between the source domain and the target domain. As shown in Table 1, the proposed WiADG with domain

Table 1:   Gesture recognition accuracy under environmental dynamics.

Approach	True postive rate (%)	
	Conference room → Office	Office → Conference room
WiG (Source Only)	48.8	36.1
WiAG (Source Only)	49.7	35.4
WiADG (Source Only)	50.7	49.3
WiADG (Domain Adaptation)	83.3	66.6

adaptation achieves tremendous improvement of accuracy by over 30% compared to WiADG (source only) which only leverages deep neural networks for supervised learning. Compared with other approaches, WiADG is still the state-of-the-art approach for this transfer scenario benchmark. We can discover that the major improvement for this scenario is attributed to Steps 2 and 3 that aim to train the target encoder to generate source-like feature distributions. These results provide compelling evidence to demonstrate that adversarial domain adaptation with deep neural networks enables robust WiFi-based gesture recognition under environmental dynamics.

6. Potentials for WiFi-based Sensing

Deep neural networks learn robust features and domain adaptation helps transfer knowledge to a new environment, which empowers robust and adaptive WiFi-based gesture recognition application. As WiFi infrastructures have been deployed almost everywhere in the commercial regions and smart homes, WiFi-based sensing will play a vital role in the upcoming sensing technology. The fast development of deep models further enables more WiFi-based sensing applications and practical scenarios. Here we would like to elaborate some potential directions with regard to how recent deep learning research can be leveraged in WiFi-based sensing.

6.1 *Transfer learning for robust feature extractor*

The reason why we transfer knowledge from source domain to target domain is that the feature extractor (encoder) can only describe the CSI sample from the source domain but fails to extract effective features in a new environment. If we aim to train a robust CSI feature extractor that is suitable for most circumstances, a large scale of CSI samples are required. For example, in computer vision field, one essential factor is the collection of ImageNet that includes 100 million labeled images for 1000 visual categories. If a model is aimed at visual recognition, pre-training it on ImageNet [23] becomes extremely necessary. In WiFi-based sensing, the scales of CSI data for most applications are small for deep learning model. Though new large-scale datasets such as Widar [24] were proposed recently, a general network architecture with pre-trained parameters for CSI-based sensing is still demanded.

6.2 *Few-shot learning for real-world scenario*

A large scale of CSI samples are hard to collect, which requires many volunteers conducting specified actions. Therefore, when we adopt traditional deep neural networks based on supervised learning, it is quite labor-intensive and time-consuming to obtain the training data. Few-shot learning aims to tackle the problem of lack of annotated data [25]. Suppose we only have several CSI gesture samples for training. Few-shot learning is able to learn effective features based on more powerful metric-based or similarity-based losses. There have been some works that explore this direction for few-shot gesture recognition [17] that proposes a Siamese neural network for efficient learning. More few-shot learning techniques could be integrated with CSI-based sensing.

6.3 *Cross-modality supervision for fine-grained recognition*

In smart sensing field, images and wireless signals are two kinds of modalities. They have their own characteristics that decide their functionalities. Images have high granularity and are suitable for high-level tasks such as visual tracking and semantic segmentation. However, they suffer from the privacy issue in smart homes, which is not a problem for WiFi-based sensing. Different from images that can be directly understood and then annotated in pixel level, WiFi CSI data can only be annotated with simple labels that refer to activity or gesture type.

To empower high-level task such as skeleton extraction using wireless signals, cross-modality supervision aims to leverage visual ground truth for training wireless signal, which empowers pose estimation based on wireless signals [26] as shown in Fig. 12.

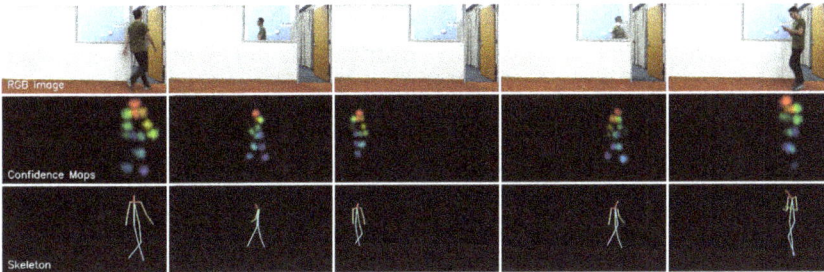

Figure 12: Pose estimation using wireless signals [26].

6.4 *Multi-modality learning for comprehensive sensing*

If camera-based method is integrated with wireless signals that can detect activities through walls, two modalities learn from other's strength and close the sensing gap together. Deep multi-modal learning contributes to more sensible ways of combining and connecting two kinds of features. For example, Zou *et al.* integrate both visual and CSI data for accurate multi-modality activity recognition [27]. In the future, comprehensive sensing will be achieved by the fusion of more modalities such as IMU data or radar data.

6.5 *Model compression for edge computing*

Another problem for WiFi-based sensing is how to deploy the algorithm in the edge devices such as WiFi routers. Smart routers have some computational resources but cannot support real-time inference of deep networks. To accelerate the speed and decrease the computations, model compression comes into existence. It mainly includes three ways including network pruning, quantization and distillation. Network pruning aims to reduce the number of parameters by pruning some connections of neurons or CNN kernels [28]. Quantization methods quantize all the weights to integers. Knowledge distillation transfers knowledge from a deep network to a shallow network, which has been explored for CSI-based sensing [29].

7. Summary

In this chapter, we introduce how deep learning enables accurate WiFi-based smart sensing technology. To begin with, CSI is illustrated theoretically and intuitively by visualizations, and we also investigate how CSI is recorded by two kinds of tools and how CSI reflects human activities. Then we present the detailed CNN that learns robust features from CSI data. When we apply the trained-once model to a new environment, its performance degrades due to domain shift. To adapt the model to a new environment, we further employ adversarial domain adaptation to obtain a target feature extractor that maps the CSI data in the new environment to the original feature space. These approaches are evaluated in the real-world settings to demonstrate their effectiveness and robustness. At last, we summarize the potentials of integrating recent advanced deep learning research with WiFi-based sensing, which will enable more WiFi-based sensing applications in more practical scenarios.

References

1. B. Spencer Jr, M. E. Ruiz-Sandoval, and N. Kurata, "Smart sensing technology: opportunities and challenges," *Struct. Control Health Monitor.* vol. 11 no. 4, pp. 349–368, 2004.

2. J. Yang, H. Zou, H. Jiang, and L. Xie, "Device-free occupant activity sensing using wifi-enabled iot devices for smart homes," *IEEE Internet Things J.* vol. 5, no. 5, pp. 3991–4002, 2018.

3. D. Halperin, W. Hu, A. Sheth, and D. Wetherall, "Tool release: Gathering 802.11n traces with channel state information," *ACM SIGCOMM Comput Commun. Revi.* vol. 41, no. 1, pp. 53–53, 2011.

4. Y. Xie, Z. Li, and M. Li, "Precise power delay profiling with commodity WiFi," in *Proc. 21st Annu. Int. Conf. Mobile Computing and Networking*, ACM, 2015, pp. 53–64.

5. H. Zou, Y. Zhou, J. Yang, H. Jiang, L. Xie, and C. J. Spanos, "Deepsense: Device-free human activity recognition via autoencoder long-term recurrent convolutional network," in *2018 IEEE Int. Conf. Communications (ICC)*, IEEE, 2018, pp. 1–6.

6. H. Zou, J. Yang, Y. Zhou, L. Xie, and C. J. Spanos, "Robust WiFi-enabled device-free gesture recognition via unsupervised adversarial domain adaptation," in *2018 27th Int. Conf. Computer Communication and Networks (ICCCN)*, IEEE, 2018, pp. 1–8.

7. S. J. Pan, Q. Yang, *et al.*, "A survey on transfer learning," *IEEE Trans. Knowledge and Data Engi.* vol. 22, no. 10, pp. 1345–1359, 2010.

8. E. Tzeng, J. Hoffman, K. Saenko, and T. Darrell, "Adversarial discriminative domain adaptation," in *Computer Vision and Pattern Recognition (CVPR)*, vol. 1, 2017, p. 4.

9. W. Wang, A. X. Liu, M. Shahzad, K. Ling, and S. Lu, "Device-free human activity recognition using commercial WiFi devices," *IEEE J. Selected Areas Commun.* vol. 35, no. 5, pp. 1118–1131, 2017.

10. H. Wang, D. Zhang, J. Ma, Y. Wang, Y. Wang, D. Wu, T. Gu, and B. Xie, "Human respiration detection with commodity WiFi devices: do user location and body orientation matter?" in *Proceedings of the 2016 ACM Int. J. Conf. Pervasive and Ubiquitous Computing*, 2016, pp. 25–36.

11. Z. Yang, Z. Zhou, and Y. Liu, "From RSSI to CSI: Indoor localization via channel response," *ACM Comput. Surv. (CSUR)* vol. 46, no. 2, p. 25, 2013.

12. J. Gjengset, J. Xiong, G. McPhillips, and K. Jamieson, "Phaser: Enabling phased array signal processing on commodity WiFi access points," in *Proc. 20th ann. Inte. Conf. Mobile Computing and Networking*, ACM, 2014, pp. 153–164.

13. H. Zou, Y. Zhou, J. Yang, W. Gu, L. Xie, and C. Spanos, "Freedetector: Device-free occupancy detection with commodity WiFi," in *2017 IEEE Int. Conf. Sensing, Communication and Networking (SECON Workshops)*, IEEE, 2017, pp. 1–5.

14. H. Zou, Y. Zhou, J. Yang, W. Gu, L. Xie, and C. Spanos, "Freecount: Device-free crowd counting with commodity WiFi," in: *GLOBECOM 2017–2017 IEEE Global Communications Conference, IEEE*, 2017, pp. 1–6.

15. J. Yang, H. Zou, H. Jiang, and L. Xie, "Fine-grained adaptive location-independent activity recognition using commodity WiFi," in: *Wireless Communications and Networking Conference (WCNC), 2018 IEEE*, 2018, pp. 1–6.

16. J. Yang, H. Zou, H. Jiang, and L. Xie, "CareFi: Sedentary behavior monitoring system via commodity WiFi infrastructures," *IEEE Tran. Vehi. Technol.*, 2018.

17. J. Yang, H. Zou, Y. Zhou, and L. Xie, "Learning gestures from WiFi: A siamese recurrent convolutional architecture," *IEEE Internet Things J.*, vol. 6, no. 6, pp. 10763–10772, 2019.

18. H. Zou, Y. Zhou, J. Yang, W. Gu, L. Xie, and C. Spanos, "WiFi-based human identification via convex tensor shaplet learning," in *AAAI*, 2018.

19. Y. LeCun, Y. Bengio, and G. Hinton, "Deep learning," *Nature*, vol. 521, no. 7553, pp. 436–444, 2015.

20. I. Goodfellow, J. Pouget-Abadie, M. Mirza, B. Xu, D. Warde-Farley, S. Ozair, A. Courville, and Y. Bengio, "Generative adversarial nets," in *Advances in neural information processing systems*, 2014, pp. 2672–2680.

21. W. He, K. Wu, Y. Zou, and Z. Ming, "WiG: WiFi-based gesture recognition system," in *2015 24th Int. Conf. Computer Communication and Networks (ICCCN)*, IEEE, 2015, pp. 1–7.

22. A. Virmani, and M. Shahzad, "Position and orientation agnostic gesture recognition using WiFi," in *Proc. 15th Ann. Int. Conf. Mobile Systems, Applications, and Services*, 2017, pp. 252–264.

23. J. Deng, W. Dong, R. Socher, L.-J. Li, K. Li, and L. Fei-Fei, "Imagenet: A large-scale hierarchical image database," in *2009 IEEE Conf. Computer Vision and Pattern Recognition*, IEEE, 2009, pp. 248–255.

24. Y. Zheng, Y. Zhang, K. Qian, G. Zhang, Y. Liu, C. Wu, and Z. Yang, "Zero-effort cross-domain gesture recognition with Wi-Fi," in *Proc. 17th Annu. Int. Conf. Mobile Systems, Applications, and Services*, 2019, pp. 313–325.

25. J. Snell, K. Swersky, and R. Zemel, "Prototypical networks for few-shot learning," in *Advances in neural information processing systems*, 2017, pp. 4077–4087.

26. M. Zhao, T. Li, M. Abu Alsheikh, Y. Tian, H. Zhao, A. Torralba, and D. Katabi, "Through-wall human pose estimation using radio signals," in: *Proc. IEEE Conf. Computer Vision and Pattern Recognition*, 2018, pp. 7356–7365.

27. H. Zou, J. Yang, H. Prasanna Das, H. Liu, Y. Zhou, and C. J. Spanos, "WiFi and vision multimodal learning for accurate and robust device-free human activity recognition," in *Proc. IEEE Conf. Computer Vision and Pattern Recognition Workshops*, 2019, pp. 426–433.

28. S. Chen, W. Wang, and S. J. Pan, "Cooperative pruning in cross-domain deep neural network compression," in *Proc. 28th Int. Joint Conf. Artificial Intelligence, AAAI Press*, 2019, pp. 2102–2108.

29. J. Yang, H. Zou, S. Cao, Z. Chen, and L. Xie, "MobileDA: Towards edge domain adaptation," *IEEE Internet of Things J.*, vol. 7, no. 8, pp. 6909–6918.

https://doi.org/10.1142/9789811218842_0005

Chapter 5

Deep Learning for Device-free Human Activity Recognition Using WiFi Signals

Linlin Guo*,‡, Hang Zhang†,§,¶, Weiyu Guo†, Jian Fang*, Bingxian Lu*, Chenfei Ma†, Guanglin Li†, Chuang Lin†,¶ and Lei Wang*,‖

*School of Software Technology,
Dalian University of Technology
†CAS Key Laboratory of Human-Machine
Intelligence-Synergy Systems, Shenzhen Institutes
of Advanced Technology(SIAT), Chinese Academcy
of Sciences(CAS), China
‡linlin.teresa.guo@gmail.com
§hang.zhang1@siat.ac.cn
¶chuang.lin@siat.ac.cn
‖lei.wang@dlut.edu.cn

Abstract

WiFi-based device-free human activity recognition applications have been popular for a decade in smart-environment sensing domains. Existing researches extract features of human activity as the input of classic classification methods to train the learning model and then leverage the trained model to recognize activities. The propagation characteristics of WiFi signals are variable for individuals under different place conditions even in the same environment with time. Depending on fixed features, data cannot effectively represent human activity and adapt to a dynamic indoor environment. As deep learning has demonstrated its effectiveness in device-free sensing domains in recent years, deep learning methods have been explored to address difficulties faced by WiFi signals. In this chapter, we focus on how to weaken the accuracy differences among individuals on human activity recognition and improve the robustness in a single indoor environment. Based on this, we design a novel deep learning model called LCED which consists of one LSTM-based Encoder, features image representation and one CNN-based Decoder. LSTM-based Encoder can learn time-sequence representation and encode it to an equal-length vector. Each equal-length vector is represented using features of image representation to compress and keep key details. CNN-based Decoder provides better recognition accuracy by capturing the local effective information of the signal image based on the spatial distribution. Experimental results show that the average accuracy of sixteen activities is high, 95%. The accuracy difference among individuals on activity recognition averagely decreases by 3%.

Keywords: deep learning, device-free, human activity recognition, WiFi signals, signal processing, indoor environment

1. Introduction

Human activity recognition is an important research topic in the human–computer interface domain which explores human activity recognition as the initial step to analyze user's behavior for providing effective guides [1], detect abnormal behavior for elder's safety monitoring, recognize activities for controlling household appliances and identifying user identity. Currently, there are four modalities regarding human activity recognition in different applications according to the related position between objects and sensing devices [2]. (i) Body-worn: worn by the user to describe the body movements such as smartphone, watch, band accelerometer and gyroscope, which is commonly used in health products and is a mature technique for human activity recognition. (ii) Object: built-in objects to capture objects' movements such as intelligent electric appliances, RFID, accelerometer on cup, etc. (iii) Ambient: applied in an environment to capture user behavior and implement user interaction such as sound, camera, WiFi, Bluetooth, etc., which focuses on applying to Smart-Homes like remoting control household appliances and real-time monitoring of the state change of plants using special sensors. (iv) Hybrid: cross-sensor boundary such as a combination of types, often deployed in a smart environment. We focus on four keywords including device-free, deep learning model, human activity recognition, and WiFi signals in an indoor environment. With increasing coverage of WiFi signals in indoor environments, ambient-based modality can leverage changes of WiFi signals to record and recognize human activity.

WiFi signals originally deliver information to implement communication function. Due to the sensitivity of propagated signals on the ambient, WiFi signals can sense changes in indoor environments, named the sensing function. With the rapid development of wireless devices and ubiquitous deployment of WiFi infrastructure in public places and home environments, WiFi-based sensing has attracted increasing academia and industry attention. Compared with wearable sensors, WiFi sensing can provide non-intrusive and continuous sensing of human activities without attaching any devices over the air. WiFi signal using through-wall and wide propagation ability can offset dead-zone and occlusion issues with which vision-based human activity recognition and monitoring applications are confronted. Received signal strength indicator (RSSI) and channel state information (CSI) are two measurements of WiFi signals. RSSI can roughly locate targets [3] and detect human behavior like daily activity or gestures [4]. During the propagation process of WiFi signals, existing multipath effect leads to unstable and unreliable performance for RSSI-based human sensing. CSI can distinguish multiple propagated paths using time-frequency information [5] and implement fall detection [6], sleep monitoring [7], gait recognition [8–10], gesture classification [11–13] and fine-grained vital signs monitoring [14, 15] in indoor environments. The principle behind WiFi-based human activity recognition is that a persons behavior affects wireless signal propagation and causes a unique signal variation pattern. This is a pattern matching problem. The literature [4, 16, 17] extract RSSI waveforms to represent gestures, construct fingerprint profile to recognize human activity and use dynamic time warping (DTW) to determine the periodicity of sleeping. However, the pattern matching method cannot guarantee stable performance of human activity recognition due to time-varying property of WiFi signals in indoor environments. The learning-based methods use features of activity data to train a model and leverage the trained model to infer the corresponding activity.

Recent learning-based researches are divided into two classes: one class [18–20] is to extract effective features from raw activity data to recognize activities using classic classification algorithms; another class [21–23] is to design deep learning models to capture more effective information of raw activity data for improving the robustness of human activity recognition in different scenarios. Table 1 lists related researches on WiFi-based sensing in

Table 1: Summation of existing works regarding wireless sensing.

Reference	Recognition methods	#Person/Scenes	#Activity	#Samples	Accuracy
WiFall [6]	LibSVM (ML)	4/3	Fall, Fall-like	100pkts/s, --	87%
RT-Fall [30]	v-SVM (ML)	6/4	Fall, Fall-like	100pkts/s,5500	91%, 92%
E-eyes [16]	EMD, DTW	4/2	Walking, One-place activities	20pkts/s, --	96%
CARM [18]	HMM+Activity-CSI model	25/3	Daily activities (8)	2500pkts/s,1500	96%
HuAc [19]	SVM	10/3	Activities and Gestures (16)	30pkts/s, 24000	93%
DFLAR [20]	Softmax regression algorithm	--/2	Localization and Activities (4)	--	90%
DFLAGR [21]	Softmax+Deep learning	--/2	Activities and Gestures	--	88%
Shi et al. [26]	DNN+SVM	11/2	Activities	1000pkts/s,3336	94%,91%
Chen et al. [23]	ABLSTM (DP)	13/3	Activity (6+7)	1000pkts/s,--	95%
Niu et al. [25]	Adversarial network	11/6	Activity (7)	200pkts/s	--
WiSLAR [31,32]	Temporal Unet	--	Activity (6)	1394	95%,88.6%
Wi-sleep [7]	Pattern recognition	--	Sleep	20pkts/s	--
Liu et al. [33]	PSD-based K-means clustering	6/3	Sleep	--	--
HSU et al. [27]	DNN	10/8	Sleep	100 nights	--
Zhao et al. [28]	CNN+RNN	25/--	Sleep	100 nights, EEG	79.8%
Smokey [17]	Pattern recognition	--/3	Smoking	30pkts/s	97.6%
WiDance [34]	Light-weight pipeline	30/--	Motion direction	1024pkts/s, 10000	92%
WiDir [35]	Fresnel zone model	--/3	Walking direction	500pkts/s,1289	10°
BodyScan [36]	Light-weight classifier	7/2	Activity, vital sign	269 mins	72.3%,60%
Zhang et al. [29]	Fresnel model, diffraction model	11/--	Activity	200pkts/s	--
WiGest [4]	Pattern recognition	3/2	Gestures (5)	50pkts/s	96%
WiSee [13]	Pattern-matching	5/2	Gestures (9)	--	94%
WiFinger [37]	PCA+MD-DTW	5/2	Finger gestures (8)	20pkts/s	93%

Table 1: (*Continued*)

Reference	Recognition methods	#Person/ Scenes	#Activity	#Samples	Accuracy
WiAG [38]	Translation funciton+k-NN	10/1	Gestures (6)	100pkts/s, 1427	91.4%
Widar3.0 [39]	One-fit-all deep model	16/3	Gestures	1000pkts/s,--	92.7%,92.4%
Zou *et al.* [40]	JADA model	1/2	Gestures (6)	100pkts/s,400	98.75%
QGesture[10]	LEVD	5/3	Gesture (distance, direction)	2500pkts/s,--	15°,3.7cm
WiflU [8]	SVM+RBF	28/1	Gait	--, 2800	93.05%
WiGait [9]	--	25/1	Gait (velocity, stride length)	--	99.8%,99.3%
WiHear [41]	MCFS+DTW	4/6	Lip	100pkts/s	91%
WiKey [42]	DTW+kNN	10/1	Keystroke	2500pkts/s,--	96.4%
Vital-Radio [14]	--	14/--	Breath and heartbeat	448000	99%
BreathLive [43]	LR, SVM, MLP	16/--	Heart	7800pkts/s	91.7%
EQ-Radio [44]	11-SVM	30/5	Emotion recognition	130000	87%,72.3%
Zhang *et al.* [45]	First Fresnel Zone (FFZ)	8/--	Respiration	20pkts/s	98%
FullBreathe [15]	Pattern-based and model-based	8/--	Respiration	100pkts/s	2.4m
FarSense[46]	CSI-ratio model	12/2	Respiration	100pkts/s, 197 hours	8m
WiVit [47]	HMM	5/3	Vitality monitoring	200pkts/s	94.2%
Liu *et al.* [48]	DA, SVM, KNN, RF	--/2	Vital sleep	20pkts/s	--
AutoID [49]	C^3SL	20/3	Human Identification	700pkts/s	91%
Li *et al.* [50]	Spatio-temporal attention module	30/10	Activity (35)	Two datasets	87.7%,93.3%
Ohara *et al.* [51]	DNN+LSTM+HMM	--/3	State changes	1000pkts/s	--
Akbari *et al.* [52]	DNN+Stochastic features	--	Activity	Four datasets	--

terms of recognition methods, testing person/scenes, activity types, sample size and accuracy. In a single indoor environment, researches [19,20,24] using classification algorithms can achieve satisfying results on human activity recognition using extracted time-domain and frequency-domain features which contain widely used CSI amplitude variance, CSI phase difference, signal entropy, Doppler frequency shift and frequency components. The mentioned features use manual-based way which faces restrictions of researchers' subjectivity with respect to collected activity data. Moreover, the trained model on a specific individual in a specific indoor environment typically does not work well when being applied to predict another individual's activity recorded in another indoor environment. To study the two problems, researchers exploit deep learning algorithms [25–29] which can learn time-domain features and spatial information of raw activity data to recognize activity. For example, the proposed EI framework [25] uses deep-learning model to remove the environment and individual of specific information contained in the activity data and extract environment/individual-independent features shared by the collected activity data on different individuals under different environments. High-quality activity data obtained using precision devices is key for the above research, however, the limitation hinders extensive use of WiFi-based researches in indoor environments. The proposed diffraction-based sensing model [29] quantitatively determines the signal change of an individual's movements and explains the corresponding relationship in terms of signal propagation model.

Deep learning model not only captures more enriched data representation of human activity, but also improves the accuracy of WiFi-based human activity recognition according to the above analysis. We want to design a deep learning model to reduce the accuracy differences among individuals on human activity recognition. Compared with the work [25], we focus on individuals while it focuses on environments. On the technical side, the difficulty of our problem is easier but this problem is very important for the home-level indoor environment. We do not remove interference factors as well as the previous work did, but focus on the shared features for various activities using the temporal and spatial models to keep stable recognition accuracy of individuals. The challenge to solve the problem is how to select and determine the problem corresponding to the deep learning structure required in spatial-temporal features learning. In this chapter, we propose an LSTM-CNN Encoder–Decoder model named LCED to reduce the accuracy differences among individuals on human activity recognition. The LSTM-based encoder encodes a time-sequence of WiFi signals into a fixed dimension of gray representation. The inconsistent problem of input and output lengths and time-learning dependencies are solved. The CNN-based decoder decodes the different grayscale representations into the classes of the activities. The encoder and the decoder of the LCED model are jointly trained to maximize the conditional probability of a vector of target labels given source labels.

2. Related Work

The research aims to design a deep learning model to reduce accuracy difference among individuals on human activity recognition using WiFi signals in single indoor environments. Referring to Table 2 summarized by Ma *et al.* [53], the related work according to ways of data representation in the process

Table 2: Summary of WiFi sensing algorithms [53].

Model: $Y = f(x)$, X: CSI measurements, Y: detection, recognition, or estimation results.	**Algorithm**: To find the mapping function $f(\cdot)$ to detect, recognize, or estimate Y given X
Algorithm Type	Examples
Modeling-based: (1) Modeling X by theoretical models based on physical theories or statistical models based on empirical measurements; (2) Inferring $f(\cdot)$ by the model of X; (3) Predicting Y by the modeled function $f(\cdot)$ and measurements of X, sometimes assisted by optimization algorithms.	**Theoretical Models**: Fresnel Zone Model, Angle of Ar-rival/Departure, Time of Flight, Amplitude Attenuation, Phase Shift, Doppler Spread, Power Delay Profile, MultiPath Fading, Radio Propagation (Reflection, Refraction, Diffraction, Absorption, Polarization, Scattering); Statistical Models: Rician Fading, Power Spectral Density, Coherence Time/Frequency, Self/Cross Correlation; Algorithms: MUSIC, Thresholding, Peak/Valley Detection, Minimization/ Maximization
Learning-based: (1) Training: learning $f(\cdot)$ by training samples of X' and Y'; (2) Inference: predicting Y by the learned function $f(\cdot)$ and measurements of X.	**Learning Algorithms**: Decision Tree, Naive Bayes, Dynamic Time Wrapping, k Nearest Neighbor, Support Vector Machine, Self-Organizing Map, Hidden Markov Models, Convolutional/Recurrent Neural Network, Long- Short-Term Memory
Hybrid: Modeling-based $g(\cdot) \rightarrow$ learning-based $f(\cdot)$(1) Modeling the problem by $Y = f(g(X))$; (2) Getting $f(\cdot)$ and $g(\cdot)$ by modeling-based or learning-based algorithms; (3) Predicting Y by the modeled or learned function $f(g(\cdot))$ and measurements of X.	**Modeling-based** $g(\cdot) \rightarrow$ **learning-based** $f(\cdot)$: (1) Extracting mobility data by Doppler Spread $\rightarrow$ recognizing gestures by k-Nearest Neighbor; (2) Estimating position and orientation features by Channel Frequency Response $\rightarrow$ recognizing gestures by k-Nearest Neighbor.

of learning and recognition is divided into pattern-matching-based activity recognition, learning-based activity recognition and model-based activity recognition.

2.1 *Pattern matching-based activity recognition*

Pattern matching method finds unique information corresponding to one class and maps unique information to one to detect, recognize and estimate the target object. The early researches use pattern matching methods to recognize human behavior [4, 7, 13, 15–17, 41]. WiGest [4] extracts three basic primitives (rising edges, pauses and falling edges) from raw signals data and converts the three primitives to a string sequence: rising edges to positive signs, falling edges to negative signs, and pauses to zeros. The extracted string sequence is then compared with the predefined gesture templates to find the best match. WiSee [13] computes Doppler frequency shifts of reflected signals caused by gestures as a unique pattern to match one gesture as speed change of human activity does not change the pattern of positive and negative shifts. Increasing positive Doppler frequencies correspond to gesture acceleration and decreasing positive Doppler frequencies correspond to gesture deceleration. Doppler frequencies profiles corresponding to gestures may consist of patterns with only positive Doppler shifts, patterns with only negative Doppler shifts and patterns with both positive and negative Doppler shifts. Smoking is a rhythmic composite behavior that contains six motions (holding, putting up, sucking, putting down, inhaling and exhaling) in a certain order which makes smoking behavior distinguishable from daily activities such as putting arms up or down. Based on the character, Smokey system [17] leverages the rhythm/order information to recognize smoking behavior.

According to the analysis of the above researches, the challenges of pattern matching methods are to find and extract unique pattern information which can efficiently represent target objects. However, due to the time-varying property of WiFi signals, multipath effect, frequency selective fading, the recorded received signals reflected off human activity cannot keep stable conditions. Therefore, finding and extracting more effective data representation of human activity is necessary for WiFi-based human activity recognition.

2.2 *Learning-based activity recognition*

Learning-based activity recognition is divided into shallow learning (classification algorithms) and deep learning according to the learning character and the requirement of data scale. The shallow learning algorithms using a manual-based way extract effective features from raw activity data to train a model and then leverage the trained model to recognize activities. The deep learning algorithms can automatically learn efficient data representation of raw activity data to detect, recognize and estimate human activities. The learning algorithms used in WiFi-based activity recognition have Decision Tree (DT), Naive Bayes (NB), Dynamic Time Warping (DTW), Random Forest (RF), k-Nearest Neighbor (kNN), Support Vector Machine (SVM), Hidden Markov Models (HMM), Convolutional Neural Network (CNN), Recurrent Neural Network (RNN), Long Short-Term Memory (LSTM) and Generative Adversarial Networks (GANs) [6, 8, 25, 33, 30, 36, 34, 42]. WiFall system [6] extracts seven statistical features in the time-domain to detect single fall using one-class SVM algorithm. The developed RT-Fall system [30] considers effective information of both the timefrequency domain to detect and recognize fall behavior. Due to the propagation characteristics of WiFi signals, WiFi-based activity recognition using shallow learning cannot transfer the designed system to a new individual and new environment. Researchers attempt to leverage the powerful studying of deep learning models to improve the robustness of activity recognition. Several researches [20, 22, 23, 54, 31, 25] explore how to select neural network structures to construct a novel network model for the robust sensing system. These novel deep models help researchers solve simultaneous localization and activity recognition [20], increasing the sample scale [22], constructing body skeleton using wireless signals to sense through-wall objects [54], improving the ability to sense granularity from sample level to data point [31] and weakening differences of indoor environments [25]. Moreover, the combination of shallow learning algorithms and deep learning algorithms as a novel learning perspective attracts the attention of several researchers. For example, Shi *et al.* [26] extract representative features from CSI measurements as inputs of DNN to accurately identify each individual. The above researches provide proof of WiFi-based human activity recognition using deep learning models being superior to pattern-matching methods.

2.3 *Model-based activity recognition*

Model-based human activity recognition is to discover the relationship between the signal pattern and activity by theoretical models based on physical theories [45, 29, 55, 56, 39] or statistical models based on empirical measurements [18, 55]. Theoretical models of WiFi signals contain fresnel zone model [45, 29], multipath fading [56], and phase shift [39] which can analyze the attributes of human behavior and then recognize human behavior. By using diffraction model within Fresnel zone and reflection model without Fresnel zone, the research [29] can quantify the principles of signal propagation in terms of distance, multi-path and dynamic level. However, the research only explains how change of signal is caused by simple behavior in stable indoor environment. For statistical models based on empirical measurements, it discovers and determines the laws of data change by statistical metrics. CARM system [18] aims to construct statistical models based on empirical measurements to reduce the cost of training samples. It established CSI-speed model to estimate the correlation between CSI change and human movement speeds and established CSI-activity model to determine the relationship between the movement speeds of human body parts and specific activity. The two models can infer and build the statistical model between CSI change and specific activity. However, the statistical model needs high-quality data which cannot be satisfied using commercial WiFi devices.

In summary, shallow learning algorithms only can achieve simple effective information to satisfy the basic demand. Deep learning model can excavate more effective information to satisfy the high demand for WiFi sensing in practical applications, however, there is no effective explanation for better performance compared with shallow learning algorithms. Theoretical models can use the principle of signal propagation to understand and extract key information for sensing activity but need precision devices and high-quality data. This chapter focuses on the novel designed deep learning model to improve the robustness of human activity on recognition accuracy.

3. Preliminary

3.1 *WiFi signals*

Wireless signals covered in indoor environment not only provide communication function but also have the sensing function which has been attracting attention from academic and industry researchers inspired by the development of the artificial intelligence domain. WiFi signals can record dynamic changes occurring in the indoor environment during the propagation process. Figure 1 shows the basic propagation model of WiFi signals in indoor environments. During signal propagation process, WiFi signals face the signal attenuation with increasing distance and the time-varying property due to the multipath effect. WiFi signals sent by one transmitter propagate to one receiver through multiple paths, which consist of one Line-of-Sight path (LoS path) and several reflected paths. The LoS path denotes the

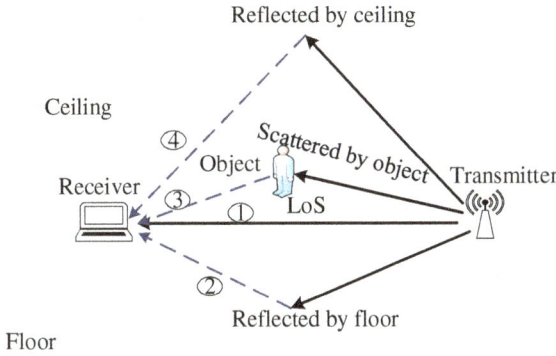

Figure 1:   Signals propagation model.

shortest distance between the receiver and the transmitter. The reflected paths contain static-based reflected paths and dynamic-based reflected paths. Static-based reflected paths caused by static objects like wall, ceiling and floor have stable distance and quantity within a short period in one indoor environment. Dynamic objects change the length of propagating paths, named dynamic-based reflected paths, which can precisely describe objects' behavior. Research institutions exploit WiFi measurements in terms of coarse-grained RSSI and fine-grained CSI to analyze and estimate human behavior using commercial wireless devices. The details of RSSI and CSI are as follows.

3.1.1 *Received signal strength indicator*

RSSI is a sum of signal energy from multiple propagation paths which include a direct path (LoS path) between the transmitter and the receiver, and multiple reflected paths caused by walls, furniture and people in the macro-view. Commodity wireless devices can record and extract RSSI data like a router and smartphone. Figure 2(a) shows RSSI change of three antennae caused by one high-throw behavior. RSSI denotes the received signal power in decibels (dBm) [5]:

$$RSSI = 10\log_2 (\| V\|^2), \tag{1}$$

where V denotes signal voltage. Due to the multipath effect, it is hard to distinguish signal components produced by different paths in an indoor environment. We treat RSSI as coarse-grained sensing information of WiFi signals

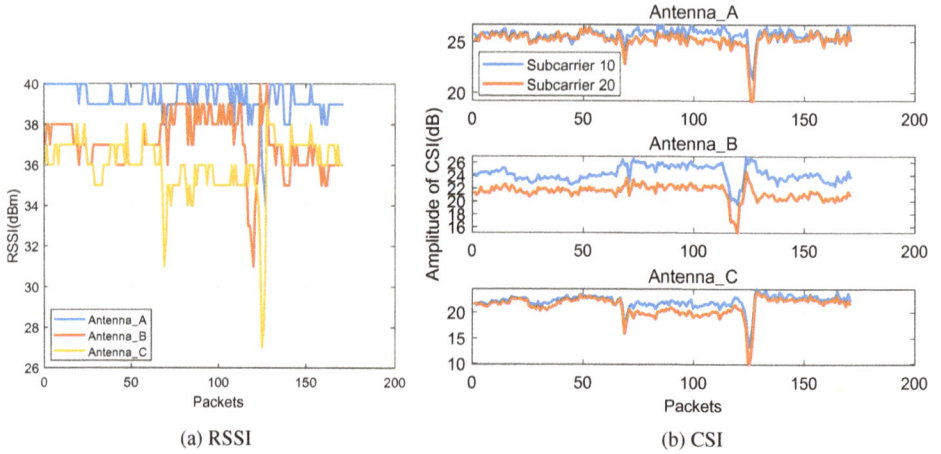

Figure 2: WiFi signal changes caused by one high-throw behavior. (a) Differences between three antennas regarding RSSI changes caused by one high-throw behavior. (b) CSI difference of adjacent subcarriers caused by one high-throw behavior.

to analyze and roughly detect dynamic changes such as human movements. In this chapter, we use RSSI to analyze dynamic changes in a meeting room. Once the variance of RSSI change exceeds the threshold value which depends on a static meeting room, we need to consider the interference of the indoor environment on human activity recognition performance.

3.1.2 *Channel state information*

CSI describes how signals propagate in the wireless channel combining the effect of time delay, energy attenuation and phase shift [6]. Compared with RSSI, Channel Impulse Response (CIR) can describe multipaths. CIR is denoted as follows:

$$h(\tau) = \sum_{i=1}^{N} a_i e^{-j\theta_i} \delta(\tau - \tau_i), \tag{2}$$

where a_i, θ_i, τ_i are the amplitude attenuation, phase shift and time delay of the ith propagation path, respectively. N is the number of propagation paths and $\delta(\tau_i)$ is Dirac delta function. During WiFi signals propagating process, multipath propagation has a property of frequency selective fading on the frequency domain. Leveraging the off-the-shelf Intel 5300 NIC with a modified driver, a group of sampled versions of Channel Frequency Response (CFR) within the WiFi bandwidth is revealed to upper layers in the format of channel state information [57]. CSI of one subcarrier is represented in the following mathematical formula:

$$H(k) = \|H(k)\| e^{j\angle H(k)}, \tag{3}$$

where $H(k)$ is a CSI of the kth subcarrier. $\|H(k)\|$ and $\angle H(k)$ are CSI amplitude and CSI phase, respectively. CSI can capture more fine-grained changes like gestures, breaths and heartbeats. Thirty subcarriers have various sensitivities to the same activity due to existing frequency selective fading. This characteristic is utilized to explore the relationship between signal patterns and activities. As shown in Fig. 2(b), CSI has a unique signal pattern produced by one high-throw behavior. Although the trends of CSI changes for thirty subcarriers are slight different, the trends of adjacent subcarriers are stable.

3.2 CNN and LSTM

3.2.1 Convolutional neural networks

CNN can select a set of convolutional kernels to obtain enrich feature maps in an optimal way and automatically extract spatial features to represent objects. The structure of CNN can be divided into multiple learning stages: convolutional layers, activation function, pooling layer and fully connected layer [58,59]. The convolution operation can extract useful features of local correlation to learn the information related to the task using a series of convolutional kernels [60]. Then, the output of the convolutional kernels is passed to the activation function, which will contribute to learning abstract representation and embedding the nonlinear in the feature space. The nonlinear transformation can learn semantic differences in the image using diversity activation functions such as *sigmoid, tanh, ReLU* and *ELU*. Pooling layer is helpful to compress data and makes the input robust to the geometric deformations [60,61]. Finally, the output of these networks is processed by one or more fully connected layers that interpret what has been read and map this internal representation to a class value. The important attributes of CNN are hierarchical learning, automatic feature extraction, multi-task processing, and weight sharing [62].

- Convolutional layer consists of a set of convolutional kernels that can divide the image into a small region (receptive field). The kernel works by convolving on the image with a specific set of weights. The operation of convolution can be expressed as follows:

$$F_l^k = (I_{x,y} * K_l^k),\qquad(4)$$

where, $I_{(x,y)}$ represents the input image, (x, y) represents the position of specific pixel points, and K_l^k represents the kth convolutional kernel of the lth layer.

- Activation function as a decision function can help in learning complex pattern representation. Choosing an appropriate activation can speed up the learning process. The activation function can be used like the following formula:

$$T_l^k = f_A(F_l^k),\qquad(5)$$

where F_l^k is the output of a convolution operation, which is assigned to the activation function to perform nonlinear transformation. The f_A is a nonlinear operation, which can return the lth layer of transformation output T_l^k.

· Pooling layer preserves key features. The convolution output of the feature patterns may appear at different positions of the image. We only need to preserve the relative position of the features. Pooling layer sums up the similar information of the receptive field and outputs the critical response [63],

$$Z_l = f_p(F_{x,y}^l),\qquad(6)$$

where Z_l represents the lth output feature image, $F_{x,y}^l$ is the lth input feature image after the activation function, and $f_p(*)$ defines the function of the pooling operation.

3.2.2 *Long short-term memory*

LSTMs can avoid gradient explode and gradient vanish problems by designing forget gate, input gate and output gate [64]. The key to LSTMs is the cell state corresponding to one rectangular box A as shown in Figure 3. The details of each gate of LSTM are supported by the related techblog [65].

- Forget gate layer determines what information to be given up from the cell state.

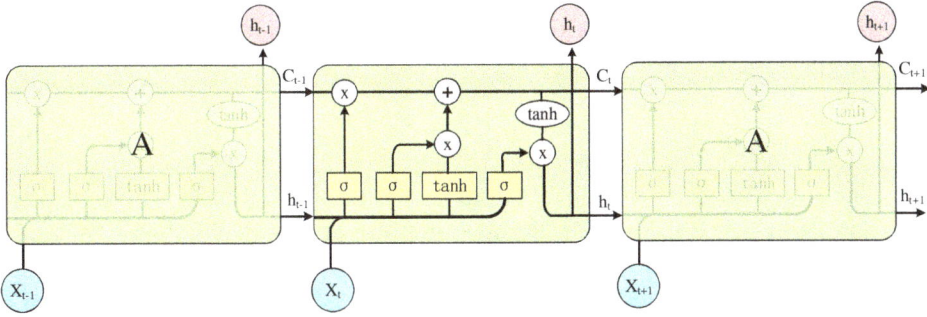

Figure 3: Structure diagram of LSTM.

$$f_t = \sigma\,(W_f \cdot [h_{t-1}, x_t] + b_f), \qquad (7)$$

where h_{t-1} denotes the output of the previous cell state and x_t denotes current input data. The forget gate layer outputs f_t which is between 0 and 1 for each number in the previous cell state C_{t-1} where 1 represents completely keeping this while 0 represents completely getting rid of this.

- Input gate layer decides what new information we are going to store in the cell state. There are two parts. First, a sigmoid layer called the input gate layer decides which values we update.

$$i_t = \sigma\,(W_i \cdot [h_{(t-1)}, x_t + b_i]). \qquad (8)$$

Next, a tan*h* layer creates a vector of new candidate values, $\tilde{C}_t$, that could be added to the cell state.

$$\tilde{C}_t = \tan h(W_C \cdot [h_{(t-1)}, x_t] + b_C). \qquad (9)$$

Then, we combine the two parts to create an update to the cell state.

$$C_t = f_t * C_{t-1} + i_t * \tilde{C}_t. \qquad (10)$$

Meanwhile, the old cell state, C_{t-1}, is updated into the new cell state C_t. The previous steps already decided what to do, we just need to follow them. We multiply the old state C_{t-1} by f_t and forget the things we decided to forget earlier. Then we add $i_t * \tilde{C}_t$. This is the new candidate values, scaled by how much we decided to update each value of cell state.

- Output gate layer decides what we are going to output. First, we run a sig-moid layer that decides which parts of the cell state we are going to output.

$$o_t = \sigma \left(W_o \left[h_{t-1}, x_t \right] + b_o \right). \tag{11}$$

Then, we put the cell state through tan$h$ and multiply it by the output of the sigmoid gate. Finally, we only output the parts we decided to.

$$h_t = o_t * \tanh \left(C_t \right) \tag{12}$$

3.3 *Motivation and solution*

The previous works utilize machine learning algorithms and deep learning algorithms to improve the accuracy of human activity recognition. The accuracy of human activity recognition exists with instability for different individuals and even the same individual in one indoor environment. Based on the problem, we design a deep learning model based on CNN and LSTM to improve the accuracy stability of individuals on human activity recognition. Note that we emphasize the differences among persons regarding the accuracy of human activity recognition not the differences between indoor environments in this chapter.

- The accuracy difference of the same individual. The collected activity data of the same activity with respect to signal patterns have a slight change at different times due to signal propagation. The slight change of signal pattern caused by an activity increases the instability of the activity on human recognition accuracy. The accuracy difference of the same person's activity recognition using machine learning algorithms is between 5% and 10%. The similar research [66] leverages deep learning algorithms reduce the difference between 3% and 7%. The result cannot be up to the demand for practical applications in daily life. To solve the problem, we design the LSTM-CNN encoder–decoder model to reduce generalization error about the accuracy of activity recognition. Compared with similar research [66], we increase three convolution layers and two full-connection layers to learn more effective information from the raw activity data.
- The accuracy differences among different individuals. Each individual has a unique behavior habit during the process of operating an activity. The behavior habit consists of speed, strength and the characteristics of the human body. The speed and strength factors affect signal pattern in terms of frequency, the speed of signal change and signal fluctuation except for the trend of signal change. In other words, the key to recognizing the same activity

operated by different individuals is to depend on the trend of signal change. Based on the above analysis, we design a novel network structure to achieve global characteristics in time-sequence data, not local characteristics.

To solve the above two problems, LSTM-based encoder and CNN-based decoder are comprise of the proposed LCED. The similar research [67] has developed a autoencoder long-term recurrent convolutional network, which focused on using compressed inherent temporal features to human activity recognition, but it ignores the spatial features of the data. Considering the CSI signal is a superposition of the space and time, we use LSTM to extract time features, it can efficiently extract long dependency and compress the data. Besides, we also utilize the spatial feature for model learning by adopting CNN to capture local information, and it can eliminate the noises reflected from the obstacles in the space to some extent. Thus, combing the spatial and temporal features can lessen the environmental noise to reduce individual recognition differentiation. In this chapter, we mainly design a novel neural network model named LCED for learning time-sequence signals data. The LCED model consists of LSTM-based encoder and CNN-based decoder to enhance the learning ability and weaken accuracy differences among individuals in human activity recognition.

4. LCED Neural Network

To improve the robustness and universality of the multi-class activities' recognition under the diverse individuals, we propose a novel encoder-decoder model called LCED. The architecture of the LCED model is described in Figure 4, which has 12 layers including the LSTM layer, convolutional layer

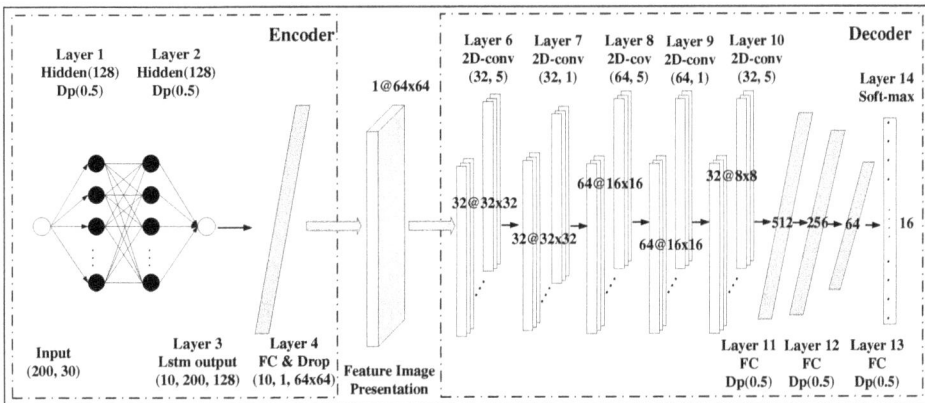

Figure 4: LCED model: LSTM-based encoder and CNN-based decoder

and fully connected layer. First, the raw signals data are treated as inputs of the encoder to extract the fixed featured vector. Then, the vector is transformed into a featured image. Finally, the spatial features of the image are extracted and classified by CNN. The idea to design the LCED model is based on the following principles: (i) The diverse activities have the distinct clusters in the space according to the visualization of PCA technique, and the featured image presentation produces diverse intensity distribution among the activities. (ii) The encoder part of the LCED is capable of learning the time-series representation and can encode it to a fixed-length vector using nonlinear transformation. What is more, the fixed-length feature vector is automatically extracted by the proposed model from the raw signals compared with the original methods based on the manual way, which can adaptively extract features and compress the key information corresponding to the differences among the individuals and activities. (iii) The decoder part is based on CNN, which can effectively capture the local features of the image based on the spatial distribution and has a better classification accuracy. In general, the proposed model has better robustness and can be more generalized due to the adaptive spatio-temporal feature extraction.

4.1 *The encoder-LSTM*

Our encoder has three layers, and it is designed by the LSTM to extract the fixed featured vector. The input to the encoder consists of a 200-length sequence of the 10 batch size with 30 channel subcarriers. The first two hidden layers are LSTM block layers, each of which consists of 128 hidden units with dropout 0.5. The last hidden layer is a fully connected layer that can compress the output of the LSTM into a fixed feature vector of size 64 * 64. In this encoder, the time t is from 1 to 200, and the last output is used for feature extraction and reduction by the FC (Full Connected) layers. The output of the featured vector output is as follows:

$$h_t = \psi(x_t, c_{t-1}, h_{t-1}), \tag{13}$$

$$y_t = \sigma(W_y h_t + b_y), \tag{14}$$

where x_t is channel state information (CSI) after filtering and y_t denotes the output of LSTM, respectively. The c_t represents the status of a cell unit and h_t is the hidden vector for each layer. The σ is an activation function and *logsoftmax* is regarded as the activation function. The W_y is the matrix weight needed to be optimized. The parameter b_y is a bias in the activation function. After passing through the LSTM layers, the LSTM output is a fixed vector and then feeds into an FC layer for features extraction and reduction. Finally, we can get the output with a 64 * 64-dimensional featured vector.

4.2 *The decoder-CNN*

For the designed decoder, there are nine layers. The output of the encoder is transformed to feature image presentation of size (64 * 64) regarded as the input of the decoder. The first five hidden layers are combined with convolutional layers and locally connected layers. The convolutional layers consist of 32 filters of 5 * 5 with stride 1 and 64 filters of 5 * 5 with stride 1, which can extract the spatial feature distribution corresponding to the image intensity and the local feature representation. The two locally connected layers consist of 32, 64 non-overlapping filters with 1 * 1 added after each convolutional layer. The convolution layers operation can be equal to the downsampling, which can learn more local features and extract high-level abstract feature representation. The locally connected layers (1 * 1) can be equivalent to the nonlinear transformer using fewer parameters to augment the ability of the model to learn more complex features [68]. The next three layers are FC and composed of 512, 256, 64 units with dropout 0.5, respectively. The decoder ends up with an FC with a *logsoftmax* function to classify 16 action categories. The detailed calculation can be expressed as follows:

$$\sigma_m^l(j) = b^l(j) + \sum_{i=1}^{i \le N_{kernel}^i} (I_{i,j}^{l-1} k_1(l,m,i) + I_{i,j+1}^{l-1} k_2(l,m,i)), \qquad (15)$$

$$I^l = [\sigma_1^l(j), \sigma_2^l(j), \sigma_3^l(j), \ldots, \sigma_m^l(j)]^T \qquad (16)$$

$$out\_class = \max (FC(\sigma * W + b)) \qquad (17)$$

where $b^l(j)$ refers to the bias in the *l* layer for the neuron. The N_{kernel}^l is the number of the kernel used in the layer. The $k_1(l, m, i)$ is the filter used in the neural network. The scalar of *l*, *m*, *i* and *w* represent the layer, the map, the *i*th kernel and the position in the kernel, respectively. $I_{i,j}^l$ represents the *I*th element which the *j*th neuron in *i*th kernel of the features extraction maps after convolution from the previous layers or the initial elements. Then, *logsoftmax* function is applied to maximize the probability distribution of the last output layer $FC(\sigma * W + b)$ to get the classification category corresponding to the given signals.

In the proposed model, the batch normalization (BN) is adopted after each convolutional layer, which can effectively avoid the phenomena of the internal covariate shift and optimize the parameter settings of the model [69]. Similarly, the *ReLu* function is added after each decoder layer to improve the learning ability of the model and it enhances the robustness among the diverse individuals.

5. Experiments and Implementation

5.1 *Experimental setting*

We aim to design a novel deep learning model to reduce the accuracy of differences among individuals in human activity recognition. Based on this thought, we select one meeting room with several furniture as the main experimental scenario in which the length and width of the meeting room are 10 m and 6 m, respectively. Figure 5 shows the details of the meeting room in terms of furniture layout, device location and the size. We collect activity data from the CSI-TOOL [70] which can be widely used for WiFi-based applications. The CSI-TOOL platform consists of two T400 laptops within Intel WiFi Link 5300 (wl5300) 802.11n NICs and use a monitoring model to guarantee the stability of collecting activity data. One T400 laptop with one antenna is treated as a transmitter to transmit signals data and another one with three antennae as a receiver receives the propagated signals data from different reflected paths. The transmitter and the receiver form one wireless link and run on 5GHz to get free from crowded 2.4 GHz

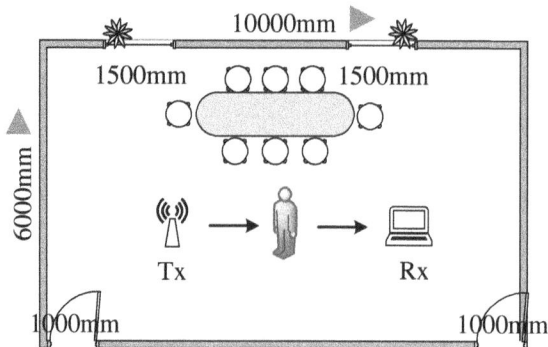

Figure 5: Experimental environment.

interference as commerical routers that are inexpensive only support 2.4 GHz. The transmitter and the receiver install the same CSI-TOOL codes using Matlab 2010 and Ubuntu system 12.04. The packet delivery fraction is 30 pkt/s, which is widely used in WiFi-based applications. The packet delivery fraction can keep sensory ability and does not affect normal network communications and applications. The collect platform can not only receive real-time signals data but also accomplish real-time signals pattern display in the process of doing activities. It can help us to discover the low-quality signal data without delay and improve the quality of the collected signals.

5.2 *Activity data*

We describe details of activity data and leverage PCA technique to estimate the quality of the collected WiFi activity data. We use activity data of 10 volunteers as the first group activity data to evaluate performance of LCED model on reducing accuracy difference of individuals. Ten volunteers, comprising four females and six males whose ages are from 18 to 30 and the heights are from 160 cm to 190 cm are studied. The distance between transmitter and receiver is 4 m and the height between the floor and antennas is 0.8 m. Note that a volunteer locates in the middle of the two laptops. The second group activity data steps from collecting 16 activities of one volunteer with three heights in the meeting room. The heights are 0.6 m, 0.9 m and 1.2 m, respectively. The third group activity data consists of 16 activities of one volunteer with three distances collected in the indoor lobby. The three distances are 1 m, 3 m and 6 m, respectively. The 16 activities are Horizontal arm wave, High arm wave, Two hands wave, High throw, Draw x, Draw tick, Toss paper, Forward kick, Side kick, Bend, Hand clap, Walk, Phone call, Drink water, Sit, and Squat, respectively. Each activity has 30 samples and each sample has time-sequence signal data which is extracted in the format of a 30 * packets * 3 matrix. The format describes 30 subcarriers, the length of time-sequence signal data, and three antennas.

As we know, the data error brought by the collected activity data offsets the differences among the individuals' activities. To explore the accuracy difference of a single person and multiple people regarding activity recognition, we first ensure the quality of collected activity data and reduce the influence of activity dataset on human activity recognition. We use eigendecomposition-based PCA to analyze the distribution of each activity in activity dataset and roughly evaluate the quality of activity data. Figure 6 demonstrates the

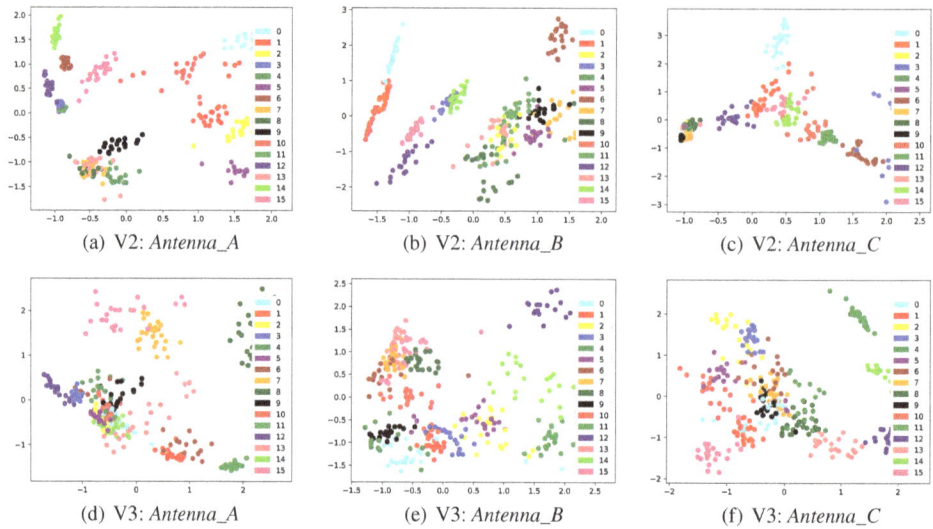

(a) V2: Antenna_A   (b) V2: Antenna_B   (c) V2: Antenna_C

(d) V3: Antenna_A   (e) V3: Antenna_B   (f) V3: Antenna_C

Figure 6: The visualization of the two volunteers' spatial category distribution in 2D-based PCA.

activity distribution of each antenna for randomly selecting two volunteers. The activity data of volunteer2 is superior to volunteer3's data because the former activity data has a distinct boundary of each class. However, the latter cannot distinguish some activities from the 16 activities data due to the overlap of some activities. Once the bad/low quality of activity data occurs, we analyze the reasons and design two ways to deal with them.

- Bad-quality data caused by device faults. The device faults include the poor contact between antennas and the mainboard, reading an issue of the wireless network card and high drop rates caused by a bad network environment. Device faults decrease the analyzability of the collected activity data due to large empty packets or irregular change of time-sequence data corresponding to an activity. For the bad quality activity data caused by these reasons, we directly ignore these activity data.
- Low-quality data caused by individual behavior. Individual behavior issues stem from non-standard activity, body shaking and velocity change of human activity. The non-standard activity and body shaking can change the corresponding signal pattern and the corresponding frequency. Human activity with high/slow speed can change the corresponding frequency but cannot change the trend of the signal pattern. Compared with the bad-quality activity data, the low-quality activity data have a chance to be recovered using the data analysis technique.

5.3 *Pre-processing*

Pre-processing aims to get high-quality activity data by removing outlier and reducing noise data. During this work, we design an LSTM-CNN-based deep learning model which directly uses filtering signal data to reduce the accuracy difference of human activity recognition caused by individuals. To make a comparison between classification algorithms and the proposed deep learning model regarding human activity recognition, we need to deal with raw activity data to meet the demands of the data format for the above classification algorithms. Note that the emphasis of this work still focuses on the proposed deep learning model. The common step of the two ways is to utilize a low-pass filter technique to remove high-frequency signals corresponding to noises and remain with low-frequency signals corresponding to the true activity. After that, we design a variance-based moving window to determine the start and the end of time-sequence activity data corresponding to the activity. The remaining data is treated as the goal of the pre-processing phase and directly as the input of LCED. However, we still extract features as the input of classification algorithms.

- Classification-based pre-processing needs to extract effective features in a manual-based way. We extract statistical features of the time-sequence signals corresponding to an activity as inputs of classification algorithms such as KNN and SVM. Before extracting statistical features, we use an iteration method to determine the optimal length of the sliding window to smooth the time-sequence signals. The statistical features include CSI mean, CSI variance and CSI standard deviation.
- LCED-based pre-processing leverages variance-based moving window function and adaptive threshold to determine the start and the end of the time-sequence signal corresponding to an activity. And then LCED model needs to obtain equal-length sample by cutting out the rest of the more than 200 packets and be filled with less than 200 packets for given time-sequence signals. Due to the difference of activities in the time cost, we set a size of 200 packets for each activity sample to meet the need of data format for the proposed deep learning model.

5.4 *Featured image presentation*

In this work, we design a novel activity recognition method based on featured image presentation. The output of the encoder which is the 4096-dimensional vector is converted into an image (64 * 64) by linear transforming. Once

| (a) *Activity* 1: FIP | (b) *Activity* 4: FIP | (c) *Activity* 6: FIP | (d) *Activity* 8: FIP |

| (e) *Activity* 1: FIP | (f) *Activity* 4: FIP | (g) *Activity* 6: FIP | (h) *Activity* 8: FIP |

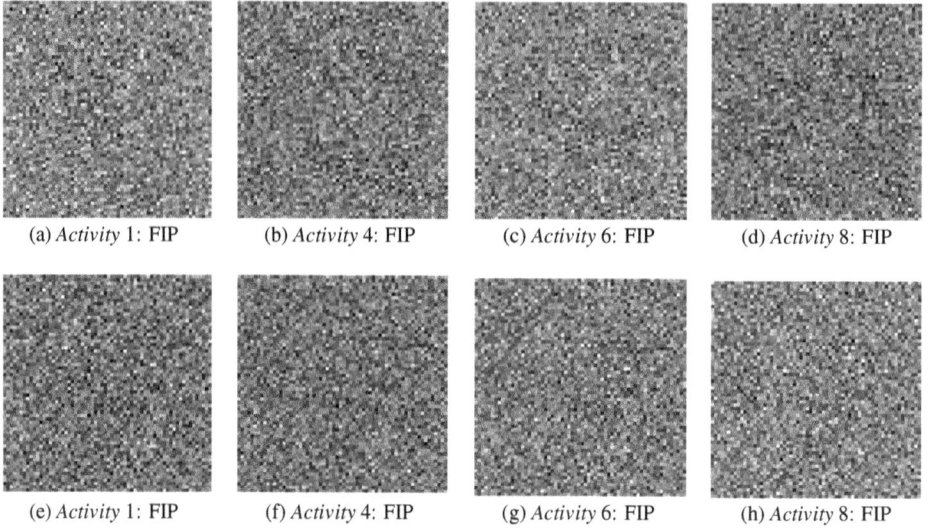

Figure 7: FIP of five activities on the *Antenna_A* for two volunteers.

time-sequence signal data is transformed into an image, we directly use image-based techniques to analyze and extract key information of signal data for high-stability performance. Figure 7 describes the Featured Image Presentation (FIP) of five activities on the *Antenna_A* for two volunteers. FIP records the spatial distribution of reflected signals caused by human activity. In this way, it can integrate the spatial with temporal characteristics of the raw signals. Besides, the max–min is applied to normalize the pixel of the image for improving the training speed.

5.5 *Training of LCED*

We evaluate the model for seven individuals. The implementation of the model is based on Pytorch framework, and the model is trained in one NVIDIA GeForce GTX 1080Ti GPU with approximately 27,335MB. In the processing to train the LCED model, the datasets are divided with a ratio of 3 to 1, and the batch size is set as 10 for batch gradient derivation. We apply 1000 trials for model training to find the minimum error point. Regraded as the optimizer, the *Adam* is used to learn adaptive gradient for different weights, which is initialized with $lr = 0.001$ and $\beta_1 = 0.9$, $\beta_2 = 0.999$, $\varepsilon = 10^{-8}$, and the *CrossEntropyLoss* function is applied to calculate the error loss between the prediction and the real to backpropagation derivation. Moreover, the dropout with 50% is adopted after each fully connected layer to prevent overfitting. After completing each training epoch, the LCED model is set as test mode, and then we experiment with the entire testing set to evaluate the performance to fine tune the parameters.

5.6 *Reference models*

To verify the stability and accuracy of individuals in human activity recognition, we compare the proposed LCED model with the diverse reference models in the time-series classification problem. The details of the contrast methods are described as follows:

- KNN is widely used in a multi-class classification problem. We implement the KNN program using the scikit-learn library. Firstly, we manually extract the mean value of 30 subcarriers of each antenna as features of WiFi signals and normalize the processed signals. Next, the PCA is applied to reduce the feature dimension to 3D space. Lastly, the KNN model is built with 3 neighbors to train and test the entire set. For the evaluation of the performance, we calculate *Accuracy, F1Score* and *Recall,* which are served as general evaluating metrics.
- SVM algorithm is often used on nonlinear binary classification tasks. The SVM uses the same data preprocessing and model training samples as the KNN. For the performance evaluation of human activity recognition, we calculate the accuracy and plot the confusion matrix for 16 activities. The confusion matrix can provide a clear relationship among these activities and determine similar activities.
- sRNN consists of 128 single hidden layers. The same preprocessing method for KNN is applied to divide the datasets with equal length. For hyperparameters, the sRNN is initialized in the same way as the proposed LCED. Besides, the K- Average operation is utilized to estimate the accuracy of the LCED performance.

6. Evaluation

The section mainly analyzes and estimates the performance of WiFi-based human activity recognition using the proposed LCED model in terms of different individuals, different distances between one transmitter and one receiver, and different heights between wireless devices and the floor. The details of these analyses are as follows.

6.1 *Performance of individuals*

The research goal is to analyze the impact of individuals' differences on the accuracy of human activity recognition. We use the proposed LCED model to evaluate the performance of individuals in human activity recognition using WiFi CSI. More-over, we also leverage three methods: KNN, SVM, sRNN as contrast methods to compare with the proposed LCED. Table 3 shows the performance of 16 activities of *Volunteer* 2 using KNN. We take *Antenna_A* as an example and analyze recognition performance of 16 activities in terms of *Precision*, *Recall* and *F1-score*. *Precision* is the number of the right samples from all samples which are predicted an activity. *Recall* denotes the number of right samples from 30 samples of an activity. The precision of *Activity* 1 is 1.00 and *Recall* is 0.83. The above two metrics show that several samples of *Activity* 1 are misidentified as another activity. *Activity* 3, *Activity* 4, *Activity* 12, *Activity* 13 and *Activity* 15 achieve recognition accuracy of 100%. However, only *Activity* 4, *Activity* 13 and *Activity* 15 achieve recognition accuracy of 100% for the same volunteer using *Antenna_B's* activity data. By combining with results of the two antennas on human activity recognition, average

Table 3: KNN performance for one individual with two antennae.

	Antenna_A			*Antenna_B*		
	Precision	Recall	F1-score	Precision	Recall	F1-score
0	1.00	0.83	0.91	0.62	0.83	0.71
1	0.89	1.00	0.94	0.88	0.88	0.88
2	1.00	1.00	1.00	0.56	0.90	0.69
3	1.00	1.00	1.00	0.79	1.00	0.88
4	0.50	0.62	0.55	0.62	0.38	0.48
5	0.91	0.91	0.91	0.57	0.73	0.64
6	1.00	1.00	1.00	1.00	0.73	0.84
7	0.57	0.80	0.67	—	—	—
8	0.43	0.43	0.43	1.00	0.71	0.83
9	0.77	0.91	0.83	1.00	0.27	0.43
10	1.00	0.91	0.95	0.91	0.91	0.91
11	1.00	1.00	1.00	0.88	0.88	0.88
12	1.00	1.00	1.00	0.83	1.00	0.91
13	1.00	0.43	0.60	0.45	0.71	0.56
14	1.00	1.00	1.00	1.00	1.00	1.00
15	1.00	0.71	0.83	0.70	1.00	0.82

accuracy of 16 activities using KNN achieves 80% except for *Activity* 5 and *Activity* 8 due to low-quality data. In summary, KNN can roughly detect and recognize daily activity using WiFi signals in a simple indoor environment.

Differences of individuals find expression in recognition accuracy of each activity in the same indoor environment using the same classification algorithm. We select confusion matrix of activity recognition accuracy to analyze relationship of each activity. Figure 8 shows the confusion matrix of two volunteers for human activity recognition using SVM algorithm. For *Volunteers* 2's confusion matrix, the *Antenna_A*'s performance is superior to the *Antenna_B* and the *Antenna_C* as the confusion matrix of *Antenna_B* occurs with few misjudgement samples. Moreover, there are some same activities with low accuracy in three antennae like *Activity* 5 because other activities are mistaken for *Activity* 5, *Activity* 2, *Activity* 12, *Activity* 13, *Activity* 15, *Activity* 16 of the 16 activities achieve good performance for *Volunteer* 2 and *Activity* 1, *Activity* 2, *Activity* 12 and *Activity* 13 also achieve better performance for *Volunteer3*. The similar activities keep similar performance for KNN and SVM such as *Activity* 12 and *Activity* 13. According to the above analysis, we know that recognition algorithms depend on extracted features to recognize human activity, which leads to huge difference in accuracy of recognition of the same activity for the same volunteer.

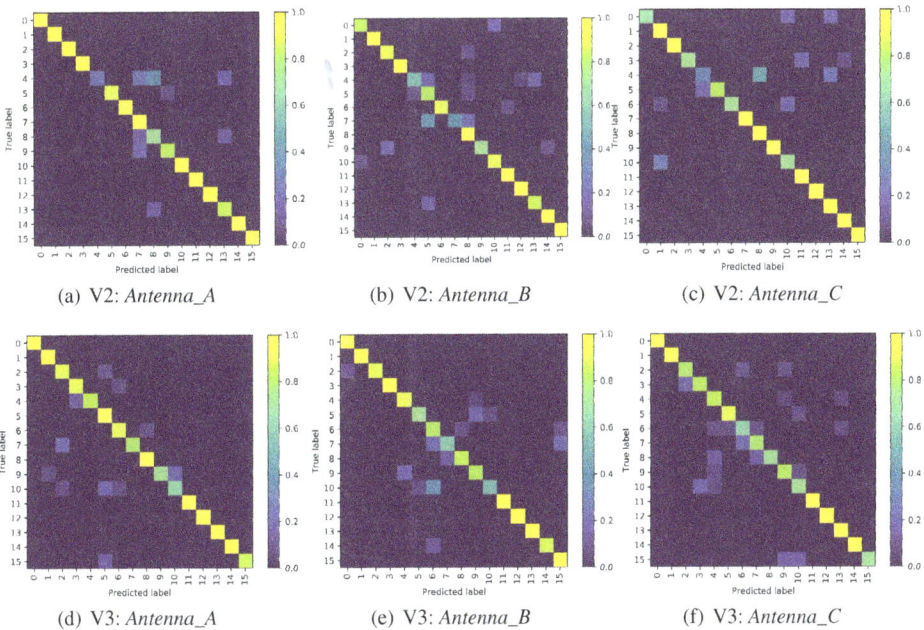

| (a) V2: *Antenna_A* | (b) V2: *Antenna_B* | (c) V2: *Antenna_C* |
| (d) V3: *Antenna_A* | (e) V3: *Antenna_B* | (f) V3: *Antenna_C* |

Figure 8: Two volunteers and three antennae: the accuracy of 16 activities using SVM.

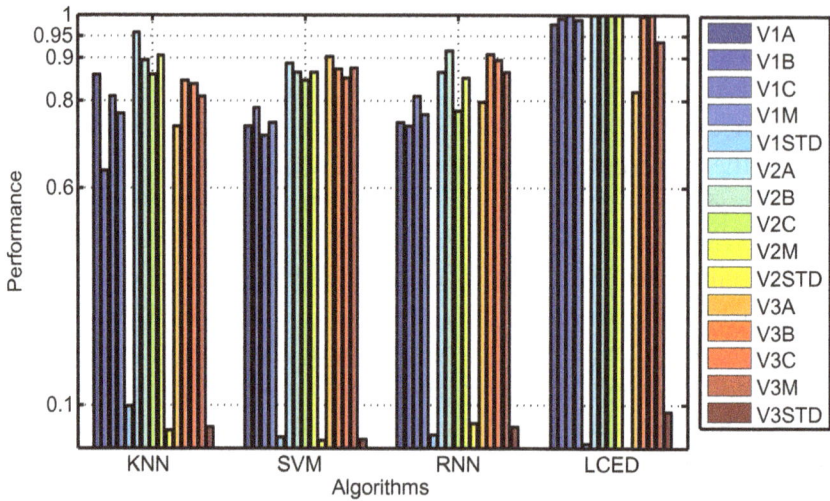

Figure 9:   Performance analysis of four algorithms. *V1A* is the accuracy of *Volunteer* 1's activity data on *Antenna_A*. *V1M* denotes the average accuracy of *Volunteer* 1's activity data on three antennas (*Antenna_A*, *Antenna_B*, *Antenna_C*). *V1STD* is the standard deviation of *Volunteer* 1's activity data on three antennas.

We randomly select activity data of three volunteers as analysis objects to estimate the influence of three antennas on activity performance using four recognition algorithms. The experimental results are shown in Figure 9. According to the comparison of three volunteers regarding recognition accuracy using KNN, *Volunteer* 2 achieves an average accuracy of 90% and *Volunteer* 3 shows high 80% recognition accuracy. The worst accuracy is *Volunteer* 1 with average 77%. *Antenna_A* of *Volunteer* 1 and *Volunteer* 2 on activity recognition accuracy achieve 5% higher than other antennas. The stability of *Volunteer* 3 on recognition accuracy using KNN is superior to *Volunteer* 1 and *Volunteer* 3. Although the accuracy of *Volunteer* 1 is still lower than the other two individuals using SVM, RNN, the average stability of recognition accuracy on three antennas is higher than the other two individuals. Compared with KNN, SVM and RNN, the accuracy of activity recognition increases 10% on average using the proposed LCED model. Moreover, the average stability of three volunteers on the accuracy of human activity recognition is superior to the other three algorithms. In summary, the important performance achieved by the proposed LCED not only reduces the difference of recognition accuracy on three antennas but also enhances the robustness of activity recognition on individuals.

We focus on the average difference of each antenna on human activity recognition for individuals using WiFi signals. Figure 10 shows performance

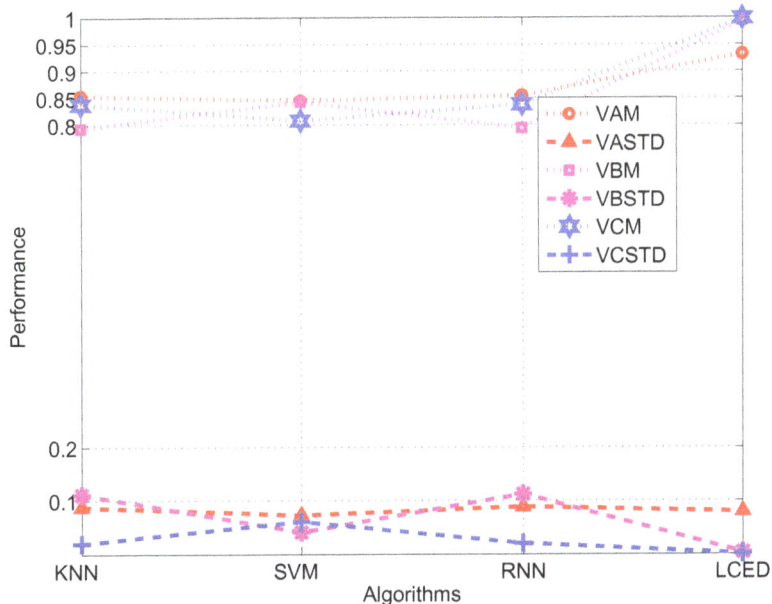

Figure 10: Performance analysis of three antennas under four algorithms. *VAM* is the average accuracy of all *Antenna_As*; *V ASTD* is the standard deviation of all *Antenna_As*.

difference of each antenna for seven volunteers under the four recognition algorithms. We take *Antenna_As* of seven individuals as an example to analyze their performance. The average accuracy of seven volunteers on *Volunteer_A* is superior to *Antenna_B* and *Antenna_C* except for LCED's result. However, the stability of *Antenna_A* is lower than the other two antennas according to the *VASTD*. In other words, the difference of recognition accuracy of the seven volunteers on *Antenna_A* is huge. The three antennas have stability accuracy with a high of 80% average accuracy, respectively, and up to 95% accuracy by using LCED model. The standard deviation of the three antennas is less than 0.1 for KNN, SVM and RNN. Compared with the above, the standard deviation of the *Antenna_B* and *Antenna_C* using LCED model is less than 0.01.

We discuss difference of three antennas of individuals under the prosed LCED. Figure 11 lists recognition accuracy of three antennas of seven individuals using LCED model. The average accuracy of three antennas regarding activity recognition is higher than 93% and the standard deviation is less than 0.02 except for the performance of *Volunteer* 3. Compared with the above-mentioned three recognition algorithms, accuracy of *Volunteer* 3's three antennas is superior.

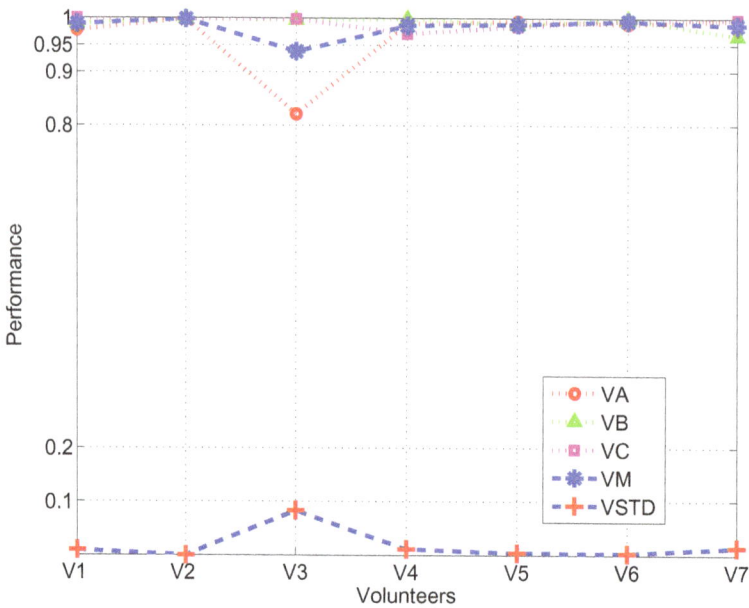

Figure 11: Performance analysis of individuals. *VA* is the accuracy of a *Antenna_A*; *VM* is the average accuracy of three antennas; *VSTD* is the standard deviation of three antennas.

6.2 *Performance of different distances*

Different distances between transmitter and receiver produce important influence on reflected signals by human activity and directly affect the recognition accuracy of human activity in indoor environments. We use the proposed LCED model to study the impact of three levels of distances on human activity recognition. We first discuss why we select 1 m, 3 m, 6 m as examining distance. As we know, the communication distance of WiFi signals is less than 15 m using a commercial wireless router and the sensing distance with satisfying accuracy is less than 5 m. Once interference occurs or in the complex indoor environment, the sensing distance is less than 3 m. Figure 12 shows the accuracy difference of the three distances on human activity recognition. The accuracy of human activity recognition obtained in 1 m using KNN is superior to the other two distances and is between 80% and 90%. SVM and LCED reduce the influence of distance on human activity recognition. The accuracy difference between three distances is less than 0.5% for SVM and 0.8%. Compared with KNN, SVM and RNN algorithms, although the proposed LCED model has no best performance, the accuracy stability of three antennas in activity recognition is the best, as with SVM. This performance indicator is an important factor to evaluate the applicability of the proposed LCED model in a practical indoor environment.

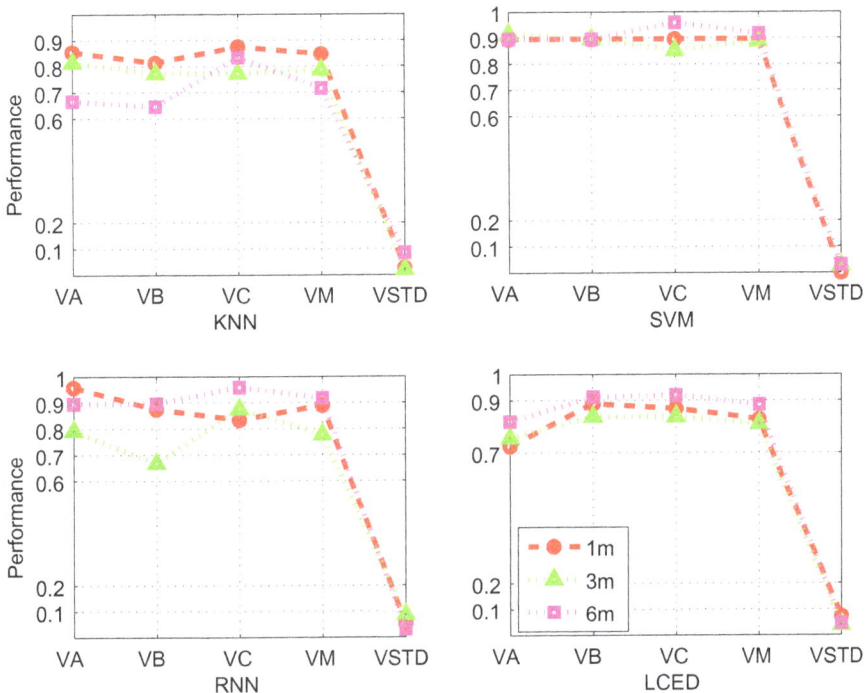

Figure 12: Effect of different distances on human activity recognition.

Compared with the accuracy of the other two distances, the accuracy of activity recognition at 6 m distance is the best under SVM and LCED and this result contradicts the previous one held by common sense. We discuss the interesting issue that occurred in experimental results. The issue is what is the limitation of sensing distance using commodity wireless devices for WiFi-based human activity recognition. From the results shown in Fig. 12, the distance of 6 m cannot be up to the limit of sensing distance [46, 71]. Therefore, the following research will explore the issue.

6.2.1 *Performance of different heights*

We collect activity data at three heights to evaluate the impact of heights on human activity recognition using the proposed LCED model. We first explain why we select 60 cm, 90 cm, 120 cm of heights as evaluated standards, respectively. The 60 cm of height corresponds to the lower-body to capture the lower-body activities like a side-kick. The 90 cm corresponds to the torso part to sense whole-body activities like running. The 120 cm of height corresponds to upper-body activities to recognize gestures like hand wave. Figure 13 shows the accuracy of three heights on human activity recognition

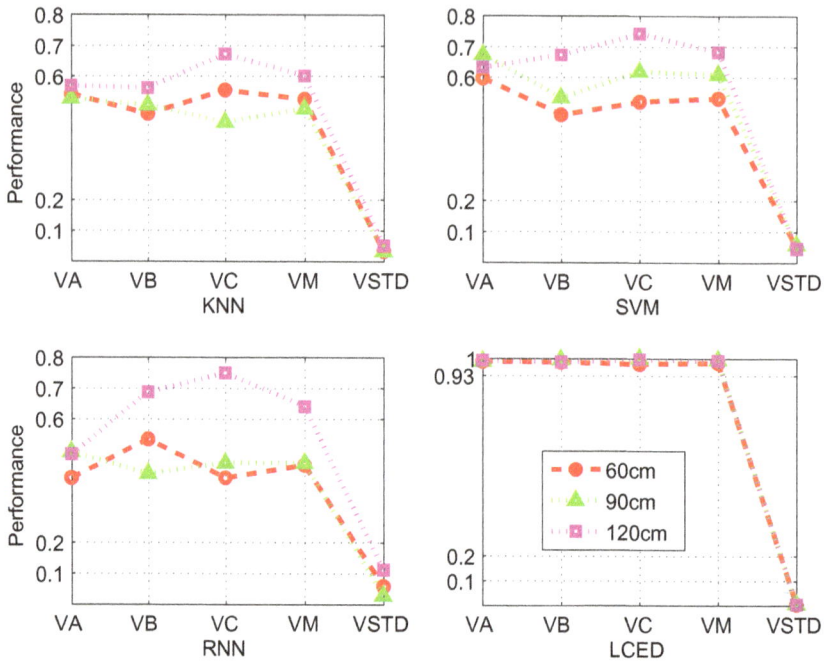

Figure 13:   Effect of different heights on human activity recognition.

using four algorithms. The results of KNN, SVM, RNN is less than 80% due to the low-quality activity data. The reason stems from additional interference caused by non-standard activities. The proposed LCED model can offset the passive effect of additional interference on activity recognition and achieve an average accuracy of 98.5%. Compared with the accuracy of the other two heights, the accuracy of activity recognition on 120 cm is best. According to the accuracy analysis of three heights on human activity recognition, we observe that the proposed LCED model can not only offset the passive effect of low-quality activity data but also keep the high stability of the accuracy on different heights.

7. Conclusion

The development of WiFi-based human activity recognition researches depends on signal processing, feature extraction and recognition methods. With increasing data scale in wireless sensing domain, the ability of recognition methods to deal with big data and capture key information is becoming increasingly important for WiFi-based human activity recognition. In this work, we compare classification algorithms with deep learning model on the performance of human activity recognition. Combining with recent researches using deep learning methods widely used in wireless sensing domain, we propose a novel deep learning model named LCED to learn the spatial and temporal features of WiFi signals and reduce the accuracy differences among individuals on human activity recognition. To be specific, we design a series of pre-processing methods to obtain high-quality activity data and use KNN, SVM, RNN as comparison methods to analyze the performance of LCED model on human activity recognition. The results show that the comparison methods achieve the accuracy of 75%–85%, and LCED model achieves the average accuracy of a high 95% with less cost. The accuracy difference in activity recognition among individuals averagely reduces to 3%. In the following research, we will focus on the accuracy difference of indoor environments on human activity recognition and attempt to design a deep mutual learning model to improve the stability of activity recognition under different indoor environments.

References

1. Y. Zeng, P. H. Pathak, and P. Mohapatra, "Analyzing shopper's behavior through WiFi signals," in *Proc. ACM WPA,* 2015.
2. J. Wang, Y. Chen, S. Hao, X. Peng, and L. Hu, "Deep learning for sensor-based activity recognition: A survey," *Pattern Recogn. Lett.* vol. 119, pp. 3–11, 2019, deep Learning for Pattern Recognition. [Online]. Available: http://www.sciencedirect.com/science/article/pii/S016786551830045X
3. P. Bahl and V. N. Padmanabhan, "RADAR: An in-building RF-based user location and tracking system," in *Proc. IEEE INFOCOM,* 2000.
4. H. Abdelnasser, M. Youssef, and K. A. Harras, "WiGest: A ubiquitous WiFi-based gesture recognition system," in *Proc. IEEE INFOCOM,* 2015.
5. Z. Yang, Z. Zhou, and Y. Liu, "From RSSI to CSI: Indoor localization via channel response," *ACM Comput. Surv.* vol. 46, no. 2, pp. 25:1–25:32, 2013.
6. C. Han, K. Wu, Y. Wang, and L. M. Ni, "WiFall: Device-free fall detection by wireless networks," in *Proc. IEEE INFOCOM,* 2014.
7. X. Liu, J. Cao, S. Tang, and J. Wen, "Wi-Sleep: Contactless sleep monitoring via WiFi signals," in *Proc. IEEE RTSS,* 2014.
8. W. Wei, A. X. Liu, and S. M, "Gait recognition using Wi-Fi signals," in *Proc. ACM UbiComp,* 2016.
9. H. Chen-Yu, L. Yuchen, K. Zach, H. Rumen, K. Dina, and L. Christine, "Extracting gait velocity and stride length from surrounding radio signals," in *Proc. ACM CHI,* 2017.

10. N. Yu, W. Wang, A. X. Liu, and L. Kong, "QGesture: Quantifying gesture distance and direction with WiFi signals," in *Proc. ACM Interact. Mob. Wearable Ubiquitous Technol,* 2018, pp. 1–23.

11. S. Sigg, S. Shi, and Y. Ji, "Teach your WiFi device: Recognize simultaneous activities and gestures from time-domain RF-features," *Int. J. Ambient Comp. Intell.,* vol. 6, no. 1, pp. 20–34, 2014.

12. P. Melgarejo, X. Zhang, P. Ramanathan, and D. Chu, "Leveraging directional antenna capabilities for fine-grained gesture recognition," in *Proc. ACM UbiComp,* 2014.

13. Q. Pu, S. Gupta, S. Gollakota, and S. Patel, "Whole-home gesture recognition using wireless signals," in *Proc. ACM MobiCom.,* 2013.

14. F. Adib, H. Mao, Z. Kebelac, D. Katabi, and R. C. Miller, "Smart homes that monitor breathing and heart rate," in *Proc. ACM CHI.,* 2015.

15. Y. Zeng, D. Wu, R. Gao, T. Gu, and D. Zhang, "Full Breathe: Full human respiration detection exploiting complementarity of CSI phase and amplitude of WiFi signals," in *Proc. ACM Interact. Mob. Wearable Ubiquitous Technol.,* 2018, pp. 148:1–148:19.

16. Y. Wang, J. Liu, Y. Chen, M. Gruteser, J. Yang, and H. Liu, "E-eyes: Device-free location-oriented activity identification using fine-grained WiFi signatures," in *Proc. ACM Mobicom.,* 2014.

17. X. Zheng, J. Wang, L. Shangguan, Z. Zhou, and Y. Liu, "Smokey: Ubiquitous smoking detection with commerical WiFi infrastructures," in *Proc. IEEE INFOCOM.,* 2016.

18. W. Wang, A. X. Liu, M. Shahzad, K. Ling, and S. Lu, "Understanding and modeling of WiFi signal based human activity recognition," in *Proc. ACM MobiCom.,* 2015.

19. L. Guo, L. Wang, J. Liu, W. Zhou, and B. Lu, "HuAc: Human activity recognition using crowdsourced WiFi signals and skeleton data," *Wireless Commun. Mobile Comput.,* 2018.

20. J. Wang, X. Zhang, X. M. Qinhua Gao, X. Feng, and H. Wang, "Device-free simultaneous wireless localization and activity recognition with wavelet feature," *IEEE Trans. Vehi. Technol.,* vol. 66, no. 2, pp. 1659–1669, 2017.

21. J. Wang, X. Zhang, Q. Gao, H. Yue, and H. Wang, "Device-free wireless localization and activity recognition: A deep learning approach," *IEEE Trans. Vehi. Technol.,* vol. 66, no. 7, pp. 6258–6267, 2017.

22. H. Zou, J. Yang, Y. Zhou, L. Xie, and C. J. Spanos, "Robust WiFi-enabled device-free gesture recognition via unsupervised adversarial domain adaption," in *Proc. IEEE ICCCN,* 2018, pp. 1–8.

23. Z. Chen, L. Zhang, C. Jiang, and W. Cui, "WiFi CSI based passive human activity recognition using attention based BLSTM," *IEEE Trans. Mobile Comput.,* 2018.

24. B. Wei, W. Hu, M. Yang, and C. T. Chou, "Radio-based device-free activity recognition with radio frequency interference," in *Proc. ACM IPSN,* 2015.

25. W. Jiang, C. Miao, S. Y. Fenglong Ma, Y. Wang, Y. Yuan, H. Xue, C. Song, D. K. Xin Ma, W. Xu, and L. Su., "Towards environment independent device free human activity recognition," in *Proc. ACM MobiCom.* 2018, pp. 1–16.
26. C. Shi, J. Liu, H. Liu, and Y. Chen, "Smart user authentication through actuation of daily activities leveraging WiFi-enabled iot," in *Proc. ACM Mobihoc.,* 2017.
27. C. Hsu, A. Ahuja, S. Yue, R. Hristov, Z. Kabelac, and D. Katabi, "Zero-effort in-home sleep and insomnia monitoring using radio signals," in *Proc. ACM UbiComp.,* 2017.
28. M. Zhao, S. Yue, D. Katabi, T. S.Jaakkola, and M. T.Bianchi, "Learning sleep stages from radio signals: A conditional adversarial architecture," in *Proc. IEEE ICML,* 2017.
29. F. Zhang, K. Niu, J. Xiong, B. Jin, T. Gu, Y. Jiang, and D. Zhang, "Towards a diffraction-based sensing approach on human activity recognition," in *Proc. ACM Interact. Mob. Wearable Ubiquitous Technol.* 2019, pp. 33–57.
30. H. Wang, D. Zhang, Y. Wang, J. Ma, Y. Wang, and S. Li, "RT-Fall: A real-time and contactless fall detection system with commodity WiFi devices," *IEEE Trans. Mobile Comput.,* vol. 16, no. 2, pp. 511–526, 2017.
31. F. Wang, Y. Song, J. Zhang, J. Han, and D. Huang, "Temporal Unet: Sample level human action recognition using WiFi," in *Proc. arXiv,* 2019, pp. 1–14.
32. F. Wang, J. Feng, Y. Zhao, X. Zhang, S. Zhang, and J. Han, "Joint activity recognition and indoor localization with WiFi fingerprints," in *Proc. arXiv,* 2019, pp. 1–10.
33. J. Liu, Y. Wang, Y. Chen, J. Yang, X. Chen, and J. Cheng, "Tracking vital signs during sleep leveraging off-the-shelf WiFi," in *Proc. ACM MobiHoc.* 2015.
34. K. Qian, Z. Zhou, Y. Zheng, Z. Yang, and Y. Liu, "Inferring motion direction using commodity WiFi for interactive exergames," in *Proc. ACM CHI,* 2017.
35. D. Z. Dan Wu, C. Xu, Y. Wang, and H. Wang, "WiDir: Walking direction estimation using wireless signals," in *Proc. ACM UBICOMP.* 2016, pp. 351–362.
36. B. Fang, N. D. Lane, M. Zhang, A. Boran, and F. Kawsar, "BodyScan: Enabling radio-based sensing on wearable devices for contactless activity and vital sign monitoring," in *Proc. IEEE MobiSys.,* 2016.
37. S. Tan and J. Yang, "WiFinger: Leveraging commodity WiFi for fine-grained finger gesture recognition," in *Proc. ACM MobiHoc.* 2016, pp. 201–210.
38. A. Virmani and M. Shahzad, "Position and orientation agnostic gesture recognition using WiFi," in *Proc. ACM MobiSys.* 2017, pp. 252–264.
39. Y. Zheng, "Zero-effort cross-domain gesture recognition with Wi-Fi," in *Proc. ACM Mobisys,* 2019, pp. 1–13.
40. H. Zou, J. Yang, Y. Zhou, and C. J. Spanos, "Joint adversarial domain adaptation for resilient WiFi-enabled device-free gesture recognition," in *Proc. IEEE ICMLA,* 2018, pp. 202–207.
41. G. Wang, Y. Zou, Z. Zhou, K. Wu, and L. M. Ni, "We can hear you with WiFi!" in *Proc. ACM MobiCom.,* 2014.

42. K. Ali, A. X. Liu, W. Wang, and M. Shahzad, "Keystroke recognition using WiFi signals," in *Proc. ACM MobiCom*, 2015.

43. C. Humang, H. Chen, L. Yang, and D. Zhang, "Breathlive: Liveness detection for heart sound authentication with deep breathing," in *Proc. ACM Interact. Mob. Wearable Ubiquitous Technol.*, 2018, pp. 12:1–12:25.

44. M. Zhao, F. Adib, and D. Katabi, "Emotion recognition using wireless signals," in *Proc. ACM MobiCom*. 2016, pp. 1–14.

45. F. Zhang, D. Zhang, J. Xiong, H. Wang, K. Niu, B. Jin, and Y. Wang, "From fresnel diffraction model to fine-grained human respiration sensing with commodity WiFi devices," in *Proc. ACM Interact. Mob. Wearable Ubiquitous Technol.* 2018, pp. 53:1–53:23.

46. Y. Zeng, D. Wu, J. Xiong, E. Yi, R. Gao, and D. Zhang, "FarSense: Pushing the range limit of WiFi-based respiration sensing with CSI ratio of two antennas," in *Proc. ACM UbiComp*, 2019.

47. X. Li, D. Zhang, J. Xiong, Y. Zhang, S. Li, Y. Wang, and H. Mei, "Training-free human vitality monitoring using commodity WiFi devices," in *Proc. ACM Interact. Mob. Wearable Ubiquitous Technol.* 2018, pp. 121:1–121:25.

48. J. Liu, Y. W. Yingying Chen, X. Chen, and J. Yang, "Monitoring vital signs and posture during sleep using wiFi signals," *IEEE Int. Things J.* vol. 5, no. 3, pp. 2071–2084, 2018.

49. H. Zou, Y. Zhou, J. Yang, W. Gu, L. Xie, and C. J. Spanos, "WiFi-based human identification via convex tensor shapelet learning," in *Proc. ACM AAAI*, 2018, pp. 1711–1718.

50. T. Li, L. Fan, M. Zhao, Y. Liu, and D. Katabi, "Making the invisible visible: Action recognition through walls and occlusions," *CoRR*, vol. abs/1909.09300, 2019. [Online]. Available: https://arxiv.org/abs/1909.09300

51. K. Ohara, T. Maekawa, and Y. Matsushita, "Detecting state changes of indoor everyday objects using WiFi," in *Proc. ACM Interact. Mob. Wearable Ubiquitous Technol.* 2017, pp. 88:1–88:28.

52. A. Akbari and R. Jafari, "Transferring activity recognition models for new wearable sensors with deep generative domain adaptation," in *Proc. ACM IPSN*, 2019, p. 85:96.

53. Y. Ma, G. Zhou, and S. Wang, "WiFi sensing with channel state information: A survey," *ACM Comput. Surv.* vol. abs/1506.00019, pp. 46:1–46:36, 2019. [Online]. Available: https://doi.org/10.1145/3310194

54. M. Zhao, T. Li, M. A. Alsheikh, Y. Tian, H. Zhao, A. Torralba, and D. Katabi, "Through-wall human pose estimation using radio signals," in *Proc. ACM CVPR*. ACM, 2018, pp. 7356–7365.

55. X. Wang, X. Wang, and S. Mao, "RF Sensing in the internet of things: A general deep learning framework," *IEEE Comm. Mag.*, pp. 62–67, 2018.

56. K. Niu, "Boosting fine-grained activity sensing by embracing wireless multipath effects," in *Proc. ACM CoNEXT*, 2018, pp. 1–13.

57. D. Halperin, W. Hu, A. Sheth, and D. Wetherall, "Tool release: Gathering 802.11n traces with channel state information," *Proc. ACM SIGCOMM CCR,* vol. 41, no. 1, 2011.

58. K. Jarrett, K. Kavukcuoglu, M. Ranzato, and Y. LeCun, "What is the best multi-stage architecture for object recognition?" in *IEEE 12th Int. Conf. Computer Vision,* 2009, pp. 2146–2153.

59. A. Khan and A. S. Q. Anabia Sohail, and U. Zahoora, "A survey of the recent architectures of deep convolutional neural networks," *CoRR,* 2019. [Online]. Available: http://arxiv.org/abs/1901.06032

60. Y. LeCun, K. Kavukcuoglu, and C. Farabet, "Convolutional networks and applications in vision," in *Proc. 2010 IEEE Int. Symp. Circuits and Systems,* 2010, pp. 253–256.

61. D. Scherer, A. Müller, and S. Behnke, "Evaluation of pooling operations in convolutional architectures for object recognition," in K. Dia-mantaras, W. Duch, and L. S. Iliadis *Artificial Neural Networks — ICANN 2010,* Springer Berlin Heidelberg, 2010, pp. 92–101.

62. A. Qaisar, I. M. E.A, and J. M.Arfan, "A comprehensive review of recent advances on deep vision systems," *Artif. Intell. Rev.,* vol. 52, pp. 39–76, 2019.

63. C.-Y. Lee, P. W. Gallagher, and Z. Tu, "Generalizing pooling functions in convolutional neural networks: Mixed, gated, and tree," in *Art. Intel. Stat.,* 2016, pp. 464–472.

64. Z. C. Lipton and J. Berkowitz, "A critical review of recurrent neural networks for sequence learning," *CoRR,* vol. abs/1506.00019, 2015. [Online]. Available: http://arxiv.org/abs/1506.00019

65. C. Olah, "Understanding LSTM networks," https://colah.github.io/posts/2015-08-Understanding-LSTMs/, 2015.

66. L. Guo, H. Zhang, C. Wang, W. Guo, L. Wang, and C. Lin, "Towards diversity activity recognition via LSTM-CNN encoder-decoder neural network," in *Proc. ACM IJCAI Workshop,* 2019, pp. 1–8.

67. H. Zou, Y. Zhou, J. Yang, H. Jiang, L. Xie, and C. J. Spanos, "DeepSense: Device-free human activity recognition via autoencoder long-term recurrent convolutional network," in *IEEE Int. Conf. Commun. (ICC),* 2018, pp. 1–6.

68. J. Gu, Z. Wang, J. Kuen, L. Ma, A. Shahroudy, S. Bing, T. Liu, X. Wang, and W. Gang, "Recent advances in convolutional neural networks," *Computer Science,* 2015.

69. S. Ioffe and C. Szegedy, "Batch normalization: Accelerating deep network training by reducing internal covariate shift," in *Int. Conf. Machine Learning,* 2015.

70. D. Halperin, W. Hu, A. Sheth, and D. Wetherall, "Linux 802.11n CSI Tool," http://dhalperi.github.io/linux-80211n-csitool/, 2010.

71. L. Chen, X. Chen, S. I. Lee, K. Chen, D. Han, D. Fang, Z. Tang, and Z. Wang, "WIDESEE: Towards wide-area contactless wireless sensing," in *Proc. ACM SenSys.,* 2019.

Chapter 6

Graph Convolutional Neural Network for Skeleton-based Video Abnormal Behavior Detection

Weixin Luo[*,†,‡,§], Wen Liu[*,†,‡,¶] and Shenghua Gao[*,‖]

*School of Information Science and Technology,
ShanghaiTech University, China
†Shanghai Institute of Microsystem and Information Technology,
Chinese Academy of Sciences, China
‡University of Chinese Academy of Sciences, China
§luowx@shanghaitech.edu.cn
¶Liuwen@shanghaitech.edu.cn
‖gaoshh@shanghaitech.edu.cn

Abstract

Video anomaly detection aims to detect abnormal events given only normal events where pedestrians regularly walk in surveillance videos. It is popular to leverage encoder–decoder-based reconstruction or prediction methods, to model a normal distribution upon normal data. Whereas, the background noise of reconstructed or predicted results may harm the final performance. To tackle this human-related task, we introduce a spatial temporal graph convolutional networks-based prediction network for skeleton-based video anomaly detection, which detects anomalies based on skeletons, thus, alleviating the noise from complex backgrounds. Specifically, we build a normal graph describing graph connection of joints in normal data. Then, a fully-connected layer is utilized to predict the future joints. Finally, the future joints in normal events can be well predicted while the abnormal ones lead to a large error. To our knowledge, this is the first work to apply graph convolutional networks on skeleton-based video anomaly detection. Experiments show that our proposed normal graph achieves the state-of-the-art performance, compared to those image-level reconstruction-based methods, image-level prediction-based methods, as well as skeleton-based RNN-based methods.

Keywords: graph convolutional networks, video anomaly detection

1. Introduction

Video anomaly detection aims to detect those unexpected events caused by appearance and motion, which have attracted numerous researchers to investigate them. The general setting of this task is to provide only normal data in the training set, while the testing set contains both normal and abnormal events to be detected.

Following this unsupervised setting, auto-encoder-based [1] and prediction network-based [2] methods are popular in terms of network design. Specifically, they use an encoder–decoder-based network to model a normal distribution, where abnormal data will be distinguished from the normal boundary. Also, since videos have an additional dimension of time compared to images, recurrent neural networks (RNNs) [3] and Stacking multiply frames over channel [1] can be utilized to model the time dependency.

Most of these previous methods, however, take the whole image as input, which means the irrelevant background will be also included during modeling the normal distribution. Thus, some false positives may be introduced during anomaly detection because of illumination and irrelevant moving objects in the background. Further, the burden of the models will increase in order to memorize these complex backgrounds, rather than focusing on modeling the normal distribution of foreground objects. In conclusion, it is desirable to design some mechanisms to push the models to focus only on the foreground objects such as humans, since most anomalies are caused by humans. To specifically characterize the human patterns, skeleton-based action recognition and using RNN [4, 5] or temporal CNNs [6, 7] can be leveraged. Whereas, these proposed networks flavor the image-level data, the skeletons are presented as coordinates. Recently, graph convolutional networks (GCNs) were proposed to tackle such non-grid data by introducing a graph-based affinity matrix. Some researchers adopt such a powerful tool to conduct skeleton-based action recognition. GCNs, however, have not yet been well studied for video anomaly detection.

In this chapter, we introduce a spatial temporal GCN-based prediction network for skeleton-based video anomaly detection. Specifically, spatial–temporal graph convlutional networks (ST-GCNs) [8] is adopted to automatically capture the normal patterns in terms of spatial configuration and temporal dynamics. In this graph, nodes are the human joints and edges are the spatial and temporal connections of these joints. For the spatial connection, we follow the configuration [8] which connects the joints physically. As for the temporal connection, it links the joints across time. The spatial graph convolution is first used to spread information over space, followed by a temporal convolution that fuses the information over time. Then, a fully connected layer is appended to predict the future joints. Finally, we conduct video anomaly detection based on the prediction errors of joints, where large errors denote anomalies.

1.1 *Contributions*

We summarize our contributions as follows: (i) We propose a spatial temporal GCN-based prediction network for skeleton-based video anomaly detection, where it efficiently explores the normal pattern from human joints, with a spatial graph network and a temporal one to spatially and temporally connect the joints, respectively. To our knowledge, this is the first work to apply GCNs to video anomaly detection; (ii) Our proposed method achieves the state-of-the-art performance on ShanghaiTech-Campus and CUHK Avenue.

The rest of this chapter is organized as follows: In Section 2, we introduce the related work of video anomaly detection, skeleton-based action recognition and graph neural networks. In Section 3, our proposed spatial temporal graph convolutional networks-based prediction network is introduced for skeleton-based video anomaly detection. In Section 4, extensive experiments including ablation study and visualization are conducted to validate the effectiveness of our method. In Section 5, we conclude our work.

2. Related Work

2.1 *Video anomaly detection*

The general setting of this task is to provide only normal data in the training set, while the testing set contains both normal and abnormal events to be detected. With this setting, we categorize the previous work into three classes: (i) handcrafted feature-based video anomaly detection; (ii) sparse coding-based video anomaly detection; (iii) deep learning-based video anomaly detection.

(1) **Handcrafted feature-based methods:** Previous work [9] usually leverages low-level patterns such as trajectories to conduct video anomaly detection, while it is not robust to crowd scenes due to the complexity of the background. Then, researchers adopt two robust features, namely, histogram of oriented gradient (HOG) [10] and histogram of flow (HOF) [11], to handle such a problem. For example, Adam *et al.* [12] utilize an exponential distribution to model a normal distribution with local flow features. In addition to that, Kim *et al.* [13] use a mixture of probabilistic PCA model. Besides, one-class SVM is intuitively suitable to this task that can be induced to a single class classification. These previous methods, however, are not optimal to those large-scale video anomaly detection datasets due to these suboptimal handcrafted based features and classifiers.

(2) **Sparse coding-based methods:** The other popular way to conduct video anomaly detection is dictionary learning and sparse coding, which assumes that the input feature is linearly reconstructed by the basis of a dictionary where the coefficients are constrained with sparsity. Based on that, some researchers [14, 15] setup a dictionary characterizing normal events, leading to a large reconstruction error for abnormal events. However, the optimization procedure is time-consuming. Thus, to accelerate the optimization, Lu *et al.* [16] dropped the sparse regularization and introduced multiple dictionaries without any sparsity constraint. Further, Luo *et al.* [17] proposed temporally coherent sparse coding inspired recurrent neural networks (RNNs) to map the optimization procedure into a stack of RNNs.

(3) **Deep learning-based methods:** As for deep learning-based methods, a reconstruction or a prediction network both based on an encoder–decoder, is usually adopted to model the normal distribution, where abnormal events will be reconstructed or predicted with a big error. More specifically, Xu *et al.* [34] propose Appearance and Motion DeepNet (AMDN) which utilizes deep neural networks to automatically learn feature representations. Hasan *et al.* [1] stack multiple consecutive frames in channels and feed them into a convolutional auto-encoder (AE) to reconstruct the inputs. Considering the temporal dependence of

videos, researchers [3, 18] leverage ConvLSTM to form a ConvLSTM-AE, which is a disentanglement of motion and appearance. In addition, Liu *et al.* [2] explicitly design a prediction network in terms of appearance and motion constraints for video anomaly detection.

2.2 Skeleton-based action recognition

Most actions are completed by humans, which means skeletons of humans are obviously the cues of action recognition. In addition, analysis of skeletons can reduce complexity of action recognition, where human pose estimation is well studied [19, 20]. With human poses extracted from images, the approaches of analyzing poses can be categorized into two classes, namely, handcrafted feature-based methods and deep learning-based methods. The former one usually utilizes covariance matrices of joints [21], rotation or translation of parts [22], and relative positions of joints [23]. The latter one uses RNN [4, 5] or temporal CNN [6, 7]. Furthermore, Yan *et al.* [8] leverage a spatial–temporal graph convolutional network (ST-GCN) for skeleton-based action recognition. In light of this work, we introduce a ST-GCN-based prediction network for skeleton-based video anomaly detection. Recently, a relevant work [24] also adopts a reconstruction and prediction network for skeleton-based video anomaly detection, where skeletons across time can be split in to two components, e.g., global coordinates representing global body moving and local ones representing local body posture. Further, a message-passing encoder–decoder recurrent neural network (MPED-RNN) is used to explore the combination of these two representations for improving performance. It, however, needs a model regularization which looks for a network with the smallest capacity that is still within 5% of the initial validation loss achieved by the network with high capacity. We argue that such a high capacity network is over-qualified for skeleton representation with low dimensions. Therefore, such a remedy that seeks a smallest capacity network will be suboptimal. It should be noticed that our proposed method tackles this task as a normal distribution modeling of skeleton-based graph connection across both space and time, which is significantly different from MPED-RNN.

2.3 Graph neural networks

The perspectives of graph neural networks can be explained by two aspects, namely, the spectral perspective [25–28] and the spatial perspective [29, 30]. The former one considers graph convolution from a spectral viewpoint, while the latter one just applies one dimension convolution over the neighbor of a node. In addition, Si *et al.* [31] introduce an attention enhanced graph convolutional LSTM network for skeleton-based action recognition.

3. Method

3.1 *Spatial graph convolutional networks*

To begin with, we briefly review the traditional spatial graph convolutional networks including two components, namely, nodes and edges. Such networks are beneficial for those non-grid data such as skeletons used in this chapter. After extracting N joints for each person with a pose estimator pretrained on other dataset, we introduce a graph $\mathcal{G}^l = \{\mathcal{J}^l, \varepsilon^l\}$ at lth layer, where $\mathcal{J}^l = \{j_i^l \in R^C \mid i = 1,\ldots,N\}$ is the set of joint nodes with dimension of C, and $\varepsilon^l = \{e_i^l\}$ is the set of edges describing the connection of joints. C denotes the feature dimension of a node j_i^l that can be 2 for 2D pose or 3 for 3D pose at the first layer. In addition to that, we denote the neighbor of j_i^l as $\mathcal{N}(j_i^l)$. Then graph convolution can be formulated as follows:

$$f(j_i^l) = \sigma\left(j_i^l W_0^l + \sum_{j_k^l \in \mathcal{N}(j_i^l)} \frac{1}{c_{i,k}} W_1^l j_k^l \right), \tag{1}$$

where $c_{i,k}$ is a normalization term to ensure that the aggregated features are invariant to the number of neighbors. Two trainable parameters $W_0^l, W_1^l \in R^{C \times C'}$ are used to conduct channel learning, transforming the feature dimension from C to C'. Also, σ is a nonlinear activation function.

We also introduce an adjacent matrix $\mathbf{A} \in R^{N \times N}$ where $\mathbf{A}_{m,n}$ denotes the fixed or trainable weights of connection between joint j_m^l and joint j_n^l. Here the weights of connection can be binary (e.g., 0 for disconnection and 1 for connection) or arbitrary values. In this chapter, we choose a fixed $\mathbf{A}$ for all layers. Then, Eq. (1) can be reformulated as follows:

$$GCN(\mathcal{J}^l) = \Lambda^{-\frac{1}{2}}(\mathbf{A}+\mathbf{I})\Lambda^{-\frac{1}{2}}\mathcal{J}^l W^l, \tag{2}$$

where $\Lambda \in R^{N \times N}$ is a degree matrix. I is an identity matrix describing self-connection of joints. $\mathcal{J}^l \in R^{N \times C}$ is the set of joints at layer l. $\Lambda^{-\frac{1}{2}}(\mathbf{A}+\mathbf{I})\Lambda^{-\frac{1}{2}}$ means a normalization over the adjacent $\mathbf{A} + \mathbf{I}$ which functions as $c_{i,k}$ in Eq. (1).

3.2 *Spatial–temporal graph convolutional networks*

The above formulations only consider spatial domain in GCNs. Following [8], it is straightforward to extend spatial GCN to ST-GCN by redefining the neighbor $\mathcal{N}(j_{i,t}^l)$ as $\mathcal{N}(j_{i,t}^l) \cup \{j_{l,m}^l \,||\, m - t \,| \leq \lfloor \frac{L}{2} \rfloor\}$ where L denotes window size across-time, which means the ith joint $j_{i,t}^l$ at time t and layer l not only takes the spatial neighbor as inputs but also connects the same joints over consecutive L frames. In practice, Eq. (2) is applied over a input tensor $\mathcal{J}^l$ with shape of $(\mathcal{C}, \mathcal{L}, \mathcal{N})$ to conduct a spatial graph convolution, leading to an output tensor with shape of $(\mathcal{C}', \mathcal{L}, \mathcal{N})$. Then, a 1D convolution with filter size of L is conducted over the second dimension of the output tensor, which leads to a final tensor with shape of $(\mathcal{C}', \mathcal{L}', \mathcal{N})$.

3.3 *Spatial connection configuration for skeleton*

We follow the spatial connection configuration for a skeleton in Ref. [8] including three adjacent matrices. In general, for the ith joint j_i in a skeleton, its neighbor can be defined as shown in Fig. 1(a). For instance, the neighbor of joint #1 is {#2, #3, #4). Leveraging the spatial configuration proposed in Ref. [38], the neighboring joints can be decomposed into three subsets, depending on the distance to the nose of the skeleton. More

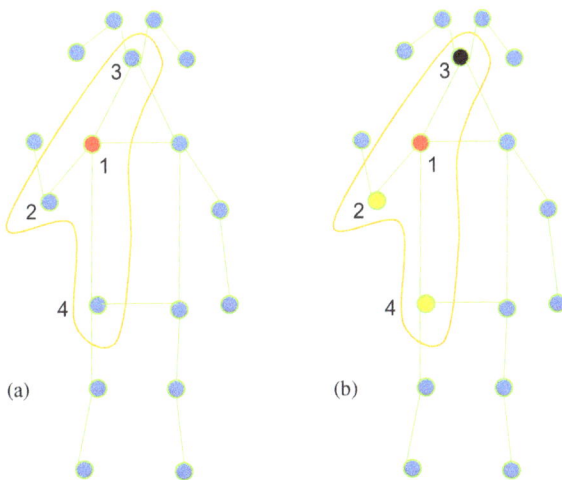

Figure 1: The visualization of the adjacent matrix. A general physical connection of # 1 joint is illustrated in (a). Another spatial connection configuration is illustrated in (b), considering the distance between each joint and the nose. It further splits the neighboring joints of #1 joint into three sets: (i) the joint #1 itself; (ii) the neighboring joints (#3) closer to the nose than the joint #1; (iii) the neighboring joints (#2, #4) farther away from the nose than the joint #1.

specifically, these three subsets can be denoted as follows: (i) the ith joint self; (ii) the neighboring joints closer to the nose than the ith joint; (iii) the neighboring joints farther away from the nose than the ith joint. After normalizing the whole adjacent matrix, we split it, namely, $(\Lambda^{-\frac{1}{2}}(\mathbf{A}+\mathbf{I})\Lambda^{-\frac{1}{2}})$ into three sub-matrices with the above criterion. Thus, Eq. (2) can be reformulated to

$$GCN(\mathcal{J}^l) = \sum_{i=1}^{3}\left(\Lambda^{-\frac{1}{2}}(\mathbf{A}+\mathbf{I})\Lambda^{-\frac{1}{2}}\right)_i \mathcal{J}^l \mathbf{W}_i, \tag{3}$$

where $\mathbf{W}_i \in R^{C \times C'}$ is still a trainable parameter of the ith subset of neighbors.

3.4 Spatial–temporal graph convolutional networks-based prediction network

Based on the above introduction of ST-GCN, we introduce a ST-GCNs-based prediction network, which handles joints of each pedestrian in a video with a predefined adjacent matrix. We name our method *Normal Graph* for simplicity. First, multiple ST-GCN are leveraged to accumulate joint information across space with some physical connections of joints as well as time. Then, we append a prediction module with a fully connected layer, which includes a trainable parameter W_p, to predict future joints. It should be noticed that we do not adopt a reconstruction module, since reconstruction is more prone than prediction to learn an identity mapping, which may result in a performance drop [2]. The whole pipeline is illustrated in Figure 2. First,

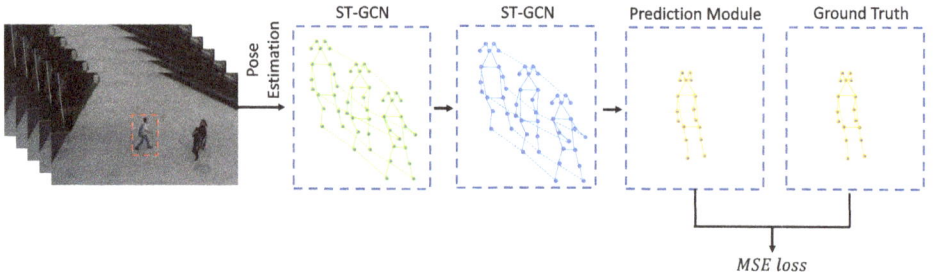

Figure 2: The whole pipeline of our proposed ST-GCN-based prediction network. With joints extracted from a video clip, we firstly stack 9 ST-GCNs and take these joints from time 1 to time L as inputs, to characterize the spatial temporal relationship between joints. Further, a prediction module with a fully connected layer takes the output tensor from the previous module as inputs and predicts the joints at time $L+1$. Also, a mean square loss will be used between the ground truth joints and the predicted joints.

a human pose estimator, namely, Alpha-Pose [19] is utilized to extract joints j_t for each person in the tth frame of a video clip with $T+1$ frames. Then, several stacking ST-GCNs take the set of joints $\mathcal{J}^l$ with the shape of (B, C, L, N), where B is batch size and C is number of channel, while L is time length and N is number of joints, as input and output a tensor with the shape of (B, C', L', N). The output tensor would be reshaped into (B, C', L', N). Finally, the prediction module with a fully connected layer takes this aggregated feature as input and predicts a future skeleton $\hat{\mathcal{J}}^l_{L+1} \in R^{B \times C \times N}$ at time $L+1$. The objective function is demonstrated as follows:

$$\min_{W^l_i, W_p} L_{T+1} = \frac{1}{2} \left\| \hat{\mathcal{J}}^l_{L+1} - \mathcal{J}^l_{L+1} \right\|^2_2, \tag{4}$$

where the parameters to be optimized in the whole network are each W^l_i at the lth ST-GCN and W_p in the prediction module.

3.5 *Skeleton-based video anomaly detection*

The prediction loss 4 is calculated for each detected person in a video clip, then the abnormal scores can be gathered by a max pooling operation upon multiple detected persons:

$$s_T = \max\{L^i_{T+1} \mid i = 1, \dots, P\}, \tag{5}$$

where P is the number of persons detected in this video clip.

4. Experiment

Our method is evaluated on two public video anomaly detection datasets, namely, ShanghaiTech Campus [17] and CUHK Avenue [16]. We compare our proposed Normal Graph with MPED-RNN [24] which leverages RNN to predict future joints and reconstruct current joints. All experiments are evaluated on the metric of frame-level AUC.

4.1 *Dataset*

The **ShanghaiTech Campus dataset** [17] is a new large-scale anomaly detection dataset, recorded on several spots in campus. The anomalies defined in this dataset can be caused by humans and vehicles. The previous work [24] splits a subset from ShanghaiTech Campus that contains only those human-related anomalies, named HR-ShanghaiTech, which will be used in this chapter. We also evaluate our method on this subset.

The **CUHK Avenue dataset** [16] is also a popular anomaly detection dataset, recorded at a single view of the metro. We also extract those human-related anomalies and name it HR-Avenue.

4.2 *Implementation details*

We utilized Alpha-Pose [19] to extract skeletons for each frame in a video. We follow the tracking algorithm used in Ref. [24], which combines sparse optical flow with skeletons. For the configuration of ST-GCN, we follow the setting in Ref. [8] which is composed of 9 ST-GCN layers. The first three layers, the following three layers and the last three layers have 64, 128 and 256 channels, respectively. The Resnet mechanism is applied on each ST-GCN.

4.3 *Comparison with state-of-the-art methods*

In this section, we compare Normal Graph with appearance-based [1, 2, 17, 32] as well as skeleton-based methods [24]. The results illustrated in Table 1 demonstrate that skeleton-based methods achieve better AUC than those with appearance-based ones, especially on the HR-ShanghaiTech dataset where anomalies are associated only with humans. We believe appearance-based methods will introduce noise from background during reconstruction [1, 17] or prediction [2]. In addition, our method performs better than MPED-RNN which models the relationship of skeletons with RNN, while

Table 1: The performance of different baselines on ShanghaiTech (ST), Human-Related ShanghaiTech (HR-ST) and CUHK HR-Avenue.

		ST	HR-ST	HR-Avenue
		AUC ↑/ EER ↓		
Appearance	Conv-AE [10]	0.704 / 0.368	0.698 / 0.375	0.848 / 0.196
	ST-AE [34]	N/A / N/A	N/A / N/A	0.809 / 0.244
	TSC sRNN [22]	0.680 / 0.371	N/A / N/A	N/A / N/A
	Pred [19]	0.728 / 0.323	0.727 / 0.321	0.862
Skeleton	MPED-RNN[23]	0.734 / N/A	0.754 / N/A	0.863 / N/A
	Normal Graph	0.741 / 0.301	0.765 / 0.292	0.873 / 0.188

our method utilizes spatial temporal graph convolutional networks to model the relationship in a graph. In conclusion, our proposed method achieves the state-of-the-art AUC and EER on the ShanghaiTech (ST), the HR–ShanghaiTech (HR–ST) and the CUHK Avenue, which demonstrates the effectiveness of our method.

4.4 *The number of ST-GCNs*

We further conduct an ablation study on the number of ST-GCNs to explore the mechanism of ST-GCNs. The output channels of 9 ST-GCN are set to 64, 64, 64, 64, 128, 128, 128, 256, 256, respectively. The number of ST-GCNs is gradually increased and a fully connected layer is always utilized for prediction. Demonstrated in Fig. 3, the performance on the CUHK Avenue dataset gradually increases when the number of ST-GCNs increases. Especially, it arrives at the best performance on the CUHK Avenue dataset when the number of ST-GCNs is 9. We assume that deeper ST-GCN models will cause difficulty in optimization, where some works [33] propose a residual mechanism to increase the layers of GCN, which is out of scope in this work.

4.5 *The effectiveness of ST-GCNs*

To further investigate the effectiveness of our proposed method, we introduce a baseline which contains only 9 fully connected layers and another baseline which includes only spatial GCNs. For fair comparison, the dimensions of the 9 fully connected layers are the same with the 9

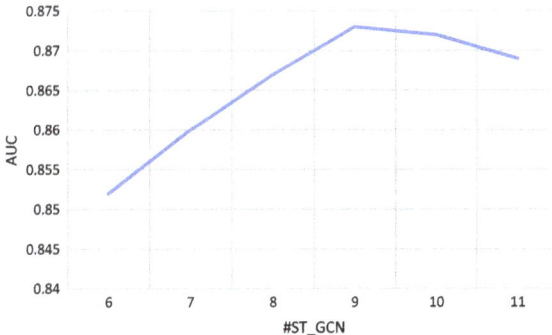

Figure 3: The performance of Normal Graph on the CUHK Avenue dataset when changing the number of ST-GCNs.

Table 2: The demonstration of the effectiveness of our proposed normal graph.

	CUHK Avenue
9 Fully connected layers	0.802
Normal Graph without temporal connection	0.857
Normal Graph	0.873

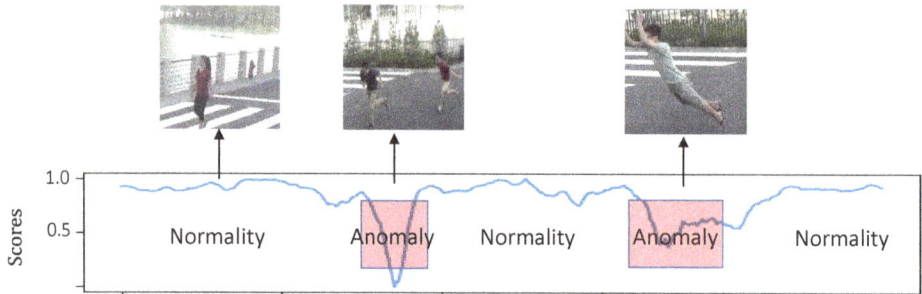

Figure 4: The visualization on the HR–ST dataset.

ST-GCN-based Normal Graph. Illustrated in Table 2, stacking only several fully connected layers results in a dramatical performance decline. It proves that fully connected layers cannot explicitly characterize the spatial and temporal relationship of human joints, while GCNs are able to model such a complex relationship with a specific adjacent matrix. Further, the GCNs with only spatial connection and without temporal connection cannot well characterize the relationship across time, resulting in a slight performance decline.

4.6 *The demo of normal graph*

To further analyze our method, we visualize scores normalized from the scores in a video clip on the HR–ST dataset. As demonstrated in Fig. 4, our proposed Normal Graph can well detect HR anomalies. Besides, we found that failure cases always happen in some subtle actions and occlusion scenes which cannot be well captured by joints.

4.7 *The visualization of anomaly detection*

We also visualize some successful detected results in both HR–ST dataset and HR-Avenue dataset, as shown in Figs. 5 and 6, respectively. The visualizations

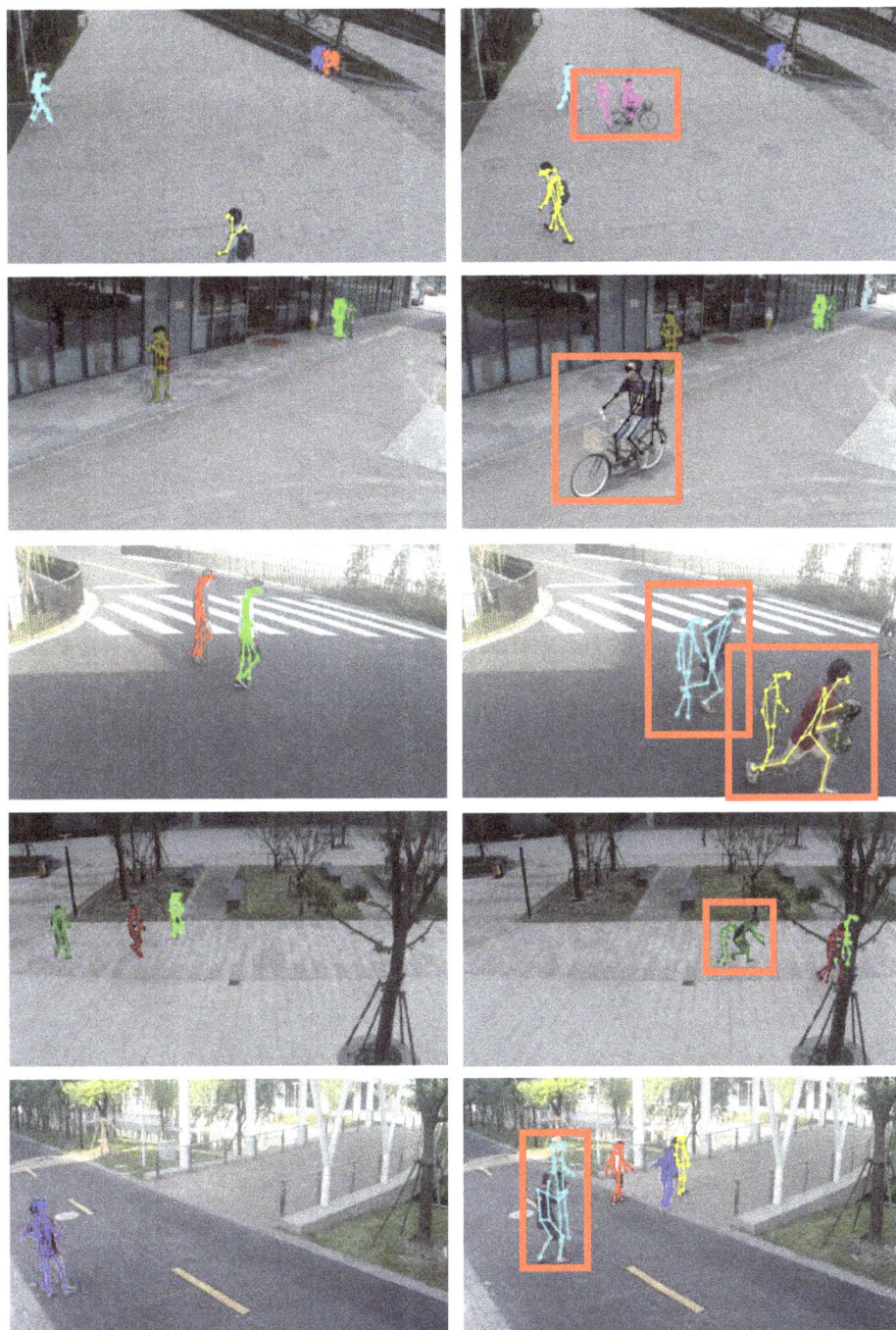

Figure 5: The detection results on the HR–ST dataset. The red boxes denote the abnormal regions. Best viewed in color.

Figure 6:   The detection results on the HR Avenue dataset. The red boxes denote the abnormal regions. Best viewed in color.

show that the pedestrian acting abnormally cannot be well predicted in terms of joint positions, while the normal cases are predicted well.

4.8 *The analysis of true positive rate and false negative rate*

Besides the measurement on AUC, we also provide the analysis of true positive rate (TPR) and false positive rate (FPR) when a certain threshold (0.5) is set to determine whether a frame is normal or abnormal. We conduct such experiments on HR–ST dataset and find that TPR is 0.76 and the FPR is 0.37.

We also find that humans who are on bikes are easily missed by our proposed skeleton-based method since they appear quite similar to those walking pedestrians. Thus, one can handle such an issue by combining the image-based solution and the skeleton-based solution. We leave it for the future.

5. Conclusion

The image-level reconstruction or prediction-based methods for video anomaly detection may introduce noise from background. To minimize the affect from background, we tackle this problem by a skeleton-based prediction network which combines several spatial temporal graph convolutional networks and a prediction module. With the model trained on only normal data in the training set, abnormal events represented by skeletons will be automatically discovered. To our knowledge, this is the first work to apply graph convolutional networks on skeleton-based video anomaly detection. Experiments show that our proposed normal graph achieves state-of-the-art performance, compared with those image-level reconstruction-based methods, image-level prediction-based methods, and skeleton-level RNN-based methods.

References

1. M. Hasan, J. Choi, *et al.*, "Learning temporal regularity in video sequences," in *CVPR*, 2016.
2. W. Liu, W. Luo, D. Lian, and S. Gao, "Future frame prediction for anomaly detection — a new baseline," in *CVPR*, 2018.
3. W. Luo, W. Liu, and S. Gao, "Remembering history with convolutional lstm for anomaly detection," in *ICME*, 2017.

4. A. Shahroudy, J. Liu, T. T. Ng, and G. Wang, "NTU RGB+D: A large scale dataset for 3D human activity analysis," in *Proc. IEEE Conf. Computer Vision and Pattern Recognition*, 2016, pp. 1010–1019.

5. J. Liu, A. Shahroudy, D. Xu, and G. Wang, "Spatio-temporal LSTM with trust gates for 3D human action recognition," in: *Euro. Conf. Computer Vision*, Springer, 2016, pp. 816–833.

6. C. Li, Q. Zhong, D. Xie, and S. Pu, "Skeleton-based action recognition with convolutional neural networks," in *2017 IEEE Int. Conf. Multimedia & Expo Workshops* (*ICMEW*). 2017, pp. 597–600. IEEE.

7. Q. Ke, M. Bennamoun, S. An, F. Sohel, and F. Boussaid, "A new representation of skeleton sequences for 3D action recognition," in *Proc. IEEE Conf. Comput. Vision and Pattern Recognition*. 2017, pp. 3288–3297.

8. S. Yan, Y. Xiong, and D. Lin, "Spatial temporal graph convolutional networks for skeleton-based action recognition," in *Thirty-Second AAAI Conf. Artificial Intelligence* (2018).

9. F. Tung, J. S. Zelek, and D. A. Clausi, "Goal-based trajectory analysis for unusual behaviour detection in intelligent surveillance," *Image and Vision Computing*, vol. 29, no. 4, pp. 230–240 (2011).

10. N. Navneet, and B. Triggs, "Histograms of oriented gradients for human detection," in *CVPR*, 2005.

11. N. Dalal, B. Triggs, and C. Schmid, "Human detection using oriented histograms of flow and appearance," in *ECCV*, 2006.

12. A. Adam, E. Rivlin, I. Shimshoni, and D. Reinitz, "Robust real-time unusual event detection using multiple fixed-location monitors," *TPAMI*, 2008.

13. J. Kim, and K. Grauman, "Observe locally, infer globally: A space-time MRF for detecting abnormal activities with incremental updates," in *CVPR*, 2009.

14. B. Zhao, F. Li, and E. P. Xing, "Online detection of unusual events in videos via dynamic sparse coding," in *CVPR*. 2011.

15. Y. Cong, J. Yuan, *et al.*, "Sparse reconstruction cost for abnormal event detection," in *CVPR*, 2011.

16. C. Lu, J. Shi, and J. Jia, "Abnormal event detection at 150 FPS in MATLAB," in *ICCV*, 2013.

17. W. Luo, W. Liu, and S. Gao, "A revisit of sparse coding based anomaly detection in stacked rnn framework," in *ICCV*, 2017.

18. Y. Chong, and Y. Tay, "Abnormal event detection in videos using spatiotemporal autoencoder," *ISNN*, 2017.

19. H. S. Fang, S. Xie, Y. W. Tai, and C. Lu, "RMPE: Regional multi-person pose estimation," in *ICCV*, 2017.

20. Z. Cao, T. Simon, S. E. Wei, and Y. Sheikh, "Realtime multi-person 2D pose estimation using part affinity fields," in *CVPR*, 2017.

21. M. E. Hussein, M. Torki, M. A. Gowayyed, and M. El-Saban, "Human action recognition using a temporal hierarchy of covariance descriptors on 3D joint locations," in *Twenty-Third Int. Joint Conf. Artificial Intelligence*, 2013.

22. R. Vemulapalli, F. Arrate, and R. Chellappa, "Human action recognition by representing 3D skeletons as points in a lie group," in *Proc. IEEE Conf. Comput. Vision and Pattern Recognition*, 2014, pp. 588–595.

23. J. Wang, Z. Liu, Y. Wu, and J. Yuan, "Mining actionlet ensemble for action recognition with depth cameras," in 2012 *IEEE Conf. Comput. Vision and Pattern Recognition*. IEEE, 2012, pp. 1290–1297.

24. R. Morais, V. Le, T. Tran, B. Saha, M. Mansour, and S. Venkatesh, "Learning regularity in skeleton trajectories for anomaly detection in videos," in *Proc. IEEE Conf. Comput. Vision and Pattern Recognition*, 2019, pp. 11996–12004.

25. M. Henaff, J. Bruna, and Y. LeCun, "Deep convolutional networks on graph-structured data," 2015, arXiv preprint arXiv:1506.05163.

26. D. K. Duvenaud, D. Maclaurin, J. Iparraguirre, R. Bombarell, T. Hirzel, A. Aspuru-Guzik, and R. P. Adams, "Convolutional networks on graphs for learning molecular fingerprints," in *Advances in Neural Information Processing Systems*. 2015, pp. 2224–2232.

27. Y. Li, D. Tarlow, M. Brockschmidt, and R. Zemel, "Gated graph sequence neural networks," 2015, arXiv preprint arXiv:1511.05493.

28. M. Defferrard, X. Bresson, and P. Vandergheynst, "Convolutional neural networks on graphs with fast localized spectral filtering," in *Advances in Neural Information Processing Systems*, 2016, pp. 3844–3852.

29. J. Bruna, W. Zaremba, A. Szlam, and Y. LeCun, "Spectral networks and locally connected networks on graphs," 2013, arXiv preprint arXiv:1312.6203.

30. M. Niepert, M. Ahmed, and K. Kutzkov, "Learning convolutional neural networks for graphs," in *Int. Conf. Machine Learning*, 2016, pp. 2014–2023.

31. C. Si, W. Chen, W. Wang, L. Wang, and T. Tan, "An attention enhanced graph convolutional LSTM network for skeleton-based action recognition," in *Proc. IEEE Conf. Comput. Vision and Pattern Recognition*, 2019, pp. 1227–1236.

32. Y. Zhao, B. Deng, C. Shen, Y. Liu, H. Lu, and X. S. Hua, "Spatio-temporal autoencoder for video anomaly detection," in *Proc. 25th ACM Int. Conf. Multimedia*, 2017, pp. 1933–1941.

33. G. Li, M. Müller, A. Thabet, and B. Ghanem, "Can GCNs go as deep as CNNs?" in *ICCV*, 2019.

34. D. Xu, E. Ricci, Y. Yan, J. Song, and N. Sebe, "Learning deep representations of appearance and motion for anomalous event detection," *BMVC*, 2015.

Part III

Deep Learning for Remote Sensing

Chapter 7

Perspective on Deep Learning for Earth Sciences

Gustau Camps-Valls

Image Processing Laboratory (IPL), Universitat de València, Spain
gustau.camps@uv.es
http://isp.uv.es and @isp_uv_es

Abstract

Machine learning and deep learning (DL) in particular have made a huge impact on many fields of science. In the last decade, advanced deep learning methods have been developed and applied to Earth data science problems extensively. Applications on classification and parameter retrieval are making a difference: methods are very accurate, can handle large amounts of data and deal with spatial and temporal data structures efficiently. Nevertheless, several important challenges still need to be addressed. Current standard deep architectures struggle to learn useful Earth feature representations in an unsupervised way, and struggle with long-range dependencies so distant driving processes (in space and time) are not captured, and they cannot cope with non-Euclidean spaces efficiently. DL models are still obscure and resistant to interpretability too and, as other data-driven techniques, they do not necessarily learn physically meaningful and, more importantly, causal relations. Advances are needed to cope with arbitrary signal structures and data relations, physical plausibility and interpretability. This chapter reviews the current approaches and discusses ways forward to develop new DL methods for the Earth sciences in all these directions.

Keywords: deep learning, learning representations, hybrid machine learning, interpretability, causality, remote sensing, geoscience, Earth system science

1. Introduction

The last 5–10 years have witnessed a dramatic increase in scientific and technical contributions to the field of machine learning from the landmarks provided by deep learning [1]. A vast amount of algorithms and network
architectures, including the well-known convolutional neural network (CNN)
[2], recurrent nets with long short-term memory (LSTM) units [3], or generative approaches like the generative adversarial networks (GANs) [4], have
been developed theoretically and are widely used in practice. Application of
deep learning has had an enormous impact on fields like natural language
processing, computer vision and speech recognition. It goes without saying
that they are now the state of the art in these fields, where distinctively there
is a clear spatial and/or temporal data structure.

The use in Earth science problems has also emerged in the last decade,
mainly focused on resolving problems of image classification and object detection from remote sensing satellite images, spatio-temporal analysis of Earth
system data cubes in the geosciences, or detection and tracking of extremes in
the climate sciences [5]. Some excellent reviews on the specific subfield of satellite Earth observation data processing have been published by Zhang *et al.* [6]
and Zhu *et al.* [7]. DL methods are now a successful tool for tackling geoscience
and climate data problems, involving the atmosphere, the land and the ocean.
Deep learning has achieved notable success in challenging problems of change/
anomaly detection, land use and cover classification [6], biogeophysical parameter retrieval [8, 9], weather forecasting [10], extreme weather detection [11]
and ENSO prediction [12], just to name a few example applications. Camps-
Valls *et al.* [17] review the recent advances in deep learning for Earth sciences.

However, despite the promising results obtained so far, important problems
need to be addressed. We identify five major challenges in the use and adoption
of DL in the Earth sciences: (1) in Earth and climate sciences, the data characteristics fulfil the "four Vs" paradigm of big data (volume, velocity, veracity,
variety), and notably the wealth of heterogeneous datasets come without labels,
hence learning feature representations in unsupervised settings is of paramount
relevance, a field that is still in its infancy; (2) standard DL architectures deal successfully with space and time data structures, but locality, stationarity and
Euclidean spaces are too rigid assumptions in the current context whenever
teleconnections, compound extreme events, non-gridded data and multiscale
processes happen in the wild; (3) DL models are excellent approximators, but
very often do not respect the most elementary laws of physics, like mass or
energy conservation, so *consistency* and (perhaps more importantly)

confidence/trustworthiness of their predictions is compromised; (4) DL models are often obscure, overparameterized models, and the learned relations are difficult to grasp, understand and visualize; and (5) finally, and more importantly, DL models cannot learn causal relations between the considered variables, so it can be argued that there is no actual learning after all.

This chapter aims to highlight the paramount importance of these fundamental challenges, providing a perspective about current models and the ongoing activities to resolve them, and lays down a road-map for the future needs in the field. Section 2 reviews the (ab)use of deep learning in Earth sciences by providing a brief review of applications and challenges. We review and further elaborate new methods and ways forward to address them in Section 3. We finish in Section 4 by some final remarks and a perspective about the future steps forward.

2. Deep Learning Challenges in Earth Sciences

The last decade has witnessed a dramatic increase in scientific and technical contributions to the field of deep learning in the analysis of Earth data. A vast amount of algorithms and network architectures have been developed and applied in remote sensing and the geosciences: (1) the well-known convolutional neural network (CNN) has been widely used for supervised, semisupervised and unsupervised remote sensing image classification [6 ,7, 13, 14]; (2) recurrent nets, and the LSTM in particular, have demonstrated good potential to deal with time series of biogeophysical parameters estimation, forecasting and target tracking [15, 16]; and (3) generative approaches, like generative adversarial networks (GANs) or variational auto-encoders (VAE), are used for data synthesis, radiative transfer model emulation, hyperspectral image classification and clustering. Excellent reviews are currently available.

Despite the large body of empirical evidence, the current application of DL methods faces several important challenges:

(1) **Learning unsupervised feature representations:** Learning feature representations from multivariate structured data, such as time series or images, is of paramount relevance for data compression, visualization and understanding. When few or no labels are available, one has to resort to learning in an unsupervised setting. A wide range of methods is available; from autoencoders, to GANs and sparse dictionary learning algorithms. However, accounting for semantically meaningful representations in Earth data is challenging because of the presence of non-stationary and

nonlinear coupled processes, different spatio-temporal scales involved, teleconnections and extremes in the wild.

(2) **Complex structures, non-stationarities and uncertainty:** Novel deep learning methods are necessary to observe data (problem) specifics: to deal with signal-to-noise relations, several structured/correlated outputs, non-stationary processes in space and time and high-dimensional spaces. New network topologies should not only account for the local neighbourhood relations at possibly different scales but also account for long-range relationships (for example, for climate teleconnections). Another important aspect to account for is that of quantifying and propagating uncertainties in an efficient manner, which will require integrating concepts from probabilistic inference into deep networks into deep networks, as it is the case of the active field of Bayesian deep learning.

(3) **Physical consistency:** Deep learning algorithms excel at fitting observations, but predictions may not be physically consistent or even plausible, due to extrapolation or observational biases. This issue calls for the integration of domain (expert) knowledge and achievement of physical consistency. It is urgent to teach models about the governing physical rules of the Earth system and provide them with meaningful constraints on top of the pure observational ones. A similar discussion is happening nowadays around the concept of symbolic-AI models in natural language processing, where semantic rules would play the role of governing physical equations in our field.

(4) **Model interpretability:** Improving the prediction accuracy is very relevant, of course, but has demonstrated to be insufficient, especially in Earth and climate applications where the far-end more ambitious goal is understanding phenomena through machine learning, as well as learning about relations and relevance of covariates. Certainly, interpretability and model understanding are key nowadays, including proper visualization of the results and decisions made by users.

(5) **Learning causal representations:** Going beyond mere fitting in the current DL dogma is a challenge in itself. Either in supervised or unsupervised DL, one learns a discriminative potentially useful feature representation from data, but the network has not necessarily learned any causal relation between the involved covariates. Learning causal feature representations should be a priority, particularly in current times where accountability, explainability and intepretability are needed in general, and in the relevant case of attribution of causes to climate change.

Addressing these challenges could lead to a remarkable difference in the Earth and climate sciences. We posit, however, that DL alone will not be able to solve all challenges: both hybrid (physics-aware) machine learning and causality will be needed to make a definite leap towards discovering relations, causes and counterfactuals beyond mere curve fitting.

3. Recent Advances and Ways Forward

3.1 *Unsupervised learning of representations*

Fields like remote sensing, computer vision or natural language processing work with the so-called structured domains, in which the original data representation has temporal and/or spatial components. From a geometrical viewpoint, data can be represented in their original coordinates, but visualizing, understanding and designing algorithms therein is not ammenable or even appropriate, mainly due to the high dimensionality and nonlinear dependencies between covariates. This is why learning alternative, typically simpler and compact feature representations of the data has captured the interest of the scientific community. This is the field of dimensionality reduction, for which one has both supervised and unsupervised algorithms.

In Earth sciences, auto-encoders have been the preferred method of choice. The use of auto-encoders in remote sensing is actually widespread in a wide range of applications, including feature extraction and image classification, spectral unmixing, image fusion, and change detection. Recent approaches have considered alternative criteria to summarize variance or correlation, and explored sparsity. Sparsity is among the properties of a good feature representation, and very useful in practice: sparse representations lead to faster, more compact and interpretable models. State-of-the-art unsupervised learning methods such as sparse restricted Boltzmann machines (RBMs), sparse auto-encoders (SAE), sparse coding (SC), predictive sparse decomposition (PSD), sparse filtering and orthogonal matching pursuit (OMP-k) have been successfully used in the literature to extract sparse feature representations. Sparsity has two faces: population and lifetime sparsity, and lately convNets have been successfully trained to achieve both. An unsupervised algorithm called "enforcing lifetime and population sparsity." (EPLS) can generate "pseudo-labels" by imposing both forms of functional sparsity [13]. The method was successfully applied to unsupervised learning of remote sensing images and parsimonious representations of multisensory data blending.

Learning a particular, eventually high-dimensional, density is at the core of many machine learning problems like regression, classification or data representation. However, the problem of PDF estimation is notoriously difficult when considering moderate and high-dimensional data. In the deep learning community, three families of methods are responsible for the majority of the progress in PDF estimation: VAEs, generative adversarial networks (GANs) and normalizing flows (NFs). In recent years, GANs have excelled in tackling the PDF estimation problem in a very efficient way in the context of computer vision and remote sensing applications [4,18,19]. GANs have been used for generating samples to improve classification algorithms as well as for adapting existing algorithms among different satellites. A plethora of GAN models and architectures are nowadays available, see for instance the GanZoo. While GANs generate data in a state-of-the-art way, several extensions have been proposed in order to restrict the generated data to have particular characteristics. For instance, the InfoGAN maximizes the mutual information between some latent space features and the target space. By doing so one has control about imposing different features when generating the data, while at the same time we can explore the feature space. When data has a distinct spatio-temporal structure, as in Earth data science, specific GANs like the 3D-GAN can be better suited. A family of GANs specifically designed to control the features of the generated images are the conditional GANs (CondGANs). This family of GANs generates samples conditioned on particular features, and one can, for instance, generate samples conditioned to a particular class of interest (e.g., types of land covers, climatic zones, or biomes). In CondGANs, we lose the concept of latent space since in this case the latent space is defined by the conditioning samples. Note that this approach is more powerful than the original GANs since we can control what type of samples we are generating. However, this approach is supervised, therefore we need to have labeled samples. In general, however, having labeled samples is often costly in remote sensing and geoscience applications. Despite the good (and surprising) extrapolation capabilities of GANs, some important caveats are identified, namely the difficulty for selecting hyperparameters in the training phase, and their quantitative evaluation. These are important issues in Earth data problems in particular.

The concept of transfer learning in DL is becoming really useful for remote sensing applications. Two main approaches exist: (1) retraining of a network for one problem departing from the frozen network obtained from solving a similar problem in, and (2) considering a part of the net as feature extractor whose latent representation learned from one data problem helps in solving another. Transfer learning is tightly related to domain adaptation, that is, learning either representations or transformations of the data or algorithm to accomodate the statistics of the new image acquisitions, a common situation happening in the everchanging natural environments. Techniques are not important to uncover if cross-domain representations can be learned, but in practice, they are useful to predict a class or biophysical parameter from what has been learned from another similar problem or variable. Possible benefits are shorter training time, as well as higher accuracy. The work exploited these techniques to transfer the architecture of a CNN trained to estimate temperature to estimate moisture. Unsupervised DL methods for domain adaptation consider either (1) adapting the inner representation attempts to minimize a statistical divergence criterion between the representations of the two domains at a given layer in the network, or (2) adapting the inputs distribution to align the input data distributions in the source and target domain before training the classifier, which is typically done either using a common latent representation (like auto-encoders) or using image-to-image translation principles. Variants of those methods, closer to real-world applications, are an active subject of research.

Recent developments learn representations that encode the underlying stable structures and shared information between different parts of high-dimensional data. This has inspired the contrastive learning framework that tries to find representations by maximizing the agreement between differently augmented views of the same data example via a contrastive loss in the latent space. Excellent results in natural (photographic) images could be eventually explored in the context of remote sensing image processing and analysis.

3.2 *Towards non-stationary extended networks*

Deep learning methodologies have been traditionally divided in spatial learning (for example, convolutional neural networks for object detection and classification) and sequence learning (for example, forecasting and prediction). Currently, though, there is a growing interest in blending these two approaches. After all, Earth data can be cast as spatial structures evolving through time: weather forecasting or hurricane tracking are clear examples. We are faced with time-evolving multi-dimensional structures, such as organized precipitating convection and vegetation states as pointed out in Reichstein *et al.* [5] Studies are starting to apply combined convolutional-recurrent deep networks for precipitation nowcasting or extreme weather forecasting, for example. Modeling atmospheric and ocean transport, wild fires, soil or ice movements as well as vegetation dynamics are other paradigmatic problems where spatiotemporal dynamics are important. This is the natural scenario where DL excels; exploiting spatial and/or temporal regularities in huge amounts of data. However, current methods still have to resolve the problem of dealing with long-range phenomena, non-stationary and non-Euclidean processes and uncertainty quantification and error propagation.

3.2.1 *New architectures and algorithms*

Even when space and time are considered altogether, the vicinity that models look at is quite limited. Convolutions and memory units are, by construction, filters with finite impulse response functions. On the one hand, recurrent and LSTM networks cannot remember longer sentences and sequences due to the vanishing/exploding gradient problem. Several architectures currently try to tackle the myopic problem of neural networks.

A promising architecture combines standard convolutional networks with human attention mechanisms in neurosciences: the so-called *attention deep learning networks* can look at the information in distant regions driving processes and activations in a local window. Attention networks have excelled in classification problems where salient patterns of a target are fired by activity in distant regions. The idea has been also developed in the field of natural language processing with networks called *transformers* [20], which efficiently tackle the problem of sequence transduction with self-attention and focused recurrences. Transformers have been limited to look at limited ranges, but recently the so-called *reformers* promise to break the long-range dependency problem.

On the other hand, and perhaps more important, is the fact that in Earth sciences one typically acquires and works with gridded data (either naturally because data comes from model simulations or by *ad hoc* interpolation), but very often this is just an idealization for the sake of convenience, as the raw data is sparse, discontinuous and not necessarily stationary. These particularities call for models able to work in non-Euclidean domains, such as graphs and manifolds. DL defined over graphs and manifolds is complicated, mainly because the non-Euclidean nature of data makes the definition of basic operations (such as convolution) rather elusive. Geometric DL [21] extends these deep learning operations over graphs, but has not been naturally used to work on graph/manifold data so abundant in Earth system science problems.

3.2.2 *Bayesian deep learning*

Features extracted from a dataset are given as point estimates, and do not capture how much the model is confident in its estimation. This is in contrast to probabilistic models, which allow reasoning about model confidence, but often at a very high computational cost. First suggested in the 1990s and studied extensively by MacKay and Neal, Bayesian deep learning (BDL) offers a probabilistic interpretation of deep learning models by inferring distributions over the models' weights. These models are robust to overfitting, offer uncertainty estimates and can learn from small-to-moderate-sized datasets. Yes, in Earth sciences, very often we encouter problems of small and sparse (yet high-dimensional) data, such as those coming from costly terrestrial campaigns (in terms of time, personnel and acquisition difficulty).

Recent advances in variational inference also promise great advances in the probabilistic treatment of deep networks; being not only mathematically solid but also computationally efficient and practical. These models, however, have not been applied extensively in the Earth sciences where, given the relevance of uncertainty propagation and quantification, they could find wide adoption. Only some applications of deep Gaussian processes for parameter retrieval are worth mentioning [22].

3.3 *Hybrid modeling: Deep learning meets physics*

Process understanding is encoded in mechanistic models, and has traditionally dominated our scienfic endevor. Actually, physics modeling and machine learning have often been considered as completely different and irreconcilable fields; scientists should adhere to either a theory-driven or a data-driven approach. Yet, these approaches are indeed complementary: physical approaches are interpretable and allow extrapolation beyond the observation space by construction, and data-driven approaches are highly flexible and adaptive to data. Their synergy has gained attention lately in the geosciences. Interactions can be diverse. We firmly argue that advances in machine learning along with observational and simulation capabilities within Earth sciences offer an opportunity for an integrated agenda for upcoming gesociences, and many approaches are indeed possible.

3.3.1 *Improving parameterizations*

Physical models require setting parameters that can be seldom derived from first (domain) principles. ML and DL in particular can learn such parameterizations. For example, instead of assigning vegetation parameters empirically (or sometimes even arbitrarily) to plant functional types in an Earth system model, one could allow these parameterizations to be learned from proxy covariates with ML, thus allowing some flexibility, adaptiveness, dynamics and context-dependent properties.

3.3.2 Surrogate modeling and emulation

Emulating models in geosciences, climate sciences and remote sensing is gaining popularity. Emulators are ML models that mimic the forward physical models using a small, yet representative, dataset of simulations. Once trained, emulators can provide fast forward simulations, which in turn allow improved model inversion and parametrizations. However, replacing physical models with DL models requires running expensive evaluations offline first, and alternatives exist that construct the model and choose the proper simulations iteratively. This topic is related to active learning and Bayesian optimization, which might model complex codes such as climate model components further [23].

3.3.3 Blending networks and process-based models

DL and physics can be fully blended in several ways, as suggested in Reichstein *et al.* For example, including knowledge through extra regularization can be seen as a form of inductive bias. Another option is to learn emulators which are then combined with purely data-driven algorithms for model inversion. Finally, one can also consider a fully coupled net where layers that describe complicated and uncertain processes feed physics-layers that encode known relations of intermediate feature representation with the target variables. The integration of physics into DL models not only allows us to achieve improved generalization properties but, more importantly, endorses DL models with physical consistency. In turn, the hybridization process has an interesting regularization effect as physics discards implausible models and promotes simpler, sparser structures.

3.4 Interpreting deep neural networks

Deep learning models are highly nonlinear, generally overparameterized models. They excel in prediction accuracy, but such complexity hampers interpretability and trustworthiness. Predictive accuracy is important but often insufficient, and interpreting what the models learned becomes very important, especially in problems with economical, societal or environmental implications. Interpretability has been identified as a potential weakness of deep neural networks, in particular for the geosciences. The lack of interpretability has become a main barrier of deep learning in its wide-spread applications. Interpretability is the degree to which a human can understand the cause of a decision. In this context, Explainable AI (XAI) has emerged as an important field in machine learning: XAI is about developing methods and techniques that can be understood by humans.

Explanations are important to understand how a system works to debug or improve it, to anticipate unforeseen circumstances, to build up trust in the technology, to understand the strengths and limitations, to audit a prediction/decision, to facilitate monitoring and testing, and to guide users into actions or behaviors. Depending on the goal, different types of explanations may be suitable. Explanation techniques may be (1) model-agnostic and use surrogate functions to explain the predictions, (2) perturbation-based testing the model's response, (3) efficient propagation-based explanation techniques which leverage a model's structure, and (4) methods for meta-explanations of the model behavior.

To attain the previous goals, a plethora of techniques have been developed to gain insight from a model: (1) feature visualization to characterize the network architecture (e.g., how redundant, outlier-prone or adversarial-sensitive the network is); (2) feature attribution to analyze how each input contributes to a particular prediction); and (3) model distillation that explains a neural network with a surrogate, simpler (often linear or tree-based) model. Several families of methods can be identified.

In order to interpret what a deep network learned, one could of course use *general purpose model-agnostic methods,* such as permutation analysis or partial dependence plots. However, since neural networks have encoded knowledge in their particular architecture, it makes more sense to look at the basis functions in the hidden layers with tailored techniques. Several techniques have been developed in the last years: (1) *feature visualization* to visualize and characterize the network architecture (e.g., how redundant, outlier-prone or adversarial-sensitive the network is); (2) *feature attribution* to analyze how each input contributes to a particular prediction), and (3) *model distillation* that explains a neural network with a surrogate, simpler (often linear or tree-based) model.

Several works in remote sensing and geosciences have studied interpretability of deep nets. For example, introduced an agnostic-based method through time-series permutation which allows to study memory effects of climate and vegetation affecting net ecosystem CO_2 fluxes in forests. In

Wolanin *et al.* [24] activation maps of hidden units in convolutional nets were studied for crop yield estimation from remote sensing data; analysis suggested that networks mainly focus at growing seasons and can provide a ranking of more important covariates. Recently, in Toms *et al.* [25] the method of layer-wise relevance propagation (LRP) was used to study patterns in Earth System variability.

3.5 *Deep causal learning*

Inferring causal relations from observations has been one of the most important challenges in data science for years (see, e.g., for excellent books on the topic, and for a recent perspective paper in the field of Earth and climate sciences). Causal inference from data allows for more robust modeling of real-world phenomena based on a better understanding of the processes that govern them. It permits a better understanding of the consequences of changes (interventions) to the system under study and allows us to answer questions about hypothetical circumstances (counterfactual statements). Deep learning, however, does not grasp causal relations in datasets, so it actually acts only as an excellent, sophisticated interpolation tool.

Only very recently, we have foreseen efforts towards either incorporating or understanding DL models causally. For instance, in Bengio *et al.* [26] authors implement a meta-learning objective that maximizes the speed of domain transfer, which under certain assumptions can be seen as a way to localize changes in causal mechanisms. In Louizos *et al.* [27], authors learn individual-level causal effects from observational data that can efficiently handle confounding factors. Their method builds on latent variable modeling to simultaneously estimate the latent space, summarizing the confounders and the causal effect using autoencoders. None of these methods have been tested however in real Earth observational data yet.

4. Conclusions and Perspective

This chapter discussed the main challenges for the application and wider adoption of DL methods in Earth sciences. We argue that DL will necessarily have to (1) learn useful feature representations automatically from data, guided by sensible criteria such as sparsity to account for compactness and thus fast algorithms; (2) evolve to completely new network structures beyond convolutional and recurrent nets that only exploit locality and stationarity features, as in attention-based nets in space, transformers in time and Geometric DL for arbitrary data structures; (3) incorporate physics and process understanding to achieve not only physical consistency but also gain credibility by domain experts: recent advances can learn differential equations from data, emulate radiative transfer models accurately and replace (and optimize) layers with mechanistic models; (4) become more interpretable and amenable to scrutiny, either through imposing sparse-promoting and knowledge-based priors directly, or by developing hybrid DL models that explore the subspace of most physically plausible solutions; and (5) tackle the most important grand challenge and ambitious far-end goal of learning causal representations from data.

The future of DL in Earth and Climate sciences is exciting. We have now access to operational differentiable and probabilistic programming tools that allow optimizing arbitrary networks, losses and physics-aware architectures. Besides, current methods are able to make sense of the learned latent network representations: interpretability is just the first stop: eXplainable AI (XAI) and causal inference have to guide network training. Our long-term vision is tied to open new frontiers and foster research towards algorithms capable of discovering knowledge from Earth data, a stepping stone before the more ambitious far-end goal of machine reasoning of anthropogenic climate change. However, while the field of machine/DL has traditionally progressed very rapidly, we observe that this is not the case in tackling such grand challenges. Cognitive barriers are still on our pathway: domain knowledge is elusive and difficult to encode, interaction between computer scientists and physicists is still a barrier and education in synergistic concepts will be a hard task in the upcoming years. The ways forward exposed in this chapter, namely, hybrid DL, interpretability and causal discovery, definitely call for a strong and continuous interaction between domain knowledge experts and computer scientists.

Acknowledgment

This research was financially supported by the European Research Council (ERC) Consolidator Grant SEDAL (Statistical Learning for Earth Observation Data Analysis) project under Grant Agreement 647423.

References

1. I. Goodfellow, Y. Bengio, and A. Courville, *Deep Learning*, MIT Press, 2016.
2. Y. LeCun, L. Bottou, G. Orr, and K. Müller, "Efficient backprop," in *Neural Networks: Tricks of the Trade*, Springer Berlin, pp. 9–50, 1998.
3. S. Hochreiter and J. Schmidhuber, "Long short-term memory," *Neural Comp.*, vol. 9, no. 8, pp. 1735–1780, 1997.

4. I. Goodfellow, J. Pouget-Abadie, M. Mirza, B. Xu, D. Warde-Farley, S. Ozair, A. Courville, and Y. Bengio, "Generative adversarial nets," in *Advances in Neural Information Processing Systems*, pp. 2672–2680, 2014.

5. M. Reichstein, G. Camps-Valls, B. Stevens, M. Jung, J. Denzler, N. Carvalhais, and Prabhat, "Deep learning and process understanding for data-driven earth system science," *Nature*, vol. 566, no. 7743, pp. 195–204, Feb. 2019. doi: 10.1038/s41586-019-0912-1. URL https: //doi.org/10.1038/s41586-019-0912-1.

6. L. Zhang, L. Zhang, and B. Du, "Deep learning for remote sensing data: A technical tutorial on the state of the art," *IEEE Geosci. Remote Sens. Magaz.*, vol. 4, no. 2, pp. 22–40, 2016.

7. X. X. Zhu, D. Tuia, L. Mou, G.-S. Xia, L. Zhang, F. Xu, and F. Fraundorfer, "Deep learning in remote sensing: A comprehensive review and list of resources," *IEEE Geosci. Remote Sens. Magaz.*, vol. 5, no. 4, pp. 8–36, 2017.

8. I. Ali, F. Greifeneder, J. Stamenkovic, M. Neumann, and C. Notarnicola, "Review of machine learning approaches for biomass and soil moisture retrievals from remote sensing data," *Remote Sensing*, vol. 7, no. 12, pp. 16398–16421, 2015.

9. D. Malmgren-Hansen, V. Laparra, A. A. Nielsen, and G. Camps-Valls, "Statistical retrieval of atmospheric profiles with deep convolutional neural networks," *ISPRS J. Photogram. Remote Sens.*, vol. 158, pp. 231–240, 2019.

10. S. Xingjian, Z. Chen, H. Wang, D.-Y. Yeung, W.-K. Wong, and W.-c. Woo, "Convolutional LSTM network: A machine learning approach for precipitation nowcasting," in *Advances in Neural Information Processing Systems*, pp. 802–810, 2015.

11. E. Racah, C. Beckham, T. Maharaj, S. E. Kahou, M. Prabhat, and C. Pal, "Extremeweather: A large-scale climate dataset for semi-supervised detection, localization, and understanding of extreme weather events," in *Advances in Neural Information Processing Systems*, pp. 3402–3413, 2017.

12. Y.-G. Ham, J.-H. Kim, and J.-J. Luo, "Deep learning for multi-year ENSO forecasts," *Nature*, vol. 573, no. 7775, pp. 568–572, 2019.

13. A. Romero, C. Gatta, and G. Camps-Valls, "Unsupervised deep feature extraction for remote sensing image classification," *IEEE Trans. Geosci. Remote Sens.* vol. 54, no. 3, pp. 1349–1362, 2015.

14. N. Kussul, M. Lavreniuk, S. Skakun, and A. Shelestov, "Deep learning classification of land cover and crop types using remote sensing data," *IEEE Geosci. Remote Sens. Lett.*, vol. 14, no. 5, pp. 778–782, 2017.

15. S. Besnard, N. Carvalhais, M. A. Arain, A. Black, B. Brede, N. Buchmann, J. Chen, J. G. W. Clevers, L. P. Dutrieux, F. Gans, *et al.*, "Memory effects of climate and vegetation affecting net ecosystem CO_2 fluxes in global forests," *PloS One*, vol. 14, no. 2, e0211510, 2019.

16. M. Ru-wurm and M. Körner, "Multi-temporal land cover classification with sequential recurrent encoders," *ISPRS Int. J. Geo-Inform.*, vol. 7, no. 4, p. 129, 2018.

17. G. Camps-Valls, D. Tuia, X.X. Zhu, and M. Reichstein, *Deep Learning for the Earth Sciences: A Comprehensive Approach to Remote Sensing, Climate Science and Geosciences*, Wiley & Sons, 2021.

18. D.P. Kingma and M. Welling, "An introduction to variational autoencoders," arXiv preprint arXiv:1906.02691, 2019.

19. J.E. Johnson, V. Laparra, M. Piles, and G. Camps-Valls, "Gaussianizing the earth: Multidimensional information measures for earth data analysis," arXiv preprint arXiv:2010.06476, 2020.

20. M. Jaderberg, K. Simonyan, and A. Zisserman, "Spatial transformer networks," *Advances in Neural Information Processing Systems*, pp. 2017–2025, 2015.

21. M.M. Bronstein, J. Bruna, Y. LeCun, A. Szlam, and P. Vandergheynst, "Geometric deep learning: Going beyond Euclidean data," *IEEE Signal Processing Magazine*, vol. 34, no. 4, pp. 18–42, 2017.

22. D.H. Svendsen, P. Morales-Alvarez, A.B. Ruescas, R. Molina, and G. Camps-Valls "Deep Gaussian processes for parameter retrieval and model inversion," *ISPRS Journal of Photogrammertry and Remote Sensing*, vol. 166, pp. 68–81, 2020.

23. D.H. Svendsen, L. Martino, and G. Camps-Valls, "Active emulation of computer codes with Gaussian processes — Application to remote sensing," *Pattern Recognition*, vol. 100, no. 107103, pp. 1–12, 2020.

24. A. Wolanin, G. Mateo-García, G. Camps-Valls, L. Gómez-Chova, M. Meroni, G. Duveiller, Y. Liangzhi, and L. Guanter, "Estimating and understanding crop yields with explainable deep learning in the Indian Wheat Belt," *Environmental Research Letters*, vol. 15, no. 2, pp. 024019, 2020, IOP Publishing.

25. B.A. Toms, E.A. Barnes, and I. Ebert-Uphoff, "Physically interpretable neural networks for the geosciences: Applications to earth system variability," *Journal of Advances in Modeling Earth Systems*, vol. 12, no. 9, pp. e2019MS002002, 2020, Wiley Online Library.

26. Y. Bengio, T. Deleu, N. Rahaman, R. Ke, S. Lachapelle, O. Bilaniuk, A. Goyal, and C. Pal, "A meta-transfer objective for learning to disentangle causal mechanisms," arXiv preprint arXiv:1901.10912, 2019.

27. C. Louizos, U. Shalit, J.M. Mooij, D. Sontag, R. Zemel, and M. Welling, "Causal effect inference with deep latent-variable models," *Advances in Neural Information Processing Systems*, pp. 6446–6456, 2017.

Chapter 8

Accurate Detection of Built-Up Areas in Remote Sensing Image via Deep Learning

Yihua Tan*, Shengzhou Xiong[†] and Pei Yan[‡]

National Key Laboratory of Science &
Technology on Multi-spectral Information Processing,
School of Artificial Intelligence and Automation,
Huazhong University of Science and Technology,
Wuhan 430074, China
*yhtan@hust.edu.cn,
[†]xiongshengzhou@126.com,
[‡]yanpei@hust.cn

Abstract

In recent years, with the rapid development of deep learning, deep models have become the mainstream in many research branches and application fields of machine learning. As one of the most important targets in remote sensing data, built-up area detection is an important practical field of machine learning. To demonstrate the capabilities of deep models, a large dataset was collected and labeled from data based on GaoFen-2 remote sensing satellite. In the following sections, several deep learning methods will be discussed and experiment on the dataset, including DSCNN, LMB-CNN and a FCN model based on LMB-CNN. It can be seen from the experimental results that the deep models bring great performance improvement compared with traditional algorithms.

Keywords: deep learning, remote sensing, built-up area, DSCNN, LMB-CNN, FCN

1. Introduction

The built-up area is the main area of human activity which is represented as
the aggregation area of buildings and it has always been one of the focuses of
remote sensing image analysis. There are many fields that benefit from accurate regional scope of built-up areas, for instance, accurate built-up area is
advantageous for the analysis of the urbanization process [1]. Accurate positioning of built-up areas can provide convenience for damage assessment [2].
Based on analysis of remote sensing images covering 35,000 km², a global
human settlement layer (GHSL) that has been constructed for that built-up
area is an essential component of earth observation [3, 4]. However, manual
extraction of built-up areas is really a huge consumption of time and human
resources. Thus, automatically extracting the built-up areas from remote
sensing data is crucial for utilizing accurate built-up area. To automatically
extract built-up areas via machine learning methods is challenging because
built-up areas appear with various styles and types, where a building's spectral
variation and confusion, environmental diversity, illumination changes, and
plant covering in different seasons are hard to describe in uniform features.
Different from the general single targets, built-up areas are a notion of
closed-shape regions that primarily include buildings but also elements of
no-built-up areas like lawns, small pools and hills, which makes the classification task quite challenging. In addition, the vast area to be processed means
large amounts of remote sensing images, so the speed of the algorithm is
essential for practical purposes.

There are a variety of remote sensing data sources that are available, data
sources with different characteristics also derive various algorithms. The
synthetic aperture radar (SAR) images [5] and polarimetric SAR (PolSAR)
images [6] are popular research data, because the active imaging mode of
radar can provide all-day and all-weather data. Multi-spectral images contain several different spectral segments, and the differences of spectral characteristics between ground objects become the basis of various spectral
indexes [7–9]. Further, hyperspectral images provide a more detailed division of the spectrum, and the spectral indexes were extended to them [10].
In addition, panchromatic images based on visible light imaging are more
similar to data sources of general computer vision tasks, which can benefit
more from research results of computer vision [11, 12], and pseudocolor
images with multisource fusion take it a further step [3, 13, 14].
Furthermore, some data sources are less frequently used because of the difficulty to obtain them, such as stereo imagery that incorporates height
information [15], road network data [16, 17], nighttime light data from

VIIRS and DMSP-OLS [18, 19], and considering the temperature differences among regions of different types, thermal infrared remote sensing is also adopted [20–22]. Panchromatic images and multi-spectral images are the most commonly used data, for their complementarity, panchromatic ones usually have high spatial resolution without color information, and the multi-spectral ones do the opposite. Panchromatic images and multi-spectral images are readily available, as there are many satellites providing them at the same time, such as series of WorldView, GeoEye-1, QuickBird, IKONOS and series of GaoFen. The following experiment will be carried out on the data of GaoFen-2 satellite.

The extraction of built-up area has attracted the attention of many scholars for a long time, there are many traditional algorithms applied to this task, including both supervised algorithms and unsupervised ones. These algorithms include almost all common machine learning methods, such as SVM [23, 24], CRF [25], MRF [26, 27], MeanShift [28], graph method [29, 30], however, due to the limitation of artificial features like spectral indexes [9], Harris points [29, 30] and famous SIFT [12, 31], these algorithms have limited effectiveness. So deep models attract greater attention, deep methods succeed because their abstract deep features have a strong generalization ability, which is suitable for complex built-up area extraction. Based on a large labeled dataset from GaoFen-2 satellite, three deep models will be described in detail and tested for performance that contains DSCNN, LMB-CNN and an FCN model. The DSCNN can utilize both panchromatic images and multi-spectral images, LMB-CNN is a lightweight and fast model that needs panchromatic images only, and the FCN is an image segmentation model that can be applied to large size remote sensing images. The experiment results show that deep methods can improve the performance of built-up area extraction obviously.

2. Related Works

Both unsupervised and supervised algorithms commonly used in machine learning are involved in the built-up area extraction algorithms proposed in recent years. Among the unsupervised methods, the threshold method is the most used one, because various index features are widely adopted, among which spectral index is the most common. For instance, Varshney selected a kernel-based threshold method to verify the performance of several spectral indexes [9]. Fuzzy threshold method was used to classify the texture-derived built-up presence index (PanTex) proposed in Ref. [32], and further research in another paper compared independent optimization threshold for each scene with the single threshold [3]. A dual-threshold filtering was proposed to integrate the information of morphological building index (MBI) and morphological shadow index (MSI) [33]. The method based on graph theory is also one of the mainstream, Liu *et al.* [64] adopted graph cut to segment their Bayes-based salient graph. Tao *et al.* [29] selected spectral clustering and graph cut method to deal with the candidate areas of built-up areas for more accurate results. Li *et al.* [30] handled the proposed corner density map with the Cauchy graph embedding optimization theory. Besides the threshold method and graph theory, there are some other algorithms being adopted, such as Edison-Meanshift and K-means algorithm are combined to extract built-up area [28].

As for supervised methods, SVM, CRF and MRF are widely accepted. Li *et al.* [23] used SVM for feature mapping and combined it with data mining and database technology. LS-SVM classifier is applied for target classification and achieved good result on a proposed new feature [30]. Tao *et al.* further applied the multi-kernel learning method to combine multiple features and achieved a better performance [31, 35]. Zhong *et al.* [25] used CRF as the basic classifier to generate group CRFs via changing the input features and fused the results. Smits *et al.* [26] applied MRF to update the land-cover maps from high-resolution imagery. Combining MRF and Bayesian model, Yu *et al.* [27] extracted built-up area with map data as the auxiliary information. Besides the methods mentioned above, Shackelford *et al.* [36] used maximum likelihood method and hierarchical fuzzy classification in their paper and Weizman *et al.* selected voting theory [37, 38].

The traditional built-up area extraction algorithms usually have limited performance and poor generalization ability, which is mainly caused by the limitation of features. It is worth noting the deep features have been proven to be superior to handcrafted ones in almost all computer vision tasks. A number of scholars have tried to apply deep learning theory to remote sensing image interpretation. Based on deep convolutional neural network and high-resolution Google Earth images, Zhang *et al.* [39] proposed a method to extract suburban buildings. Zhou *et al.* [40] discussed the performance of CNN in the target recognition task of POLSAR data. Via comparative experiments, Yang *et al.* [65] verified the performance of three novel CNN structures in detecting buildings. Li *et al.* [41] proposed two key modules, the L-shaped module is designed to extract complementary information hierarchically and the V-shaped module with nested connections integrates these features to obtain the salient objects. Cheng *et al.* [42] utilized measure learning to improve the CNN performance of remote sensing scene interpretation. Chen *et al.* [43] and Xu *et al.* [44] applied CNN to fuse several different remote sensing data, such as hyperspectral images and LIDAR images. As for built-up area extraction task, Li *et al.* [45] proposed a CNN-based fusion model of multi-scale features and they dealt the task on a high-resolution SAR image. As a type of built-up area, urban villages are of great significance as the product of China's urbanization development. Li *et al.* [31] combined unsupervised learning strategy and CNN model to handle the lack of labeled data. Bramhe *et al.* [46] applied Inception-V3 and VGGNet to Sentinel-2 images which are pre-trained on ImageNet. Based on high spatial resolution remote sensing images, Li *et al.* [47] adopted CNN and CapsNet to construct four experiments and made a comparative analysis.

For the detection task of built-up area, almost all classical machine learning algorithms have been employed but the benefits are limited. There are some studies in recent years demonstrating that deep learning has obvious advantage in this task. Since Alexnet gains the champion of 2012 ImageNet classification competition [48], deep learning theory has made great progress, which is accompanied by the rapid development of GPU. Profiting from the operational capability of GPU, it is possible to rapidly process large-scale remote sensing data utilized deep learning. Adopting deep methods to extract built-up areas is gradually becoming the mainstream and several instances will be described in detail in what follows.

3. Data Description

To extract built-up area from remote sensing data, panchromatic images and multi-spectral images are most commonly used and they are usually gathered synchronously, so 662,308 sample pairs are collected from 64 groups of Gaofen-2 imagery, the geographic regions corresponding to these images are distributed in 32 different provincial-level administrative regions of China, and it should be noted that multi-spectral samples contain G band, R band and NIR band data images. GaoFen-2 satellite is the first civil optical remote sensing satellite with spatial resolution reach to 1 m that was independently developed by China, which provides panchromatic images with 1-m resolution and corresponding multi-spectral images with 4-mr resolution. Consider that each sample should contain multiple buildings to have the characteristics of built-up area, each collected panchromatic sample contains 64×64 pixels and a multi-spectral sample contains 16×16 pixels. As required by the deep model training, each sample is manually labeled as city or non-city, and the dataset is named as GF-2 sample set. Follow the training rules of supervised machine learning, the dataset is divided into three parts, training set, validation set and

Table 1:   The partitioning of GF-2 sample set.

	City	Non-city	Total
Training set	67971	329418	397389
Validation set	22656	109804	132460
Testing set	22656	109803	132459

testing set. The partition ratio is 6:2:2. The following deep models are trained on this dataset or panchromatic part of it (see Table 1 and Fig. 1).

To verify the algorithms involved in the experiment, a full-frame test dataset is built, which consists of 10 full-size images. The geographic regions of the full-frame test dataset are located in five different provinces of China, including Jiangsu, Yunnan, Hebei, Jiangxi and Shandong. There is no overlapping area between GF-2 sample set and the test set. Furthermore, each panchromatic image contains 10240×10240 pixels with a manually labeled block-level ground truth. Consider the speed of compared algorithms, a portion of each testing image is cut that contains 2048×2048 pixels and made a pixel-level annotate. Table 2 shows the basic information for the test images.

4. Double-Stream Deep CNN (DSCNN)

The DSCNN model is designed with the aim to combine the information of both panchromatic image and multi-spectral image. To improve the detection accuracy by combining panchromatic data and multi-spectral data, there are two basic ideas, generate the pansharpening image by fusing panchromatic and multi-spectral data, or fuse them in feature level. DSCNN model adopts the second one on account of time consumption of pansharpening image. To the structure, simple feature fusion is performed only in the fully connected layer.

The detail of DSCNN model is shown in Fig. 2. The model contains two branches: the upper one takes panchromatic sample as inputs, which provides primary information for city sample recognition; the lower branch takes multi-spectral samples as input, which provides auxiliary color information to improve the accuracy. Consider the outstanding performance of Inception module [49, 50], the structure of both branches are based on the Inception V3 module. Although the further structure of Inception [50] has made a slight improvement in many applications, Inception V3 is more applicable in terms of the balance between performance and computational complexity. So following the design ideas of Szegedy *et al.* [49], the final DSCNN gives the Inception-based structures as in Fig. 2.

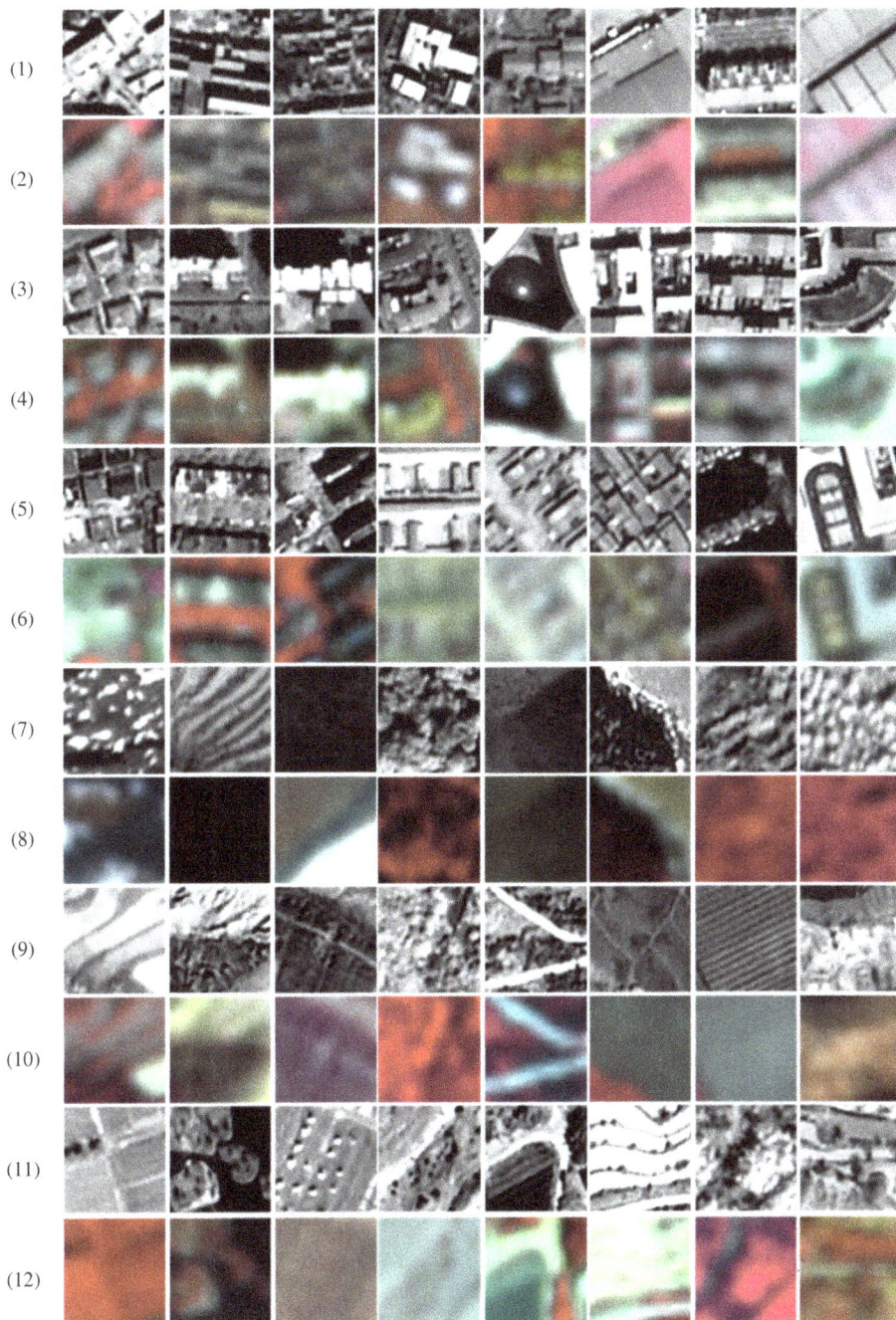

Figure 1: Part of GF-2 sample set. The rows 1, 3, 5 are the panchromatic samples from city areas, the rows 2, 4, 6 are the corresponding multi-spectral samples; the rows 7, 9, 11 are the panchromatic samples from non-city areas, the rows 8, 10, 12 are the corresponding multi-spectral samples.

Table 2:   The basic information of the full-frame test dataset.

No.	Location	Lng & Lat	Ratio of urban (10240)	Ratio of urban (2048)
1	Jiangsu	E119.36° N31.34°	0.091	0.059
2	Jiangsu	E119.39° N31.31°	0.086	0.055
3	Yunnan	E103.57° N25.03°	0.032	0.104
4	Yunnan	E103.49° N24.94°	0.090	0.161
5	Hebei	E115.22° N38.51°	0.185	0.424
6	Hebei	E115.35° N38.80°	0.107	0.152
7	Jiangxi	E116.53° N28.30°	0.212	0.057
8	Jiangxi	E116.47° N28.41°	0.047	0.038
9	Shandong	E117.31° N36.25°	0.172	0.164
10	Shandong	E117.29° N36.16°	0.113	0.128

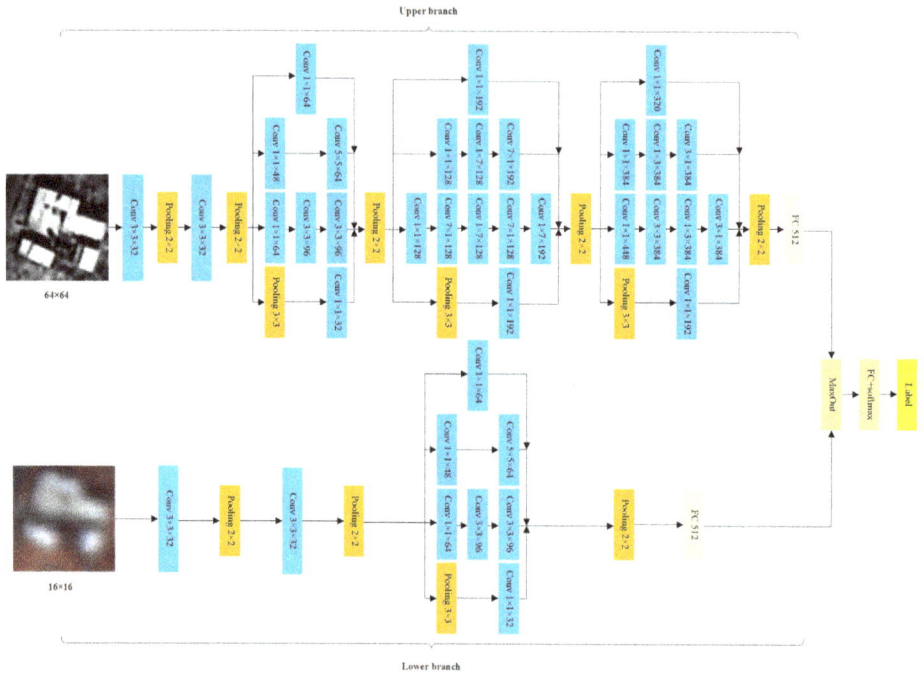

Figure 2:   The structure of DSCNN, the blue blocks mean convolutional layers, and the convolution kernel size and output channel count are marked in it.

The design of upper branch aims to extract the complex morphological features and texture features from higher resolution panchromatic images, it consists of three Inception modules. As show in Fig. 2, the output of the lower level module is fed into the next module as input with a pooling operation between them. For this branch, the input panchromatic samples contain 64 × 64 pixels and obtain a 512-dimension feature vector as the output.

The lower branch is designed to extract spectral feature from multi-spectral image, in traditional applications the data is basically adopted to calculate the simple spectral indexes, so it owns sufficient complexity through one Inception V3 module. Referring to the idea of spectral indexes, the input samples of lower branch is a pseudo color image, and just as described in the data description section, the samples are composed of G, R, NR bands of multi-spectral images. In collaboration with the input data of the upper branch, the multi-spectral samples contain 16 × 16 pixels for each, and the output of the lower branch is also a 512-dimension feature vector.

A MaxOut layer connects the two branches, and then a full connection layer and SoftMax layer are followed to identify if the input sample is built-up area or not.

For DSCNN, the model parameters are mainly derived from convolution layers and full connection layers. Each convolution layer in the model contains three successive steps: convolution, batch normalization, and activation. The calculation of convolution layer is described as follows:

$$x_j^l = f\left(\sum_{i=1}^{M} x_i^{l-1} * k_{ij}^l + b_j^l\right), \tag{1}$$

where x_j^l means the jth features map of layer l, and k_{ij}^l is convolution kernel of this layer, M is the count of feature maps that are input to the layer l, and b_j^l means the bias. The batch normalization is a very routine operation, details are available in Ref. [51]. ReLU is adopted as activation function as (2).

$$f(x) = \max(0, x). \tag{2}$$

The full connection layer can be expressed as follows:

$$y_j = \sum_{i=1}^{n} w_{ij} x_i + b_j, \tag{3}$$

where x_i means the ith dimension of input vectors and y_j is the jth dimension of output, w_{ij} and b_j are parameters that are trained by labeled samples. The activation function of full connection layer is also ReLU. In addition, the MaxOut layer outputs the larger response values of the two branches.

The optimization of model parameters is driven by cross entropy loss that is described as (4) and the Adam algorithm proposed in Ref. [52].

$$J(\theta) = -\frac{1}{m}\sum_{i=1}^{m} y^{(i)}\log(h_\theta(x^{(i)})) + (1-y^{(i)})\log(1-h_\theta(x^{(i)})). \tag{4}$$

Table 3: Comparison of the two training strategies.

	The accuracy of test part of GF-2 sample set		
	City (%)	Non-city (%)	Total (%)
Joint training	98.89	99.39	99.31
Separate training	99.05	99.79	99.66

For DSCNN model, two training strategies are proposed. Firstly, jointly training the two branches, in which the entire network is trained and optimized directly with parameters randomly initialed. Another one is training separately the two branches of DSCNN, which includes two steps: Firstly, the upper branch is trained on the panchromatic sample set singly, so is the lower branch on multi-spectral sample set. Secondly, initializing the branches with the optimal and the parameters that were obtained before, and the parameters of two branches will be frozen, then the classify head is randomly initialized and fine-tuned. The accuracy of test sample set of two training strategies has been shown in Table 3.

5. Lightweight Multi-Branch CNN (LMB-CNN)

There is always a huge amount of data to deal with for built-up area extraction, the process speed of an algorithm is critical, so a lightweight CNN structure is necessary. Some of the most popular networks such as AlexNet [48], MoblieNet [53], and Inception series [49, 50] can provide some design hints [54]. According to their successful design experience, the designed LMB-CNN will abide by the following guidelines:

(1) The model structure should be as simple as possible while satisfying the task. The experiments of built-up area classification have been done on GF-2 sample set and several common used networks are selected (AlexNet, MobileNet, Inception V3). The processing speed of AlexNet is the fastest according to experiment results, and the speed of Inception V3 is only $1/5$ of it, although the Inception structure has the best performance. It is worth noting that AlexNet achieved 98% accuracy with only five convolution layers on the training set, and for built-up area extraction task, data processing speed becomes more crucial when accuracy reaches a satisfied level. So considering both performance and model complexity, the convolutional layer of required model should be around 10 layers.

(2) The 1×1 convolutional layer brings many benefits. The 1×1 convolutional kernel was widely adopted in CNN [55, 56], this structure can be

used to increase the depth and nonlinearity, and it is the first choice to deal with output channel count adjusting. In addition, the increased consumption can be negligible.

(3) Multiple branch structure should be considered. Inception module is one of the most famous multi-branch architectures [55]. The residual structure of ResNet [56] is also viewed as a collection of network paths with different lengths. The multi-branch structure has been adopted by most deep models and its positive effects have been well documented.

(4) Batch normalization (BN) layer is a must. Batch normalization operation is proposed to solve the internal covariate shift phenomenon [51]. It is widely used for the ability of avoiding over-fitting and improving generalization ability.

The basic characteristics of the proposed LMB-CNN module is that it contains three branches. To reason the proposed model is named lightweight multi-branch convolutional neural network (LMB-CNN) is that it obeys the design guidelines. In Fig. 3, graphical detail is displayed. In addition, there are some intentional designs that follow above design guidelines: (1) adopt several blocks with similar structures to make the model concise; (2) each branch of a module has at most two convolutional layers to limit the total count of layers; (3) to reduce complexity, partially replace the normal convolution with the depthwise separable convolution; (4) three branches are combined to extract multiple diverse features; (5) To bring in competitive mechanism for different features, MaxOut strategy is selected to merge multiple branches.

The LMB-CNN mainly contains three convolutional modules, and each of them consists of three branches. The upper branch includes two depthwise

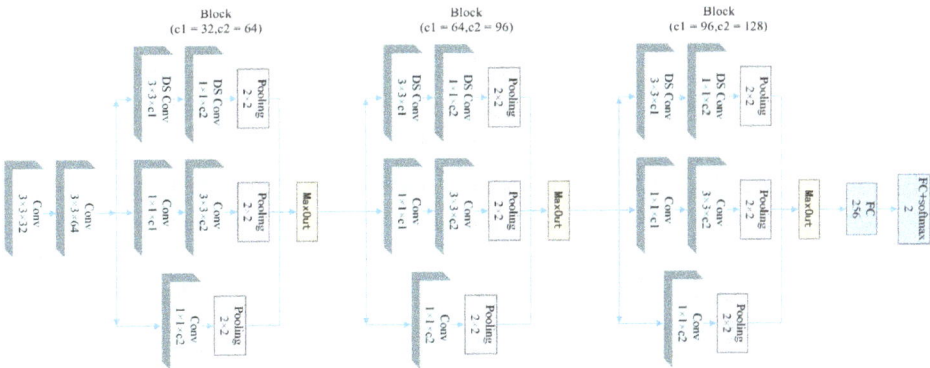

Figure 3: Structure of LMB-CNN. The kernel size and count of convolution layers are shown in the blocks. BN layers and activation layers are hidden. The "DS Conv" means depthwise separable convolution while "Conv" means standard convolution.

separable convolutions [53]; compared to standard convolution operation, depthwise convolution has fewer parameters. For depthwise separable convolution, each unit contains a 3×3 depthwise convolution followed by a 1×1 standard convolution. The depth convolution is given as follows:

$$G_{k,l,m} = \sum_{i,j} \hat{K}_{i,j,m} \cdot F_{k+i-1,l+j-1,m}, \tag{5}$$

where $F_{k+i-1,l+j-1,m}$ represents the data which locates in $(k + i - 1, l + j - 1)$ of the mth input feature map. $\hat{K}_{i,j,m}$ is the parameter that locates in $(i - 1, i - 1)$ of the mth convolutional kernel, $G_{k,l,m}$ is the data which locates in (k, l) of the mth output feature map. The middle branch that contributes the main computation is made up of two standard convolutional layers. Finally, aiming to possibly pass the low level features forward to the higher level, the last branch uses a 1×1 standard convolution to adjust the count of feature map channel just following the guidelines above. In such cases, the features from the invariable receptive field are possibly kept with the MaxOut strategy. In addition, the standard convolution is calculated as follows:

$$G_{k,l,n} = \sum_{i,j,m} K_{i,j,m,n} \cdot F_{k+i-1,l+j-1,m}, \tag{6}$$

where the variables in Eq. (6) have the same meanings as those in Eq. (5). However, there are $M \times N$ convolutional kernels, M and N refer to input channel counts and output channel counts.

After a pooling operation on the three branches, a MaxOut layer will merge their features through competitive mechanism. Therefore, the better features will be retained while the low level features also have a chance via the lower branch. Besides, the operation can fuse multiple features to maintain comparatively low computation without increasing the feature map channels.

The integrated LMB-CNN model consists of two standard convolutional layers, three convolutional blocks and two fully connected layers. To get the final feature vectors, the first fully connected layer is set to dimensions based on complexity, where higher dimension can lead to a rapid complexity increase of the entire model. The second fully connected layer is used for final classification which has two neurons followed by a softmax function. In addition, the activation function is ELU function that is proposed in Ref. [66], and is as follows:

$$f(x) = \begin{cases} x & , x > 0 \\ \alpha (\exp(x) - 1), x \leq 0 \end{cases}. \tag{7}$$

The batch normalization layer proposed in Ref. [51] has embedded after each convolutional layer as a general operation. The dimension of final feature vectors is 256 to control the computing complexity. As for training, cross-entropy loss is selected which is described as (8), and the optimization method is Adam algorithm that is proposed in Ref. [52].

$$J(\theta) = -\frac{1}{m}\sum_{i=1}^{m} y^{(i)} \log(h_\theta(x^{(i)})) + (1 - y^{(i)}) \log(1 - h_\theta(x^{(i)})). \tag{8}$$

6. Fully Convolutional Networks

The FCN has been successful in the semantic segmentation of natural scenes [57, 58], but the size of the images they work on is usually not very large. Here attempts to apply FCN to large-scale remote sensing images has been made. As shown in Fig. 4, the procedure for the proposed method can be divided into three steps:

(1) Divide the image into small blocks by checkerboard partitioning as the processing unit and extract the deep features of each block via an LMB-CNN.
(2) Rearrange the features of the blocks into multi-channel feature maps according to spatial location. The designed FCN segments the multi-channel feature maps into eight preliminary segmentation masks.
(3) Vote on the eight preliminary segmentation masks to determine the final extraction result.

It's worth pointing out that an LMB-CNN is used to extract features which are used as the inputs of an FCN. The segmentation task is essentially the classification of each local image block by considering neighboring relationships, so the classification features extracted by LMB-CNN are suitable for subsequent segmentation tasks.

The first step of the proposed algorithm divides the image into small blocks that are used to extract the deep features as the input of FCN, and the block size is 64×64 pixels considering that the resolution of the experimental data is 1 m. There are four reasons to do this other than to combine LMB-CNN with FCN to construct an end-to-end network. Firstly, treating small blocks as the processing units is more appropriate than pixels. Secondly, this step can handle the problem that a large remote sensing image causes considerable consumption of hardware resources while using FCN. More

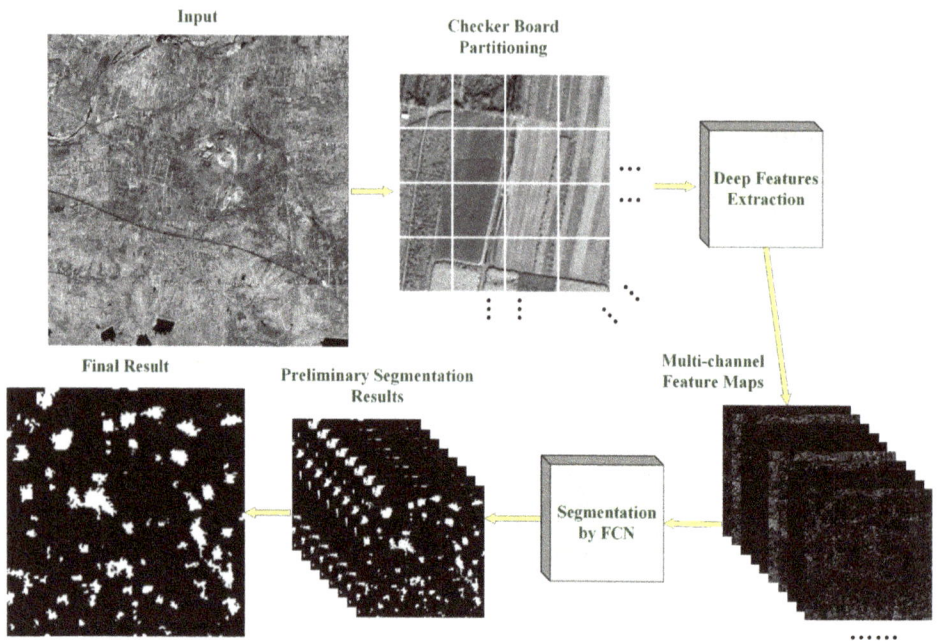

Figure 4: Main framework of the proposed algorithm.

than 20 GB of storage is required for 64 channel feature maps with a size of 10240×10240 pixels, and the actual image is usually larger. This operation reduces the input size of FCN nearly one-thousand times. Thirdly, the selected LMB-CNN model can be trained as a classification task, while the FCN model used later is trained as a segmentation task. FCN segmentation uses the block-level features extracted by LMB-CNN, combines them together sequentially in a way that both of the tasks can be learned. Finally, the block size is 64 pixels because each block can traverse two or more buildings in terms of the definition of the built-up area.

Deep features of each block are extracted by LMB-CNN, and the feature dimension was 256. Based on the spatial position of image blocks, the deep features can be rearranged into 256 channel feature maps, similar to the output of a series of convolutional layers. The first nine feature maps of the sample image are shown in Fig. 5 as examples.

The reconstructed feature maps are treated as the input of an FCN model. Considering the calculation speed, the basic structure of LMB-CNN is reused by FCN. However, the pooling operations and the full connection layers are removed to keep the feature map's size unchanged. The structure of the FCN model is shown in Fig. 6, taking an 8-kernel depthwise separable convolutional layer activated by a sigmoid function as the end of the network,

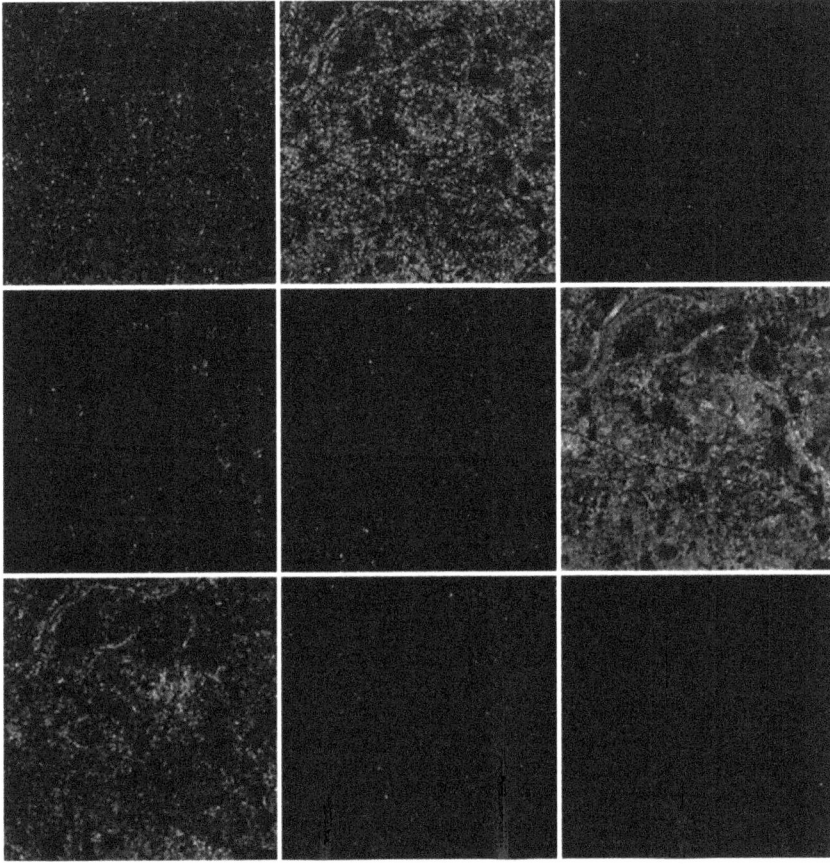

Figure 5: The first nine reconstructed feature maps with size of 160×160 (corresponding to the sample image with 10240×10240 pixels), in which the deep features are extracted by LMB-CNN.

which outputs eight channel segmentation masks to improve the segmentation reliability. Intuitively, voting on the multiple decisions is better than only one decision since different output channels consider different inner characteristics of the input features. Figure 7 gives an intuitive view of the eight outputs. Experimental results in Table 4 show that the voting strategy can reduce the false alarms and missing alarms.

To train the FCN model, considering that the extraction of built-up area is a binary classification problem, it is convenient to measure the difference between the output results of each channel and ground truth by the mean square error. It also avoids the probability calculation when using cross-entropy. Besides, for this issue the false alarms and missing alarm which attract attention are usually located on the boundary of the built-up area. The

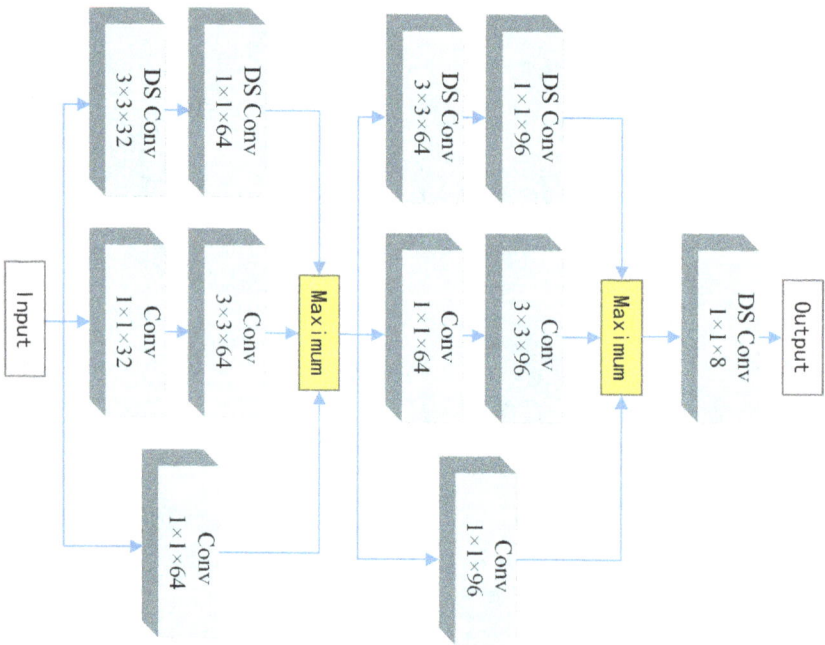

Figure 6: The structure of the FCN model.

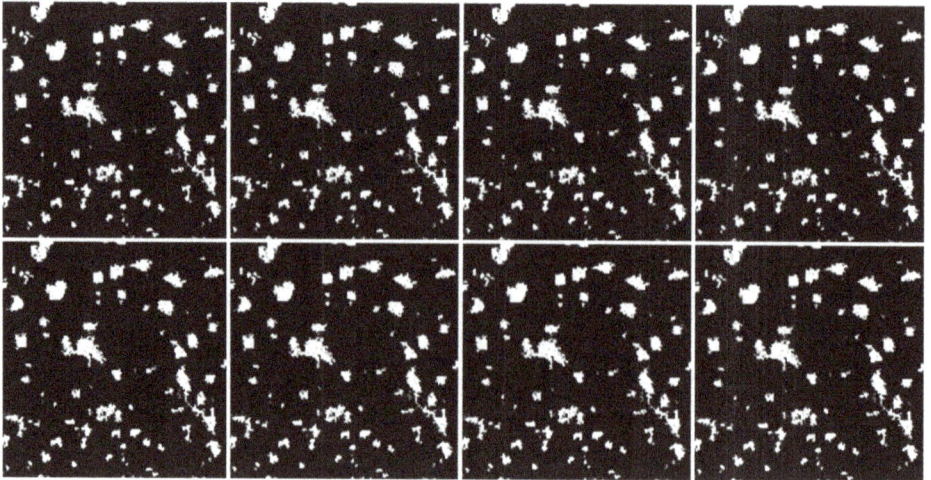

Figure 7: The eight preliminary segmentation masks that are produced by the FCN model.

sigmoid output probabilities of these samples are usually around 0.5 if one selects cross-entropy. However, the mean square error can highlight the contributions of these samples. Therefore, the loss function based on the square error is adopted for training the FCN model, but the output is divided into three

Table 4. Results of different voting thresholds.

	False alarm			Missing alarm			Time (s)		
No.	Single output	$T = 4$	$T = 7$	Single output	$T = 4$	$T = 7$	Single output	$T = 4$	$T = 7$
1	0.1224	0.0755	0.0643	0.0706	0.0988	0.1100	8.78	8.78	8.80
2	0.1097	0.0655	0.0558	0.0878	0.1046	0.1228	8.75	8.77	8.70
3	0.0887	0.0765	0.0635	0.0291	0.0485	0.0533	8.71	8.71	8.74
4	0.1450	0.0928	0.0768	0.0270	0.0340	0.0427	8.71	8.79	8.76
5	0.0264	0.0171	0.0140	0.0219	0.0316	0.0358	8.73	8.73	8.78
6	0.0653	0.0324	0.0246	0.0329	0.0394	0.0449	8.71	8.75	8.81
7	0.0696	0.0546	0.0463	0.0345	0.0414	0.0489	8.78	8.80	8.76
8	0.1509	0.1051	0.0918	0.0824	0.1073	0.1190	8.77	8.74	8.73
9	0.1064	0.0712	0.0608	0.0302	0.0454	0.0520	8.73	8.77	8.80
10	0.0700	0.0492	0.0425	0.0482	0.0541	0.0617	8.77	8.71	8.80
ALL	0.0860	0.0569	0.0479	0.0412	0.0532	0.0612	87.44	87.55	87.68

parts when calculating the loss function: correct classification, false alarm and missing alarm. The specific loss function is as follows:

$$
L(\theta) = \sum_{p^{(i)}=y^{(i)}} \left\| h_\theta\left(x^{(i)}\right) - y^{(i)} \right\|^2 + \alpha \sum_{p^{(i)}=1, y^{(i)}=0} \left\| h_\theta\left(x^{(i)}\right) - y^{(i)} \right\|^2
$$
$$
+ \beta \sum_{p^{(i)}=0, y^{(i)}=1} \left\| h_\theta\left(x^{(i)}\right) - y^{(i)} \right\|^2,
$$

(9)

where $h_\theta\left(x^{(i)}\right)$ indicates the output of the FCN model and $y^{(i)}$ is the ground truth. $y^{(i)} = 1$ represents the built-up area and $y^{(i)} = 0$ is the non-built-up area. α and β are the parameters that can be specified, and $p^{(i)}$ is the threshold discriminant result of $h_\theta\left(x^{(i)}\right)$, which is given as follows:

$$
p^{(i)} = \begin{cases} 0, & h_\theta\left(x^{(i)}\right) \leq 0.5 \\ 1, & h_\theta\left(x^{(i)}\right) \geq 0.5 \end{cases},
$$

(10)

By adjusting the weights of the loss function, the contribution of the error classification part can be increased to avoid being covered by the correct classification part. In fact, after a certain number of iterations, more than 90% of the units will be classified correctly.

The eight segmentation masks of the FCN model are shown in Fig. 7. There is an ~6% difference between each pair of masks, which indicates that each block may have varied binary values in different masks. The final segmentation result can be obtained via voting. Because each block is now a pixel in the feature maps, the voting process is converted to count the number of pixels being determined as built-up areas. The counting number is computed as follows:

$$T_i = \sum_j p_j^{(i)}, \tag{11}$$

where $p_j^{(i)}$ is the segmentation binary value in the jth mask map according to Eq. (10). According to experimental result, set the voting threshold to 4 to achieve the balance between the false alarm and the missing alarm. In practical applications, the threshold can be adjusted according to the emphasis on the false alarms and missing alarms. The range of the threshold has little effect on the overall precision.

7. Post-processing

Only a block-level result can be obtained by DSCNN, LMB-CNN or FCN compared to the pixel-level results of classical methods. In addition, built-up area refers to a region which has been exploited and constructed, it needs certain acreage larger than a threshold, and a small area of grass, pond, etc. should also be part of the built-up area. So post-processing is necessary to make the results more reasonable.

To remove the detected region where the acreage is not reached, setting the threshold of built-up area as 5-image patches is reasonable, taking the acreage of some villages into consideration. Each patch corresponds to an area of 64 m × 64 m. In addition, the lawns, pools and hills within the built-up area should be identified as parts of the built-up area, so the non-built-up area that is smaller than 5-image patches should be converted to built-up area, too.

For pixel-level results, just considering the aesthetics of the edges and simply enlarging the binary image of block-level results in the original size being enough such that the interpolation process smooths the boundary of built-up area. The results are shown in Fig. 8. To obtain more an elaborate boundary, you can follow these steps:

(1) A simple linear iterative clustering (SLIC) [59] segmentation operation is applied to the boundary blocks of built-up area, with a recommendation

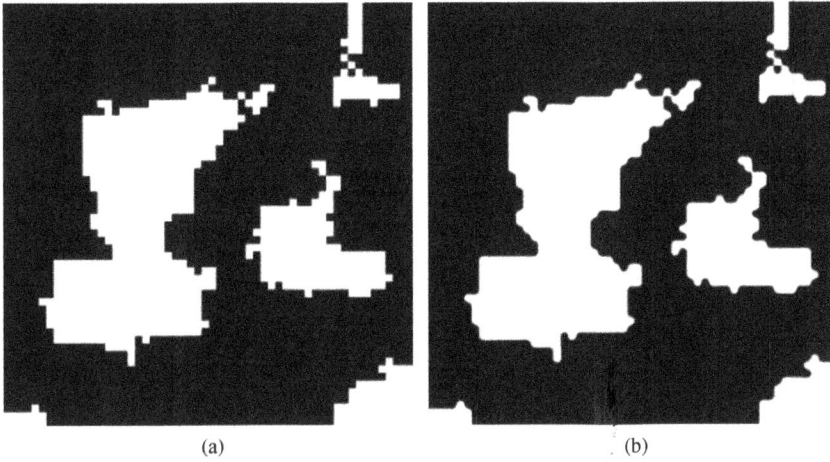

(a)   (b)

Figure 8: (a) Shows block-level result. Treat each block as a pixel and enlarge it to original size, then obtain (b).

to set the parameters spcount as 16, compactness as 20, so that each super pixel is about 16×16 pixels.

(2) Calculate the average vegetation indexes AVE_NDVI and AVE_SAVI of each super pixel, and then judge them by Eq. (14). NDVI [60] and SAVI [61] are calculated as Eqs. (12) and (13), where $Band_x$ represents the different bands of multi-spectral data, L means the adjustment coefficient of soil which ranges from 0 to 1. It has no vegetation coverage when L equals to 1, and the influence of soil background can be ignored when L equals to 0. Usually choosing $L = 0.5$ is good enough. In Eq. (14), 0 means non-built-up area while 1 means built-up area. The threshold values are the corresponding vegetation index ranges of woodland and grassland [62].

(3) Change the category of boundary super pixels which are judged as non-built-up area.

A more elaborate boundary based on super-pixel technology can be generated just like Fig. 9, but time consumption is an obvious shortcoming, as this step takes much longer than the CNN part. In practice, boundary refinement based on super-pixel basically leads to no increase in accuracy according to statistical results, therefore, the first method to make the boundary more delicate is recommended.

$$NDVI = \frac{Band_{NIR} - Band_{red}}{Band_{NIR} + Band_{red}} \qquad (12)$$

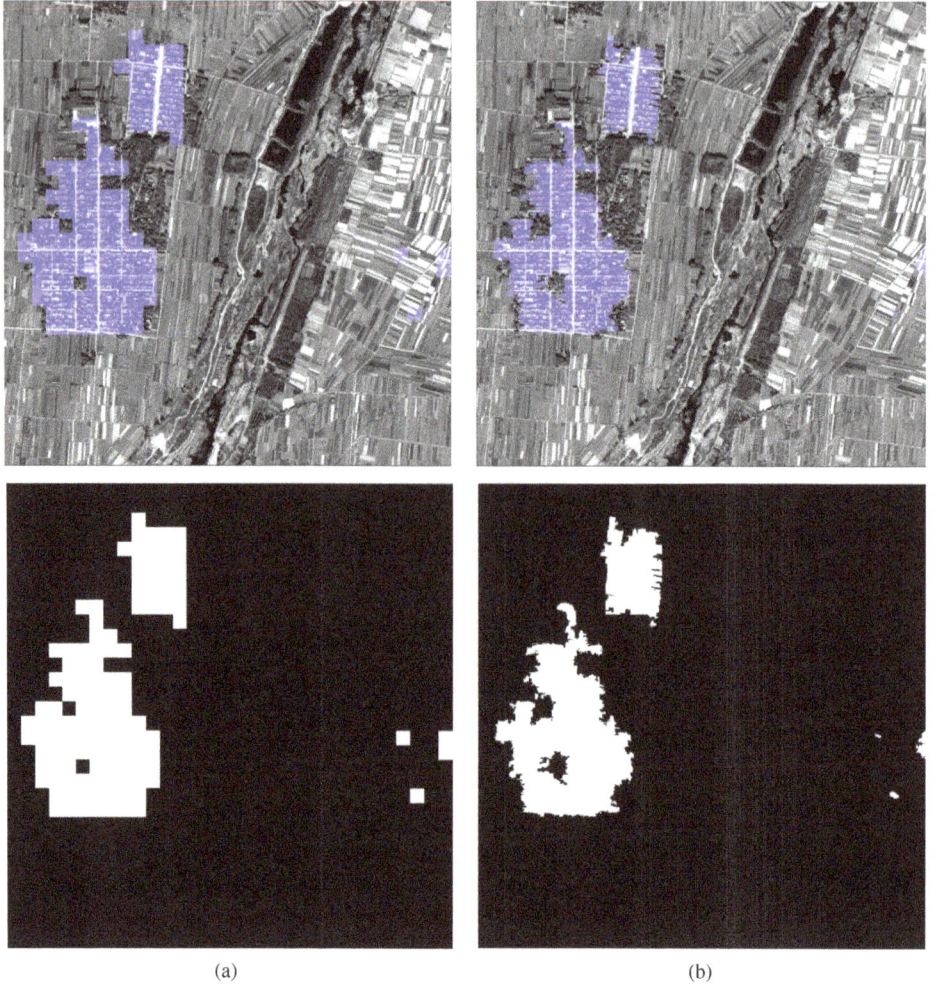

(a)   (b)

Figure 9: (a) shows block-level result of DSCNN. (b) shows result refined based on SLIC.

$$SAVI = \frac{(Band_{NIR} - Band_{red})}{(Band_{NIR} + Band_{red} + L)}(1 + L) \tag{13}$$

$$Res = \begin{cases} 1, & else \\ \\ 0, & \begin{array}{c} AVE\_NDVI > 0.29 \\ and \\ AVE\_SAVI > 0.18 \end{array} \end{cases} \tag{14}$$

8. Experimental Comparison with Classical Methods

The three depth models mentioned above have limited difference in classification accuracy, therefore, making the FCN model an example to compare with some classical methods. The performance of the algorithms is verified by four evaluation indexes, namely, User's Accuracy, Producer's Accuracy, Overall Accuracy and mean intersection-over union (IOU). The formulas of these indexes are as follows:

$$User.Acc = \frac{PT}{PT + PF}, \tag{15}$$

$$Prod.Acc = \frac{PT}{PT + NF}, \tag{16}$$

$$Overall.Acc = \frac{PT + NT}{PT + PF + NT + NF}, \tag{17}$$

$$mean\ IOU = \left(\frac{PT}{PT + PF + NF} + \frac{NT}{PF + NT + NF} \right) / 2. \tag{18}$$

The PT, PF, NT, NF are from the confusing matrix shown in Table 5. The PT represents a positive true representing the number of samples whose prediction and ground truth are both built-up areas. PF, NF and NT have similar meanings as PT according to their prediction and ground truth. User's Accuracy and Producer's Accuracy focus on the built-up area corresponding to the false alarm rate and the missing alarm rate, respectively. The other two concern the overall segmentation precision. Overall, accuracy represents the segmentation precision of the entire image. The mean IOU means the IOU averaged across the built-up area and non-built-up area. These indexes are the most commonly used in image segmentation tasks, and the first three are often used in classification tasks as well.

Table 5. Confusion matrix.

Number of pixels (probability)		Ground truth		
		City	Non-city	Total
Prediction	City	$PT(p_{11})$	$PF(p_{12})$	$PT + PF(p_{1+})$
	Non-city	$NF(p_{21})$	$NT(p_{22})$	$NT + NF(p_{2+})$
	Total	$PT + NF(p_{+1})$	$PF + NT(p_{+2})$	$n(1)$

These algorithms are verified in test images with a size of 2048×2048 which is described above considering the time consumption. Firstly, the FCN model is compared with several unsupervised built-up area extraction algorithms, and they are CGEO [30], Harris corner-based MHEC [63], anisotropic rotation-invariant textural measure (PanTex) [32], and MBI [33]. The MBI is primarily used to extract buildings, so a Gaussian smooth is added and used to extract built-up areas, where the scale parameter $s = 2 + i*5(i = 0-5)$. The PanTex, MHEC and MBI are threshold-based algorithms, and CGEO is graph-based, which utilizes superpixels to reduce complexity. In addition, the results of all the algorithms are removed in small areas below 2,500 pixels according to the actual test remote sensing image resolution. Furthermore, the results of FCN model are also demonstrated with single output, which means only one channel outputting from the FCN model. The results are shown in Fig. 10 and Tables 6 and 7.

Secondly, several supervised algorithms are selected to be compared. Weizman and Goldberger [38] presented an approach which is based on visual words (VW). Tao *et al.* [35] proposed an urban area detection method that combines multiple kernel learning and graph cuts (MKL+GC). Li *et al.* [31] adopted multi-kernel learning, multi-field integrating and multi-hypothesis voting to do this work (MMM). All the three models are trained by the same dataset as the deep model. The results are shown in Fig. 11 and Tables 8 and 9.

According to the experimental results, the proposed deep model has an obvious advantage in comparison with both unsupervised and supervised traditional algorithms. It should be noted that the resulting statistics are pixel level, but the proposed deep model is an essential block level detection. Actually, the deep model is not always optimal in all metrics on every test image, for instance, MMM algorithm obtains a highest User.Acc on image 8, and MKL + GC algorithm wins over almost the entire test set in terms of Prod.Acc. However, the MMM is only the best on image 8, its Overall.Acc on two test images may differ by more than 10%, which reflects its instability. For MKL + GC algorithm, its victory in Prod.Acc comes at the cost of an unacceptable false alarm. On the whole, the proposed deep model achieves high detection accuracy while maintaining stability, which shows that it has good generalization ability. On the other hand, benefiting from the power of GPU, the proposed deep model is tens or even thousands of times faster than the traditional algorithms, which is crucial for large-scale remote sensing image processing.

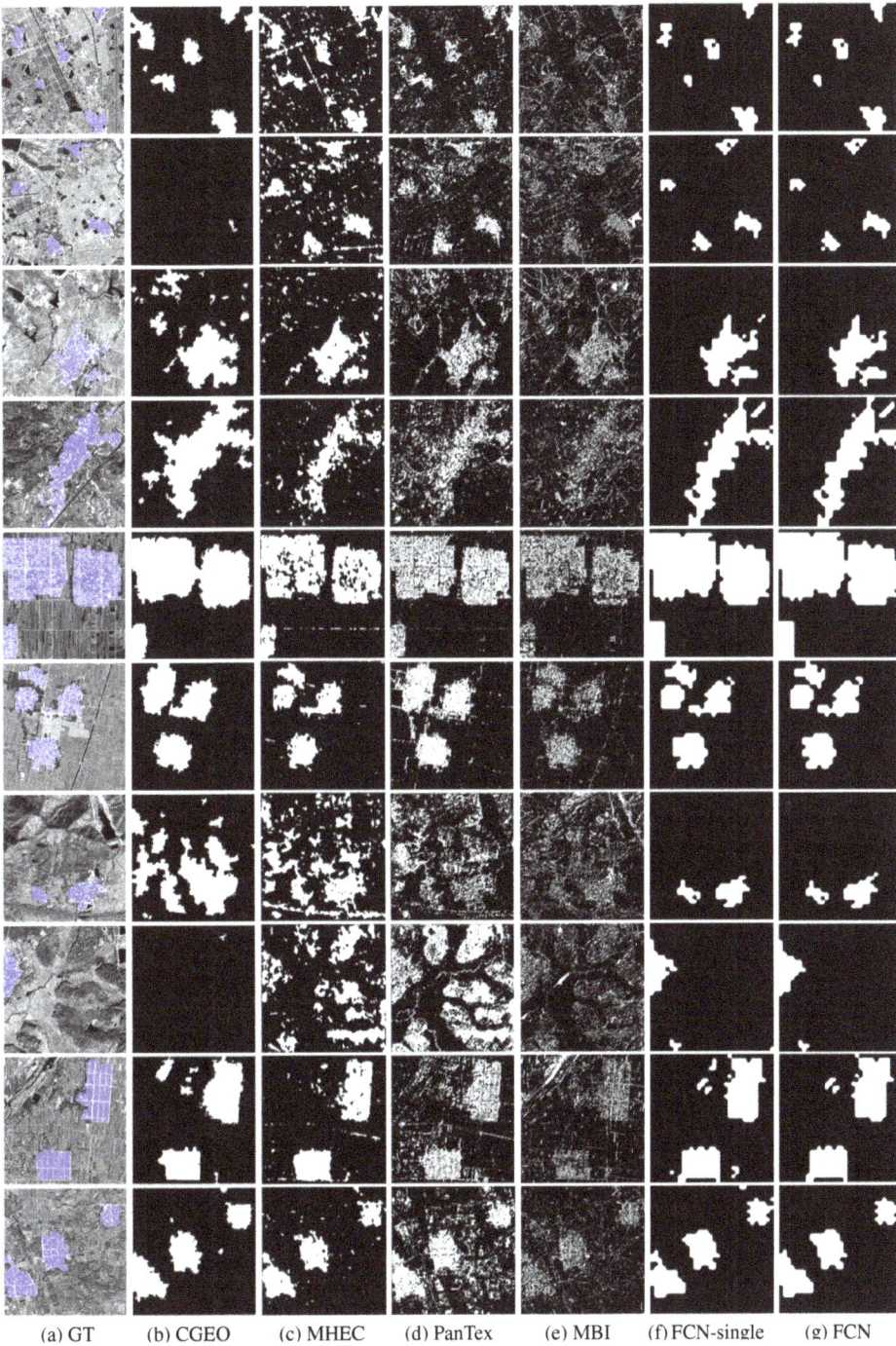

| | | | | | | |
| (a) GT | (b) CGEO | (c) MHEC | (d) PanTex | (e) MBI | (f) FCN-single | (g) FCN |

Figure 10: Results of comparing unsupervised algorithms and ours, each row corresponds to a 2048 × 2048 pixel test image.

Table 6: The evaluation indexes of unsupervised algorithms and ours.

No	CGEO	MHEC	PanTex	MBI	FCN-Single	FCN
(a) User.Acc						
1	0.4727	0.3108	0.3321	0.1768	**0.7916**	0.7640
2	**0.9631**	0.3266	0.2937	0.1641	0.8162	0.8094
3	0.4786	0.6635	0.4480	0.3388	0.7765	**0.7830**
4	0.5124	0.7645	0.4542	0.4378	**0.7879**	0.7851
5	0.9537	**0.9774**	0.9675	0.9396	0.9254	0.9186
6	0.6738	**0.9060**	0.6342	0.6922	0.8369	0.8299
7	0.1938	0.1849	0.1184	0.1303	**0.837**	0.8155
8	0.0000	0.0875	0.0658	0.0992	0.6981	**0.7052**
9	0.7291	**0.8468**	0.5508	0.4224	0.75	0.7767
10	0.7442	**0.8269**	0.4165	0.4657	0.8188	0.8086
ALL	0.6178	0.5779	0.4299	0.4022	**0.8303**	0.8285
(b) Prod.Acc						
1	**0.9851**	0.9380	0.7448	0.3657	0.9325	0.9387
2	0.0458	**0.9649**	0.7906	0.4487	0.8601	0.8669
3	**0.9789**	0.9067	0.7005	0.5514	0.9674	0.9694
4	0.9361	0.7283	0.6263	0.4035	0.9706	**0.9771**
5	0.9680	0.8202	0.7352	0.5096	0.9895	**0.9904**
6	0.9685	0.8747	0.8972	0.5129	**0.9756**	0.9749
7	**0.9890**	0.8977	0.4883	0.4272	0.8388	0.8429
8	0.0000	0.5480	0.6693	0.4079	0.9251	**0.9431**
9	0.9840	0.9438	0.7516	0.4141	0.9868	**0.9879**
10	0.9716	0.9165	0.8872	0.4653	0.9777	**0.9797**
ALL	0.9039	0.8529	0.7445	0.4693	0.9664	**0.9691**
(c) Overall.Acc						
1	0.9341	0.8733	0.8963	0.8617	**0.9815**	0.9792
2	0.9478	0.8895	0.8848	0.8452	**0.9818**	0.9816
3	0.8869	0.9425	0.8791	0.8415	0.9677	**0.9689**
4	0.8462	0.9201	0.8186	0.8204	**0.9532**	**0.9532**
5	**0.9665**	0.9158	0.8773	0.7784	0.9617	0.9587
6	0.9242	0.9673	0.9060	0.8916	**0.9675**	0.9659
7	0.7634	0.7671	0.7621	0.8035	**0.9814**	0.9800
8	0.9606	0.7640	0.6233	0.8355	0.9818	**0.9827**

Table 6: (*Continued*)

No	CGEO	MHEC	PanTex	MBI	FCN-Single	FCN
9	0.9376	**0.9629**	0.8591	0.8115	0.944	0.9516
10	0.9536	0.9647	0.8264	0.8632	0.9694	**0.9677**
ALL	0.9121	0.8967	0.8333	0.8352	0.9658	**0.9690**
(d) Mean IOU						
1	0.6997	0.5852	0.5948	0.4970	**0.8646**	0.8528
2	0.4967	0.6030	0.5761	0.4890	**0.8507**	0.8503
3	0.6739	0.7788	0.6228	0.5487	0.8604	**0.8648**
4	0.6570	0.7522	0.5778	0.5368	0.857	**0.8577**
5	**0.9339**	0.8380	0.7696	0.6054	0.9253	0.9197
6	0.7852	0.8821	0.7412	0.6501	**0.8908**	0.8863
7	0.4713	0.4677	0.4303	0.4548	**0.8508**	0.8434
8	0.4803	0.4203	0.3385	0.4598	0.821	**0.8293**
9	0.8231	**0.8811**	0.6526	0.5311	0.8379	0.8558
10	0.8377	0.8645	0.5999	0.5789	**0.8844**	0.8792
ALL	0.7398	0.7044	0.5947	0.5503	0.8857	**0.8858**

Table 7. The time consumption of unsupervised algorithms and ours.

	CGEO	MHEC	PanTex	MBI	FCN-Single	FCN
Ave time	199.7	4435.2	95387	942.3	3.0	3.0

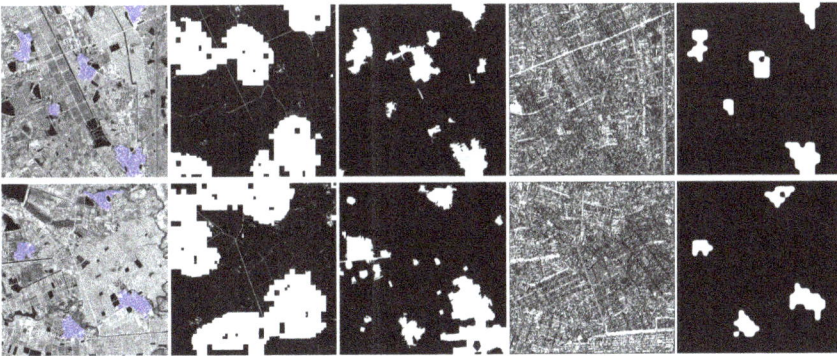

Figure 11: Results of comparing supervised algorithms and FCN, each row corresponding to a 2048 × 2048 pixel test image.

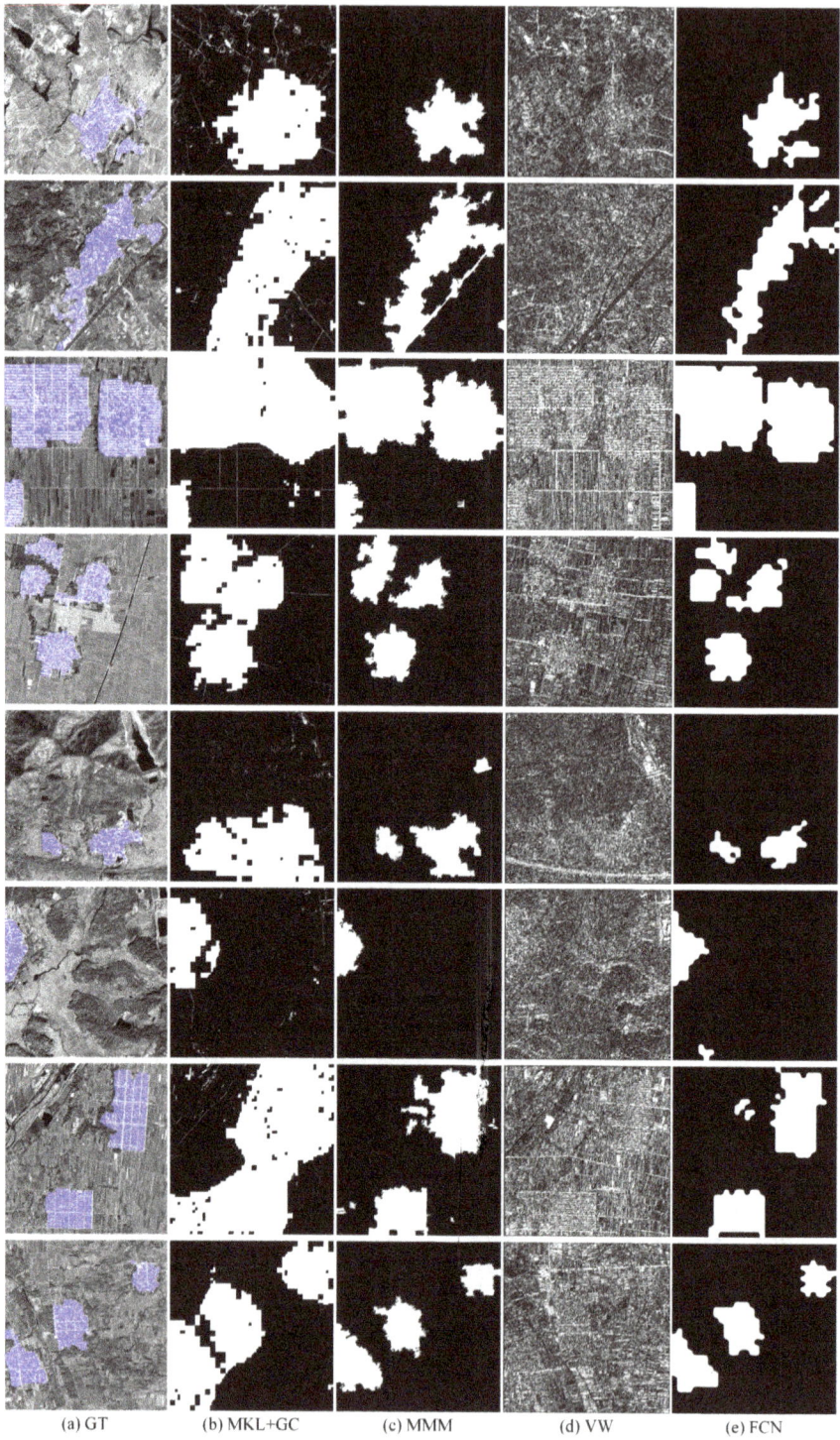

(a) GT   (b) MKL+GC   (c) MMM   (d) VW   (e) FCN

Figure 11:   (*Continued*)

Table 8: The evaluation indexes comparing supervised algorithms and ours.

No	MKL + GC	MMM	VW	FCN
(a) User.Acc				
1	0.1723	0.4850	0.0610	0.7640
2	0.1332	0.2937	0.0597	0.8094
3	0.3570	0.7390	0.1588	**0.7830**
4	0.3740	0.6177	0.1999	0.7851
5	0.7494	0.9155	0.5055	0.9186
6	0.4307	0.7437	0.2641	0.8299
7	0.2166	0.4779	0.0981	0.8155
8	0.3390	**0.8598**	0.0597	0.7052
9	0.3005	0.6682	0.2176	**0.7767**
10	0.3717	0.7423	0.1554	0.8086
ALL	0.3672	0.6866	0.1908	0.8285
(b) Prod.Acc				
1	0.9199	0.9509	0.4344	0.9387
2	**0.9925**	0.9107	0.4960	0.8669
3	**0.9977**	0.9429	0.5059	0.9694
4	**0.9950**	0.9551	0.4474	0.9771
5	0.9803	0.9641	0.6692	0.9904
6	**0.9983**	0.9550	0.5928	0.9749
7	**0.9961**	0.9861	0.5795	0.8429
8	0.9802	0.8895	0.5550	0.9431
9	**0.9988**	0.9875	0.6461	0.9879
10	**0.9976**	0.9480	0.5104	0.9797
ALL	**0.9879**	0.9577	0.5788	0.9691
(c) Overall.Acc				
1	0.7337	0.9373	0.5708	0.9792
2	0.6472	0.8756	0.5464	**0.9816**
3	0.8129	0.9594	0.6700	**0.9689**
4	0.7309	0.8976	0.6224	0.9532
5	0.8528	0.9471	0.5824	0.9587
6	0.7998	0.9433	0.6880	0.9659
7	0.7931	0.9374	0.6701	0.9800
8	0.9261	**0.9902**	0.6484	0.9827

(*Continued*)

Table 8: (*Continued*)

No	MKL + GC	MMM	VW	FCN
9	0.6195	0.9177	0.5620	**0.9516**
10	0.7838	0.9512	0.5821	0.9677
ALL	0.7700	0.9357	0.6143	0.9690

(d) Mean IOU

No	MKL + GC	MMM	VW	FCN
1	0.4440	0.7034	0.2893	0.8528
2	0.3801	0.5773	0.2805	**0.8503**
3	0.5740	0.8312	0.3516	0.8648
4	0.5264	0.7396	0.3483	0.8577
5	0.7432	0.8979	0.3889	0.9197
6	0.5973	0.8261	0.3831	0.8863
7	0.4985	0.7042	0.3320	0.8434
8	0.6299	**0.8833**	0.3142	0.8293
9	0.4228	0.7822	0.3297	**0.8558**
10	0.5617	0.8288	0.3201	0.8792
ALL	0.5502	0.7963	0.3418	0.8858

Table 9: The time consumption comparison for supervised algorithms and ours (s).

	MKL + GC	MMM	VW	FCN
Average time	7877.3	628.8	186.2	3.0

9. Conclusion

Nowadays, deep learning theory has made great progress and gained a dominant position in many subfields of machine learning, and the rapid development of GPU makes it possible to utilize deep models to process large amounts of remote sensing data quickly, so deep learning technology will become the mainstream for the extraction of built-up areas. In the comparison experiment between the FCN model and the classical algorithm, no matter the indexes about precision or time consumption, the FCN model occupies absolute superiority.

Despite the excellent performance of the deep model, the extraction of built-up area still faces many challenges. For instance, the training of deep model needs large number of labeled samples which cover as many circumstances as possible, so deep learning with few labeled samples is worthy of attention. In addition, there are many sources of remote sensing data, it is valuable to study the generalization ability of deep models across different data sources. On the other hand, generally block-level results can be obtained when applying deep methods to built-up area extraction, so refining the built-up area boundary quickly requires more attention. In conclusion, more research is needed for applying deep learning to built-up area extraction.

References

1. L. Gueguen *et al.*, "Urbanization detection by a region based mixed information change analysis between built-up indicators," *IEEE J. Sel. Topics Appl. Earth Observ. Remote Sens.*, vol. 6, no. 6, pp. 2410–2420, 2013.
2. H. Khurshid and M. F. Khan, "Segmentation and classification using logistic regression in remote sensing imagery," *IEEE J. Sel. Topics Appl. Earth Observ. Remote Sens.*, vol. 8, no. 1, pp. 224–232, 2015.
3. M. Pesaresi *et al.*, "Toward global automatic built-up area recognition using optical VHR imagery," *IEEE J Sel Top Appl Earth Obs Remote Sens.*, vol. 4, pp. 923–934, 2011.
4. M. Pesaresi *et al.*, "A global human settlement layer from optical HR/VHR RS data: Concept and first results," *IEEE J. Sel. Topics Appl. Earth Observ. Remote Sens.*, vol. 6, no. 5, pp. 2102–2131, 2013.
5. N. Li *et al.*, "A novel technique based on the combination of labeled co-occurrence matrix and variogram for the detection of built-up areas in high-resolution SAR images," *Remote Sens.*, vol. 6, pp. 3857–3878, 2014.
6. D. Xiang *et al.*, "Built-up area extraction from PolSAR imagery with model-based decomposition and polarimetric coherence," *Remote Sens.*, vol. 8, pp. 685–705, 2016.
7. S. Bouzekri, A. A. Lasbet, and A. Lachehab, "A new spectral index for extraction of built-up area using Landsat-8 data," *Ind J Remote Sens.*, vol. 43, pp. 867–873, 2015.
8. D. Kaimaris and P. Patias, "Identification and area measurement of the built-up area with the built-up index (BUI)," *Int J Adv Remote Sens.*, vol. 5, pp. 1844–1858, 2016.
9. A. Varshney and E. Rajesh, "A comparative study of built-up index approaches for automated extraction of built-up regions from remote sensing data," *Ind J Remote Sens.*, vol. 42, pp. 659–663, 2014.
10. X. Liu, A. Yue, and J. Chen, "New normalized difference index for built-up land enhancement using airborne visible infrared imaging spectrometer imagery," *J Appl Remote Sens.*, vol. 8, p. 085092, 2014.

11. J. A. Benediktsson, M. Pesaresi, and K. Amason, "Classification and feature extraction for remote sensing images from urban areas based on morphological transformations," *IEEE Trans Geosci Remote Sens.*, vol. 41, pp. 1940–1949, 2003.

12. B. Sirmacek and C. Unsalan, "Urban-area and building detection using SIFT keypoints and graph theory," *IEEE Trans Geosci Remote Sens.*, vol. 47, pp. 1156–1167, 2009.

13. G. Liu *et al.*, "A perception-inspired building index for automatic built-up area detection in high-resolution satellite images," in *Proc. IEEE Int. Geoscience and Remote Sensing Symposium, Melbourne, Australia.*, 2013, pp. 3132–3135.

14. Y. Chen *et al.*, "Built-up area extraction using data field from high-resolution satellite images," in *Proc. IEEE Int. Geoscience and Remote Sensing Symposium*, Beijing, China, 2016, pp. 437–440.

15. F. Peng *et al.*, "A new stereo pair disparity index (SPDI) for detecting built-up areas from high-resolution stereo imagery," *Remote Sens.*, vol. 9, p. 633, 2017.

16. Q. Zhou, "Comparative study of approaches to delineating built-up areas using road network data," *Trans. GIS*, vol. 19, pp. 848–876, 2016.

17. Q. Zhou and L. Guo, "Empirical approach to threshold determination for the delineation of built-up areas with road network data," *Plos One*, vol. 13, p. e0194806, 2018.

18. Y. Dou *et al.*, "Urban land extraction using VIIRS nighttime light data: An evaluation of three popular methods," *Remote Sens.*, vol. 9, p. 175, 2017.

19. Y. Yang *et al.*, "Spatial recognition of the urban-rural fringe of Beijing using DMSP/OLS nighttime light data," *Remote Sens.*, vol. 9, p. 1141, 2017.

20. P. Zhang *et al.*, "A strategy of rapid extraction of built-up area using multi-seasonal Landsat-8 thermal infrared band 10 images," *Remote Sens.*, vol. 9, p. 1126, 2017.

21. M. Tarawally *et al.*, "Comparative analysis of responses of land surface temperature to long-term land use/cover changes between a coastal and inland city: A case of Freetown and Bo town in Sierra Leone," *Remote Sens.*, vol. 10, p. 112, 2018.

22. L. Wang *et al.*, "Urban built-up area boundary extraction and spatial-temporal characteristics based on land surface temperature retrieval," *Remote Sens.*, vol. 10, p. 473, 2018.

23. J. Li and R. M. Narayanan, "Integrated spectral and spatial information mining in remote sensing imagery," *IEEE Geosci. Remote Sens.*, vol. 42, pp. 673–685, 2004.

24. C. Li *et al.*, "Texture-based urban detection using contourlet coefficient on remote sensing imagery," in *Proc. IET International Radar Conference*, Hangzhou, China, pp. 5–6, 2015.

25. P. Zhong and R. Wang, "A multiple conditional random fields ensemble model for urban area detection in remote sensing optical images," *IEEE Trans. Geosci. Remote Sens.*, vol. 45, pp. 3978–3988, 2007.

26. P.C. Smits and A. Annoni, "Updating land-cover maps by using texture information from very high-resolution space-borne imagery," *IEEE Trans. Geosci. Remote Sens.*, vol. 37, pp. 1244–1254, 1999.

27. S. Yu, M. Berthod, and G. Giraudon, "Toward robust analysis of satellite images using map information-application to urban area detection," *IEEE Trans. Geosci. Remote Sens.*, vol. 37, pp. 1925–1939, 1999.

28. D. D. Vecchi, M. Harb, and F. Dell'Acqua, "A PCA-based hybrid approach for built-up area extraction from Landsat 5, 7 and 8 datasets," in *Proc. Int. Geoscience and Remote Sensing Symposium*, Milan, Italy, 2015, pp. 1152–1154.

29. C. Tao *et al.*, "Unsupervised detection of built-up areas from multiple high-resolution remote sensing images," *IEEE Geosci. Remote Sens. Lett.*, vol. 10, pp. 1300–1304, 2013.

30. Y. Li *et al.*, "Cauchy graph embedding optimization for built-up areas detection from high-resolution remote sensing images," *IEEE J. Sel. Top Appl. Earth Obs. Remote Sens.*, vol. 8, pp. 2078–2096, 2015.

31. Y. Li *et al.*, "Built-up area detection from satellite images using multikernel learning, multifield integrating, and multihypothesis voting," *IEEE Geosci. Remote Sens. Lett.*, vol. 12, pp. 1190–1194, 2017.

32. M. Pesaresi, A. Gerhardinger, and F. Kayitakire, "A robust built-up area presence index by anisotropic rotation-invariant textural measure," *IEEE J. Sel. Top Appl. Earth Obs. Remote Sens.*, vol. 1, pp. 180–192, 2009.

33. X. Huang and L. Zhang, "Morphological building/shadow index for building extraction from high-resolution imagery over urban areas," *IEEE J. Sel. Top Appl. Earth Obs. Remote Sens.*, vol. 5, pp. 161–172, 2012.

34. Y. Li, X. Huang, and H. Liu, "Unsupervised deep feature learning for urban village detection from high-resolution remote sensing images," *Photogramm. Eng. Remote Sens.*, vol. 83, pp. 567–579, 2017.

35. C. Tao *et al.*, "Urban area detection using multiple Kernel Learning and graph cut," in *Proc. IEEE Int. Geoscience and Remote Sensing Symposium*, Munich, Germany, 2012, pp. 83–86.

36. A. K. Shackelford and C. H. Davis, "A hierarchical fuzzy classification approach for high-resolution multi-spectral data over urban areas," *IEEE Trans. Geosci. Remote Sens.*, vol. 41, pp. 1920–1932, 2003.

37. C. Unsalan and K. L. Boyer, "Classifying land development in high-resolution panchromatic satellite images using straight-line statistics," *IEEE Trans. Geosci. Remote Sens.*, vol. 42, pp. 907–919, 2004.

38. L. Weizman and J. J. I. G. Goldberger, "Urban-area segmentation using visual words," *IEEE Geosci Remote Sens Lett.*, vol. 6, pp. 388–392, 2009.

39. Q. Zhang *et al.*, "CNN based suburban building detection using monocular high resolution Google Earth images," in *IEEE Int. Geoscience and Remote Sensing Symposium*, pp. 661–664, 2016.

40. Y. Zhou *et al.*, "Polarimetric SAR image classification using deep convolutional neural networks," *IEEE Geosci. Remote Sens. Lett.*, vol. 13, pp. 1935–1939, 2016.
41. W. Li *et al.*, "Deep learning-based classification methods for remote sensing images in urban built-up areas," *IEEE Access.*, p. 1, 2019.
42. G. Cheng *et al.*, "When deep learning meets metric learning: Remote sensing image scene classification via learning discriminative CNNs," *IEEE Trans. Geosci. Remote Sens.*, pp. 1–11, 2018.
43. Y. Chen *et al.*, "Deep fusion of remote sensing data for accurate classification," *IEEE Geosci. Remote Sens. Lett.*, vol. 14, pp. 1253–1257, 2017.
44. X. Xu *et al.*, "Multisource remote sensing data classification based on convolutional neural network," *IEEE Trans. Geosci. Remote Sens.*, vol. 56, pp. 937–949, 2018.
45. J. Li, R. Zhang, and Y. Li, "Multiscale convolutional neural network for the detection of built-up areas in high-resolution SAR images," in *Proc. Int. Geoscience and Remote Sensing Symposium*, Beijing, China, pp. 910–913, 2016.
46. V. Bramhe, S. K. Ghosh, and P. K. Garg, "Extraction of built-up areas using convolutional neural networks and transfer learning from Sentinel-2 satellite images," in *Int. Arch. Photogram., Remote Sens. Spatial Inform. Sci.*, vol. XLII-3, pp. 79–85, 2018.
47. C. Li, *et al.*, "Nested network with a two-stream pyramid for salient object detection in optical remote sensing images," *IEEE Trans. Geosci. Remote Sens.*, vol. 57, pp. 9156–9166, 2019.
48. A. Krizhevsky, I. Sutskever, and G. Hinton, "ImageNet classification with deep convolutional neural network," *Neural Inform. Process. Syst.*, vol. 25, 2012.
49. C. Szegedy *et al.*, "Rethinking the inception architecture for computer vision," *IEEE CVPR*, 2015.
50. C. Szegedy *et al.*, "Inception-v4, Inception-ResNet and the impact of residual connections on learning," in *AAAI Conference on Artificial Intelligence*, 2016.
51. S. Ioffe and C. Szegedy, "Batch normalization: Accelerating deep network training by reducing internal covariate shift," *CoRR*, vol. 1511, pp. 448–456, 2015.
52. D. Kingma and J. Ba, "Adam: A method for stochastic optimization," *CoRR*, vol. 1412, 2014.
53. A. G. Howard *et al.*, "MobileNets: Efficient convolutional neural networks for mobile vision applications," *CoRR*, vol. 1704, 2017.
54. L. Smith and N. Topin, "Deep convolutional neural network design patterns," *CoRR*, 2016.
55. C. Szegedy *et al.*, "Going deeper with convolutions," in *IEEE Conf. Computer Vision and Pattern Recognition*, pp. 1–9, 2015.
56. K. He *et al.*, "Deep residual learning for image recognition," in *IEEE Conference on Computer Vision and Pattern Recognition*, pp. 770–778, 2016.

57. J. Long, E. Shelhamer, and T. Darrell, "Fully convolutional networks for semantic segmentation," in *Proc. IEEE Conf. Computer Vision and Pattern Recognition(CVPR)*, Boston, MA, USA, pp. 3431–3440, 2015.

58. K. He *et al.*, "Mask R-CNN," in *Proc. Int. Conf. Computer Vision (ICCV)*, Venice, Italy, pp. 2980–2988, 2017.

59. R. Achanta *et al.*, "SLIC superpixels compared to state-of-the-art superpixel methods," *IEEE Trans. Pattern Anal. Mach. Intell.*, vol. 34, no. 11, pp. 2274–2282, 2012.

60. J. W. Rouse *et al.*, "Monitoring vegetation systems in the great plains with ERTS," in *3rd Earth Resource Technology Satellite*, vol. 1, pp. 48–62, 1974.

61. A. R. Huete, "A soil-adjusted vegetation index (SAVI)," *Remote Sens. Environ.*, 25, pp. 295–309, 1988.

62. P. Qin and J. Chen, "A comparison between NDVI and SAVI for vegetation spatial information retrieval based on ASTER images: A case study of Huadu District, Guangzhou," *Trop. Geograp.*, vol. 28, no. 5, 2008.

63. Kovács and T. Szirányi, "Improved harris feature point set for orientation-sensitive urban-area detection in aerial images," *IEEE Geosci Remote Sens Lett.*, vol. 10, pp. 796–800, 2013.

64. Q. Liu *et al.*, "Built-up area detection based on a Bayesian saliency model," *Int. J. Wavelets Multiresolut. Inf. Process.*, vol. 15, pp. 1–16, 2017.

65. L. Yang, J. Yuan, and D. Lunga, "Toward country scale building detection with convolutional neural network using aerial images," in *IEEE Int. Geoscience and Remote Sensing Symp.*, 2017, pp. 870–873.

66. D. A. Clevert, T. Unterthiner, and S. Hochreiter, "Fast and accurate deep network learning by exponential linear units (ELUs)," *Comput. Sci.*, vol. 1511, 2015.

Chapter 9

Recent Advances of Manifold-based Graph Convolutional Networks for Remote Sensing Images Recognition

Sichao Fu and Weifeng Liu*

College of Control Science and Engineering
China University of Petroleum (East China), Qingdao, China
*liuwf@upc.edu.cn

Abstract

Due to the variety of information contained and the complexity of spatial structure for remote sensing images, graph representation learning (GRL) has aroused a wide concern because of the ability to learn more representative data features. As one of the most representative works, graph neural networks (GNNs) can take full advantage of the high-order correlation information of data with arbitrary structures to improve its learning performance when only a few labeled data are available. In recent years, an increasing number of GNN models including the spatial-domain-based GNN and spectral-domain-based GNN have been proposed and achieved excellent performance in many areas. The more typical model graph convolutional networks (GCNs) first generalizes the convolution operation of convolutional neural networks (CNNs) on Euclidean space to graphs. GCN fuses the original feature information and the complex and high-order data correlation relationships between each node by spectral graph Laplacian convolution with the one-order approximation to learn richer data features.

However, how to better preserve and acquire the locality and similarity relationships of graph structured data has always been a challenging task. In this chapter, we summarize many applications of the existing GNN. Under the framework of the combination of GCN and manifold assumptions, many GCN variants have been proposed to exactly acquire the spatial structure information of graph structured data when exploring and exploiting the local geometry structure of data manifold. Specially, we present its theory analysis in detail as well as some applications of the GCN variants, such as two-order GCN, Hypergraph p-Laplacian GCN and manifold regularized dynamic GCN.

Keywords: graph representation learning, graph neural networks, spectral domain, spatial domain, graph convolutional networks

1. Introduction

With the rapid development of satellite remote sensing techniques and sensor technology, a large number of remote sensing images with different resolutions can be easily obtained. Remote sensing images have become one of the main methods for relevant departments and agencies to monitor the ground information. At the same time, the acquired data have been greatly increased in terms of scale and complexity. How to extract the most valuable information from these massive high-dimensional data is one of the most common challenges faced by many researchers in the fields of machine learning and computer vision [1, 2].

In recent years, many popular works, such as convolutional neural networks (CNNs) [3], principal component analysis networks (PCANet) [4], canonical correlation analysis networks (CCANet) [5] and auto-encoder [6], have been proposed and widely applied in many tasks, ranging from natural language processing [7] and image [8] to audio [9] and video [10]. These existing methods can handle the data defined on the Euclidean space or regular grids efficiently. Applying these models to graphs with arbitrary space structures, such as remote sensing images, is still a challenging task. For graph structured data defined on the non-Euclidean domain, each node of graphs has a different number of neighbor nodes. In addition, with the change of space locations, its high-order data correlation information also varies. Thus, how to effectively and efficiently analyze and interpret the content of remote sensing images is vital.

To solve the above-mentioned problem, many proposed traditional (shallow) graph representation learning (GRL) [11] methods, i.e., graph embedding [12], have enjoyed tremendous success on graph structured data generated from the non-Euclidean space. The main objective of graph embedding is to find an appropriate mapping function that converts each node in the high-dimensional data into a low-dimensional potential representation. Deepwalk [13] first utilized the truncated random walks (depth-first-search) to acquire the local structure information of each node in the graphs, and then learned the latent representations of some vertices in the sentences by the skip-gram [14] model. Different from the Deepwalk, large-scale information network embedding (LINE) [15] simultaneously used the first-order and second-order proximity between the vertices to describe the local structure relationships of data distribution, i.e., breadth-first strategy. To learn more accurate structure relationships between nodes, Node2vec [16] utilized tunable parameters to control the scope of the search space of each node in the graphs by designing a random walk strategy with bias. Isometric mapping (ISOMAP) [17] aimed to obtain the globally optimal local geometry structure by using the multidimensional scaling [18] method to calculate the geodesic distance between different nodes. Locally linear embedding (LLE) [19] aimed to maintain the local geometric structure of data which utilized the local linear coordinates between any node and its neighbor nodes in the graphs to establish the final optimization objective function. Laplacian eigenmaps (LE) [20] described the similarity information between nodes by constructing a weighted graph. In other words, it utilized the graph Laplacian to preserve the manifold structure of data distribution. Hessian eigenmaps (HLLE) [21] obtained a matrix by the estimation of the graph Hessian matrix over its neighbor domain, and then acquired the local linear coordinates in the low-dimensional space by applying an Eigendecomposition to the constructed matrix. Compared with graph Laplacian, graph Hessian can describe the local geometric structure of data accurately because graph Hessian has a richer zero space by predicting higher-order derivatives, which allows its prediction results to produce corresponding smooth changes with potential manifold changes [22, 23]. Liu *et al.* [24] solved the issue of a high computing cost by introducing an approximate approximation of graph p-Laplacian, and then proposed the p-Laplacian [25, 26] regularization framework-based kernel least squares (pLapKLS) and support vector machines (pLapSVM) models for scene recognition.

With the extensive applications of deep learning in many areas, there has been an increase in the interest of researchers in generalized deep learning frameworks on the Euclidean domain including CNN, auto-encoder and generative adversarial networks (GANs) [27] to graph structured data. Recently, many popular deep GRL works, especially graph neural networks (GNNs) have obtained great success in many applications, such as computer vision [28], natural language processing [29], Internet [30], recommender systems [31], healthcare [32], chemistry [33], physics [34], etc. Now, the existing GNN can be roughly divided into two parts: the spatial-domain-based GNN and spectral-domain-based GNN.

Driven by the standard convolution operation of CNN, spatial-domain-based GNN also can make a convolution directly on the non-Euclidean data to learn a novel data representation for each node. Different from CNN, its convolution operation can be interpreted as the aggregation process of each central node with its neighbor nodes. Simonovsky *et al.* [35] utilized the edge labels to adjust the value of the filter weight, leading a dynamic spatial convolution on graphs. Geodesic CNN [36] used the local geodesic coordinates (that replaced the image patches) to acquire the local anisotropic structure of data, and then made a convolution through the slide of a window over the manifold. Compared to Geodesic CNN, anisotropic CNN [37] used the anisotropic heat kernels to construct a simpler patch operator. Diffusion CNN [38] learned a latent representation for each node in the graph structured data by introducing the diffusion scansion process of the diffusion convolution.

Spectral-domain-based GNN defines the training method of a spectral graph filter on a graph with a specific structure by introducing the convolution theorem, i.e., the Fourier transform of function convolution is the product of the function Fourier transform, to remove the noise of the input data. Different from CNN, the most typical model of the spectral-domain-based GNN, i.e., graph convolutional networks (GCNs) [39], fused the original feature information and the complex and high-order data correlation relationships between each node by the spectral graph convolution with the one-order approximation to learn richer data features. Two-order GCN [40] not only considered the direct relationships of each node, but also utilized its indirect relationships by using the two-order approximation of Chebyshev polynomials. HyperGCN [41, 42] used Hypergraph to encode the more complicated and beyond pairwise connection relationships of graph structured data. Hypergraph p-Laplacian GCN [43] extended the convolution operation of GCN to the Hypergraph p-Laplacian domain by using Hypergraph p-Laplacian to more accurately express the high-order relationships of graph structured data. Manifold regularized dynamic GCN [44] paid more attention to the optimization of graph structure information learned by introducing a regularization term about the local geometry distribution of graph structured data to drive the objective function to change over the potential sample distribution manifold.

In this chapter, we briefly introduce some related works about manifold learning (ML), and present a detailed review of the development history and its theoretical analysis of GCN. What is more important, we emphatically discuss and analyze how to exactly acquire the spatial structure information of graph structured data when exploring and exploiting the local geometry structure of data manifold for the two-order GCN, Hypergraph p-Laplacian GCN and manifold regularized dynamic GCN models. Compared to GCN, the above-mentioned methods have the following advantages:

(1) Two-order GCN acquires the structure relationships not only from the direct neighbor but also from the indirect neighbor of each node by using the two-order Chebyshev polynomial about the Laplacian.
(2) Hypergraph p-Laplacian GCN generalizes the spectral graph convolution operation of GCN in simple graphs to the Hypergraph p-Laplacian domain by using the effective combination of Hypergraph theory and p-Laplacian matrix to better preserve the high-order relationships of data.
(3) Manifold regularized dynamic GCN automatically updates the graph structure information learned by introducing a regularization term about the local geometry distribution of graph structured data to its objection function.

Extensive experimental results on remote sensing datasets demonstrate the effectiveness of two-order GCN, Hypergraph p-Laplacian GCN and manifold regularized dynamic GCN in comparison to many state-of-the-art semi-supervised learning methods. In addition, we also give some analysis of some related future woks.

2. Related Works

In this section, we briefly give a description of some related works including the local geometry structure preserving method and spectral graph convolution on graph structured data.

2.1 *Hypergraph p-Laplacian theory*

In the theory research of a simple graph, researchers often suppose that there only exists a pairwise correspondence relationship between different data objects. In the simple graph, a data object represents a vertex on the graph, and the edge between any two vertices is used to describe the correspondence relationships of different data objects. In real life, different objects are not just pairwise relationships simply, most are complicatedly diversified relationships [45], such as one-to-many and many-to-many. For the simple graph, it is difficult to accurately express the complex relationships that are common in practical applications. For Hypergraph, each edge can connect two or more vertices to describe its more complex correlation information. On the one hand, it can capture the complicated information between different data objects, on the other hand, it indirectly guarantees the integrity of data information. Figure 1 gives the difference between a Hypergraph and a simple graph.

In this chapter, we use V^h to describe a finite set of all vertices. E^h represents a family of all hyper-edges on a Hypergraph. A subset of V^h can construct a hyper-edge ($e \in E^h$) of a Hypergraph, i.e., $U_{e \in E^h} = V$. For each hyper-edge, it has a nonnegative weight w(e). W^h denotes the diagonal matrix of all hyper-edge weights. Thus, we can use $G(V^h, E^h, W^h)$ to represent any Hypergraph structure with weights. To better understand its structure for readers, we can utilize a $|V^h| \times |E^h|$ correlation matrix H to show it, i.e.,

$$h(v,e) = \begin{cases} 1, & \textit{if } v \in e, \\ 0, & \text{otherwise.} \end{cases} \tag{1}$$

The degree matrix of each vertex, i.e., the sum of all the hyper-edge weights of its connections, can be expressed in the following form.

$$d(v) = \sum_{e \in E^h} w(e) h(v,e). \tag{2}$$

For each Hypergraph, its degree matrix can be defined as the sum of all vertices or the edge weights on each hyper-edge.

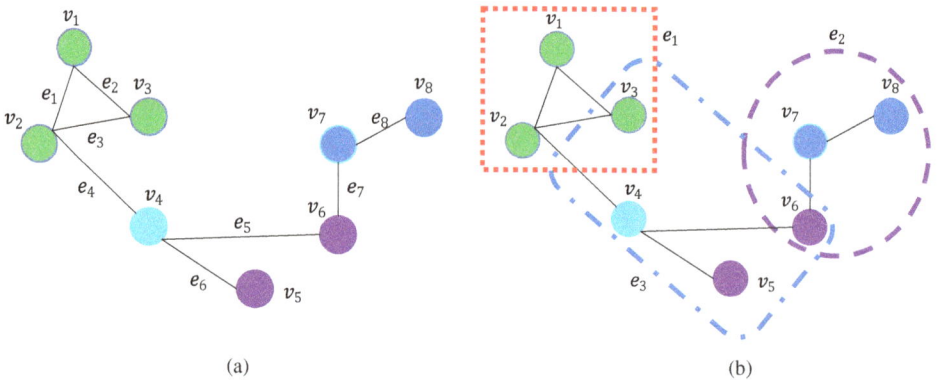

Figure 1:   The difference between a Hypergraph and a simple graph. (a) A simple graph: each edge of a simple graph has only two vertices. (b) A Hypergraph. In a Hypergraph, each edge can connect N vertices ($N \geq 2$).

$$\delta(e) = \sum_{v \in V^b} h(v,e),$$ (3)

where D_v expresses the diagonal matrix including the degree matrix of each vertex. In the past few years, many Hypergraph learning-based models have achieved great success. The first methods first convert a Hypergraph to a simple graph, and then use the spectral clustering methods that are applied to the simple graph to address the Hypergraph issue, such as clique averaging [46], Rodriguez's Laplacian [47], star expansion [48] and clique expansion [49]. The second type of methods, including Bolla Laplacian [50] and Zhou's normalized Laplacian [51], are to construct a Hypergraph Laplacian by the definition of a simple graph Laplacian. The third type of methods [52, 53] utilize a tensor to describe the Hypergraph structure. Then the Hypergraph is segmented by using the joint clustering methods.

The computational process of the adjacency matrix on the Hypergraph can be divided into four parts. Firstly, we use the K-nearest neighbor with the Euclidean distance method to acquire the adjacency matrix W^s of the simple graph. Then, we regard each central node and its K neighbor nodes as a hyper-edge. Third, we can acquire all hyper-edge weights matrix W^b of the Hypergraph. Finally, the adjacency matrix of Hypergraph can be expressed by the following definition, i.e.,

$$W^{hg} = HW^b H^T - D_v.$$ (4)

Tan *et al.* [54] gave the definition of unnormalized graph Laplacian operator Δ_2 by introducing the inner product, i.e.,

$$\langle f, \Delta_2 f \rangle_i = \frac{1}{2} \sum_{i,j=1}^{n} W_{ij} \left(f_i - f_j \right)^2,$$ (5)

where W_{ij} denotes the weight of any two nodes on the simple graph. According to the above-mentioned formula, we also can acquire the inner product form of graph p-Laplacian operator Δ_p by the choice of the parameter $p(p \geq 1)$.

$$\langle f, \Delta_p f \rangle_i = \frac{1}{2} \sum_{i,j=1}^{n} W_{ij} \left(f_i - f_j \right)^p.$$ (6)

In other words, high-order graph p-Laplacian [55, 56] is a nonlinear generalization of unnormalized graph Laplacian (one order). In addition, we can show the definition of graph p-Laplacian operator Δ_p in the unnormalized and normalized matrix forms.

$$\left(\Delta_p^{(u)}\right)_i = \sum_{j\in V} W_{ij}\varphi_p\left(f_i - f_j\right),$$ (7)

$$\left(\Delta_p^{(n)}\right)_i = \frac{1}{d_i}\sum_{j\in V} W_{ij}\varphi_p\left(f_i - f_j\right),$$ (8)

where $d_i = \Sigma_{j=1}^n W_{ij}$ is the degree function of the graph. $\varphi_{p(x)} = |x|^{p-1}\,\text{sign}\,(x)$. Compared with graph Laplacian, the upper and lower bounds on the second eigenvalue approximate the optimal Cheeger cut value well because of the existence of the tighter isoperimetric inequality on the graph p-Laplacian [57, 58]. Figure 2 shows the difference between graph Laplacian and graph p-Laplacian in the manifold structure preserving of data.

Liu et al. [26] made a generalization for the Laplacian regularized sparse coding by utilizing graph p-Laplacian to describe the manifold distribution of data. Extensive experiment results demonstrate the effectiveness of the p-Laplacian regularized sparse coding model on the human activity datasets. Luo et al. [56] introduced an efficient gradient descend to analyze the properties of its full eigenvectors on the graph p-Laplacian in detail, which can solve the multi-class clustering problems. Ma et al. [59] proposed the ensemble p-Laplacian regularization for remote sensing image recognition by the effective combination of multiple graphs to approximate the intrinsic geometry structure of data.

Recently, Ma et al. [60] proposed a Hypergraph p-Laplacian regularization-based logistic regression for remote sensing image recognition. Fu et al. [41] proposed a Hypergraph p-Laplacian GCN. These models all use the Hypergraph p-Laplacian to explore its manifold structure information. Compared with Hypergraph and graph p-Laplacian, the effective combination of Hypergraph and graph p-Laplacian (Hypergraph p-Laplacian) can better express the intrinsic structure relationships between nodes. The computational process of HL_p is divided into the following parts: First, we can compute the Hypergraph Laplacian HL according to $HL = D^{hg} - W^{hg}$ and $D_{ii}^{hg} = \Sigma_j W_{ij}^{hg}$. Second, we make an eigenvector decomposition on HL. Then, we can acquire the full eigenvalues of HL by using the gradient descend that was proposed in Ref. [56] to solve the embedding problem. Finally, we can get the approximation of HL_p by means of the above full eigenvectors and eigenvalues.

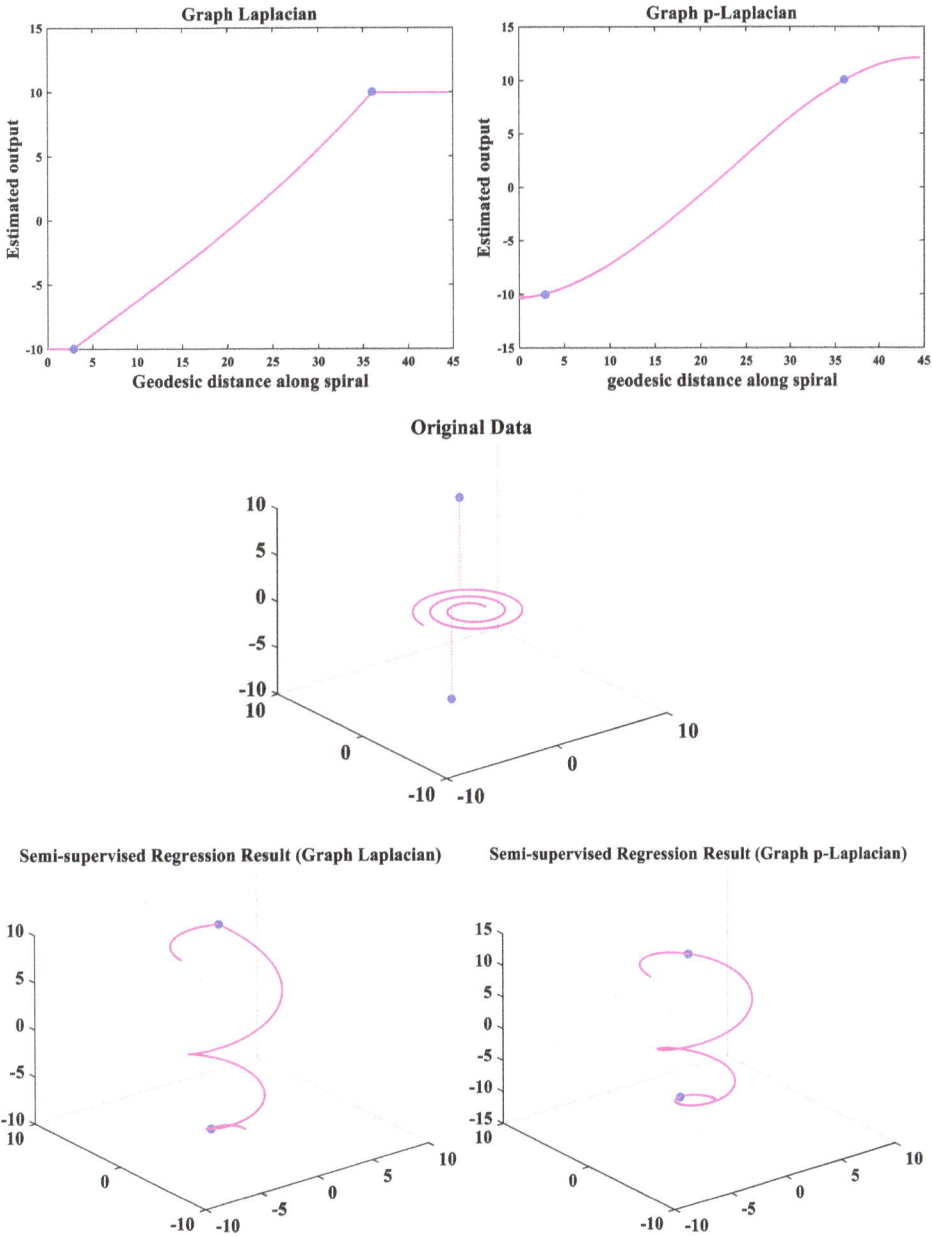

Figure 2: Differences of the semi-supervised regression for fitting two points on the one-dimensional spiral by respectively utilizing graph Laplacian and graph p-Laplacian matrix to preserve the local geometry structure relationships of data distribution. In the local geometry preserving of data, graph Laplacian always biases the solution towards a constant function and the extrapolation functions remain unchanged along the spiral for unseen data. Due to the variation smoothness of graph p-Laplacian's extrapolation function with the geodesic distance, it can fit the data properly and extrapolates perfectly to unseen data.

2.2 Spectral convolution on graphs

For graph structured data, the standard convolution of CNN is not more effective because of the irregular space structure of data [61]. To solve this problem, Bruna et al. [62] first gave the definition of convolution operation for graph structured data from the perspective of the spectral domain. It is expressed as the multiplication of the input signal $X \in R^{N \times M}$ and the filter $g_\theta(L)$ in the Fourier domain.

$$
\begin{aligned}
g_\theta(L) * X &= U\left(\left(U^T g_\theta\right) \odot \left(U^T X\right)\right) \\
&= U g_\theta(\Lambda) U^T X,
\end{aligned}
\tag{9}
$$

where $\odot$ expresses the Hadamard product and $g_\theta = \text{diag}\left(U^T g\right)$. In this definition, we can get the eigenvectors matrix U and diagonal matrix of eigenvalues Λ by the eigendecomposition of the normalized graph Laplacian $L = I_N - D^{-\frac{1}{2}} A D^{-\frac{1}{2}} = U \Lambda U^T$. A denotes the adjacency relationship matrix between different nodes that exclude self-connections. D is the degree matrix about A. This method is not applicable for large graph structured data and cannot reflect the spatial localization of the filter. Hammond et al. [62] and Defferrard et al. [63] utilized a truncated expansion with K-order Chebyshev polynomial to approximate $g_\theta(\Lambda)$ with filter coefficients θ, i.e.,

$$
g_\theta(\Lambda) = \sum_{k=0}^{k} \theta_k T_k(\tilde{\Lambda}),
\tag{10}
$$

where Chebyshev polynomial $T_k(X)$ can be acquired through the following definition: $T_0(X) = 1$, $T_1(X) = X$, $T_k(X) = 2X T_{k-1}(X) - T_{k-2}(X)$. $\tilde{\Lambda} = \frac{2}{\lambda_{max}} \Lambda - I_N$ and λ_{max} is the maximum eigenvalue of the graph Laplacian. Thus, Eq. (9) can be simplified, i.e.,

$$
\begin{aligned}
g_\theta(L) * X &= U\left(\sum_{k=0}^{k} \theta_k T_k(\tilde{\Lambda})\right) U^T X \\
&= \sum_{k=0}^{k} \theta_k T_k(\tilde{L}) X
\end{aligned}
\tag{11}
$$

Kipf et al. [39] aimed to avoid the overfitting problem in the local structure information exploiting for graph structured data by introducing the one-order polynomial about the graph Laplacian ($K = 1$ and $\lambda_{max} = 2$). In other words, Kipf et al. [39] thought that Eq. (11) can well describe the complex correlation information of data by utilizing the spectral graph convolution with the one-order approximation to consider the direct relationships between different nodes. Equation (11) is further simplified as the following form:

$$g_\theta(L) * X = \tilde{D}^{-\frac{1}{2}}(A + I_N)\tilde{D}^{-\frac{1}{2}}\theta X. \tag{12}$$

The detailed theoretical analysis can be found in Ref. [39]. We can build a deeper spectral graph convolution network (GCN) by stacking multi-layers, Eq. (12). Finally, GCN can extract more representative data information by the effective convolution fusion of the input data features and constructed structure relationships between different nodes.

3. Local Structure Preserving Methods of GCN

GCN is one of the most prominent spectral-domain-based GNN methods, which utilizes the graph Laplacian to describe the local geometric distribution of data. For GCN, how to acquire the most efficient high-order correlation information when exploring the local structure preserving methods of data is still challenging. In this section, we show and analyze the effectiveness of its local structure preserving for two-order GCN, Hypergraph p-Laplacian GCN and manifold regularized dynamic GCN in turns.

3.1 *Two-order GCN*

GCN can only capture the structural relationships between each node and its direct neighbor nodes by spectral graph convolution that uses one-order Chebyshev polynomial. Because of the lack of richer data structure informa-tion, there is insufficiency of extracted data features, and this directly reduces the classification performance of the semi-supervised classification. To solve this problem, in 2019, Fu *et al.* [40] developed a two-order graph convolu-tional network by simultaneously considering the direct and indirect relation-ships of different nodes, i.e., $\lambda_{\max} = 2$ and $K = 2$. It can acquire two-order approximation of spectral convolution on graph structured data.

$$g_\theta(L) * X = \theta_0 X + \theta_1 (L - I_N)X + \theta_2[2(L - I_N)^2 - I_N]X. \tag{13}$$

It contains three filter parameters θ_0, θ_1 and θ_2 shared on the graphs. By multi-layer convolution of Eq. (13), each node can learn a new data representation that fuses local structure information of its neighbor nodes. To further limit the filter parameters to reduce the operation of convolution layer, Eq. (13) can be rewritten as the following optimization problem:

$$
\begin{aligned}
\mathscr{g}_\theta\left(L\right)* X &= \theta\left[\left(I_N - L\right) + 2\left(L - I_N\right)^2\right]X \\
&= \theta\left[D^{-\frac{1}{2}}AD^{-\frac{1}{2}} + 2\left(D^{-\frac{1}{2}}AD^{-\frac{1}{2}}\right)^2\right]X
\end{aligned}
\tag{14}
$$

where $\theta = \theta_0 = -\theta_1 = \theta_2$. To further increase the stabilities and avoid the vanishing gradient of a deep model, Fu et al. [40] proposed three optimization methods.

$$
\mathscr{g}_\theta\left(L\right)* X = \theta\left(I_N - L\right)X = D^{-\frac{1}{2}}AD^{-\frac{1}{2}}X\theta,
\tag{15}
$$

$$
\mathscr{g}_\theta\left(L\right)* X = \theta\left[2\left(L - I_N\right)^2\right]X = 2\left(D^{-\frac{1}{2}}AD^{-\frac{1}{2}}\right)^2 X\theta,
\tag{16}
$$

$$
\begin{aligned}
g_\theta\left(L\right)* X &= \theta\left[\left(D^{-\frac{1}{2}}AD^{-\frac{1}{2}}\right)\left(I_N + 2D^{-\frac{1}{2}}AD^{-\frac{1}{2}}\right)\right]X, \\
&= D^{-\frac{1}{2}}AD^{-\frac{1}{2}}\tilde{D}^{-\frac{1}{2}}\left(2A + I_N\right)\tilde{D}^{-\frac{1}{2}}X\theta.
\end{aligned}
\tag{17}
$$

For Eq. (17), it let $I_N + 2D^{-\frac{1}{2}}AD^{-\frac{1}{2}} \rightarrow \tilde{D}^{-\frac{1}{2}}\left(2A + I_N\right)\tilde{D}^{-\frac{1}{2}}$ and $D_{ii} = \Sigma_j A_{ij}$, $\tilde{D}_{ii} = \Sigma_j(2A + I_N)_{ij}$. To evaluate the effectiveness of the above proposed convolution layer rule, Fu et al. [40] built three two-layer two-order GCNs for semi-supervised classification by stacking Eqs. (15)–(17) separately.

$$
Z = \tilde{B}\left[\sigma\left(\tilde{B}H^{(0)}W^{(0)}\right)\right]W^{(1)},
\tag{18}
$$

where σ denotes the activation function RELU. $H^{(0)} = X$ expresses the original input data features. $W^{(L)}$ are the filter parameters matrix trained. $\tilde{B} = D^{-\frac{1}{2}}AD^{-\frac{1}{2}}, \tilde{B} = 2(D^{-\frac{1}{2}}AD^{-\frac{1}{2}})^2$, and $\tilde{B} = D^{-\frac{1}{2}}AD^{-\frac{1}{2}}\tilde{D}^{-\frac{1}{2}}\left(2A + I_N\right)\tilde{D}^{-\frac{1}{2}}$ denote the constructed sample structure relationship matrix. Z denotes the extracted data features of the final convolution layer. In addition, it used Softmax classifier and optimized hyper-parameters according to the values of the cross-entropy loss function.

3.2 *Hypergraph p-Laplacian GCN*

Compared with graph Laplacian, p-Laplacian [24, 26] and Hypergraph [41, 60] theories all have demonstrated the superiority of its local structure preserving methods. Ma *et al.* [60] further provided an effective approximation of the combination of the Hypergraph and p-Laplacian matrix. In 2019, Fu *et al.* [41] first generalized graph Laplacian-based spectral convolution on simple graphs to spectral graph p-Laplacian convolution on the Hypergraph domain by introducing the Hypergraph p-Laplacian matrix to preserve the manifold distribution of data. Then it proposed the definition of its convolution operation on spectral Hypergraph p-Laplacian convolution.

$$g_\theta \left(HL_p \right) * X = \sum_{k=0}^{k} \theta_k T_k \left(\widetilde{HL}_p \right) X, \tag{19}$$

where $\widetilde{HL}_p = \frac{2}{\lambda_{max}} HL_p - I_N$. It used the computing method of the Hypergraph p-Laplacian matrix proposed in Ref. [60]. In addition, it also limited the order number of Chebyshev polynomial ($K = 1$) to cut down the computation complexity of the model. That is to say, one-order Chebyshev polynomial in the Hypergraph p-Laplacian matrix can well capture the richer sample structure relationships compared with the graph Laplacian matrix. (At the same time, it only aimed to test the effectiveness of the Hypergraph p-Laplacian matrix on spectral Hypergraph p-Laplacian convolution when comparing with other GCN variants, especially the GCN model.)

$$g_\theta \left(HL_p \right) * X = \theta_0 X + \theta_1 \left(\frac{2}{\lambda_{max}} HL_p - I_N \right) X. \tag{20}$$

In Eq. (20), $T_0 \left(\widetilde{HL}_p \right) = I_N$, $T_1 \left(\widetilde{HL}_p \right) = \widetilde{HL}_p = \frac{2}{\lambda_{max}} HL_p - I_N$ Fu *et al.* [41] recalculated the largest eigenvalue λ_{max} according to the difference of different types of databases. It can solve the issue that the limitations of $\lambda_{max} = 2$ [39] because of the difference of the constructed Hypergraph p-Laplacian matrix on the different datasets. To further avoid the overfitting problem, Fu *et al.* [41] utilized a single parameter θ to replace the two filter parameters θ_0 and θ_1, i.e., $\theta_0 = 0$ and $\theta_1 = \theta$.

$$g_\theta \left(HL_p \right) * X = \left(\frac{2}{\lambda_{max}} HL_p - I_N \right) X\theta. \tag{21}$$

If we use the optimization strategy $(\theta = \theta_0 = -\theta_1)$ that was proposed in Ref. [39], we can get the following one-order approximation of spectral Hypergraph p-Laplacian convolution:

$$g_\theta \left(HL_p \right) * X = 2 \left(I_N - \frac{1}{\lambda_{max}} HL_p \right) X\theta. \tag{22}$$

In Eq. (22), the constructed structure relationships will exist with abundant negative number. After the activation function of each convolution layer, it will remove its learned data features with negative number. Finally, the extracted sample information is not enough and cannot improve the classification performance of the model. It also indirectly demonstrated the correctness of the used optimization method proposed in Ref. [41]. According to Eq. (21), Fu et al. [41] built a convolution layer rule $f(X, W, HL_p)$. Figure 3 shows the basic framework of multi-layer Hypergraph p-Laplacian graph convolution networks for semi-supervised classification. (Note, $\tilde{C} = \frac{2}{\lambda_{max}} HL_p - I_N$.)

$$H^{(L+1)} = \sigma \left[\left(\frac{2}{\lambda_{max}} HL_p - I_N \right) H^{(L)} W^{(L)} \right]. \tag{23}$$

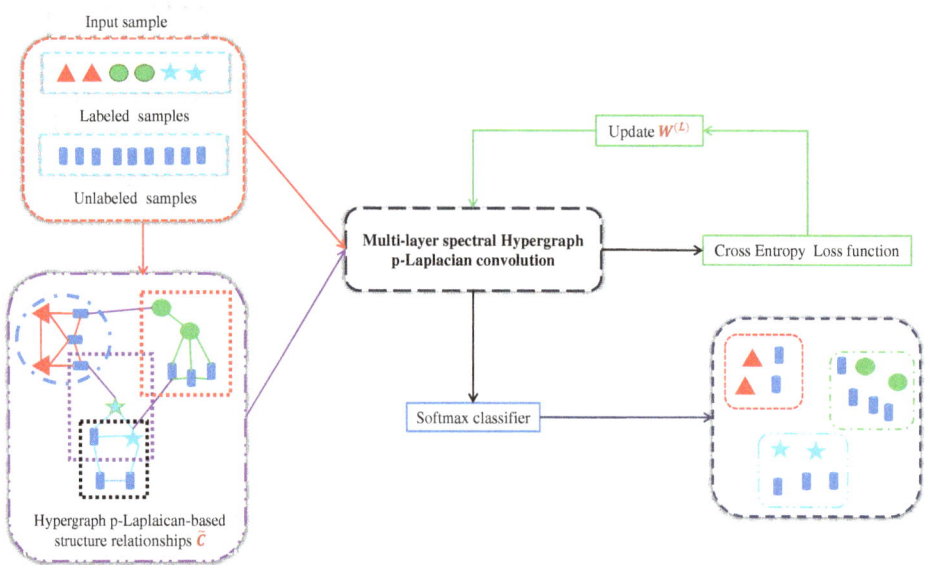

Figure 3. The basic framework of multi-layer Hypergraph p-Laplacian graph convolution networks for semi-supervised classification.

3.3 *Manifold regularized dynamic GCN*

Now, more and more GCN variants have been proposed. However, most of the proposed models only utilized a static local geometry distribution under the training iteration process. Thus, this way cannot guarantee that its captured sample structure relationships between different nodes are optimal. To generate the optimal structure information, in 2019, Fu *et al.* [44] proposed a dynamic GCN model by introducing manifold regularization framework. Compared with GCN, two-order GCN and Hypergraph p-Laplacian GCN, it simultaneously optimized the weight parameters and data structure information matrix. And then it gave a novel objective function by regarding the manifold regularization as a term of the cross-entropy loss function.

$$C(W, A) = C_1(W, A) + \beta C_2(W, A)$$

$$= -y \log Y + \beta tr(Y^T M Y), \tag{24}$$

where $C_1(W, A)$ is the cross-entropy loss function. y denotes the true label information and Y expresses the probability distribution matrix after Softamx function [41]. M denotes the matrix described in the local geometry structure of data distribution including graph Laplacian, Hypergraph Laplacian [43], graph Hessian [5], graph p-Laplacian [59] and Hypergraph p-Laplacian [60]. In manifold regularized dynamic GCN [41], it utilized graph Laplacian L to describe the adjacency relationships of data. Thus, Eq. (24) can be simplified as the following problem:

$$C(W, A) = -y \log Y + \beta tr(Y^T L Y). \tag{25}$$

For Eq. (25), Fu *et al.* [44] alternatively optimized the above-mentioned variables adjacency matrix A and weight matrix W. First, it fixed A, and then updated W by gradient descent.

$$C(W) = -y \log Y + \beta tr(Y^T L Y), \tag{26}$$

where $L = D - A$ and A of the first training iteration can be computed on the original input data by the Euclidean distance-based K-nearest neighbor method. In the following, it fixed the above optimized W, and then optimized A. Equation (25) can be denoted as

$$C(A) = -y \log Y + \beta tr(Y^T L Y). \tag{27}$$

It calculated A on the learned data features of the last convolution layer. After many times training iteration, manifold regularized dynamic GCN can learn the optimization A and W compared with the existing static GCN models.

At the same time, Fu et al. [44] further made a generalization for the one-order Chebyshev approximation based on the convolution layer rule of GCN, which aimed to solve the limitations of λ_{max} with a specific value.

$$g_\theta(L) * X = \sum_{k=0}^{k=1} \theta_k T_k(\tilde{L}) X$$

$$= \theta_0 X + \theta_1 \left(\frac{2}{\lambda_{\text{max}}} L - I_N \right) X. \tag{28}$$

In the following, it utilized a single parameter θ to cut down the convolution operation and avoid the overfitting of the model. Thus, Eq. (28) can be denoted as the following expression:

$$g_\theta(L) * X = \left(\theta_0 - \theta_1 + \frac{2}{\lambda_{\text{max}}} \theta_1 \left(I_N - D^{-\frac{1}{2}} A D^{-\frac{1}{2}} \right) \right) X$$

$$= \left(\theta_2 + \frac{2}{\lambda_{\text{max}}} \theta_1 \left(I_N - D^{-\frac{1}{2}} A D^{-\frac{1}{2}} \right) \right) X$$

$$= \left(\frac{2}{\lambda_{\text{max}}} \left(\frac{\lambda_{\text{max}}}{2} \theta_2 + \theta_1 \right) - \frac{2}{\lambda_{\text{max}}} \theta_1 D^{-\frac{1}{2}} A D^{-\frac{1}{2}} \right) X$$

$$= \frac{2}{\lambda_{\text{max}}} \left(\theta_3 - \theta_1 D^{-\frac{1}{2}} A D^{-\frac{1}{2}} \right) X$$

$$= \theta \frac{2}{\lambda_{\text{max}}} \left(I_N + D^{-\frac{1}{2}} A D^{-\frac{1}{2}} \right) X. \tag{29}$$

In Eq. (29), Fu et al. [41], let $\theta_0 - \theta_1 = \theta_2$, $\frac{\lambda_{\text{max}}}{2} \theta_2 + \theta_1 = \theta_3$ and $\theta = \theta_3 = -\theta_1$. Finally, it introduced a renormalization trick that was proposed in Ref. [39] to optimize Eq. (29), i.e., $I_N + D^{-\frac{1}{2}} A D^{-\frac{1}{2}} \rightarrow \tilde{D}^{-\frac{1}{2}} (A + I_N) \tilde{D}^{-\frac{1}{2}}$ Equation (29) can be further simplified as the following form:

$$g_\theta(L) * X = \frac{2}{\lambda_{\text{max}}} \tilde{D}^{-\frac{1}{2}} (A + I_N) \tilde{D}^{-\frac{1}{2}} X\theta. \tag{30}$$

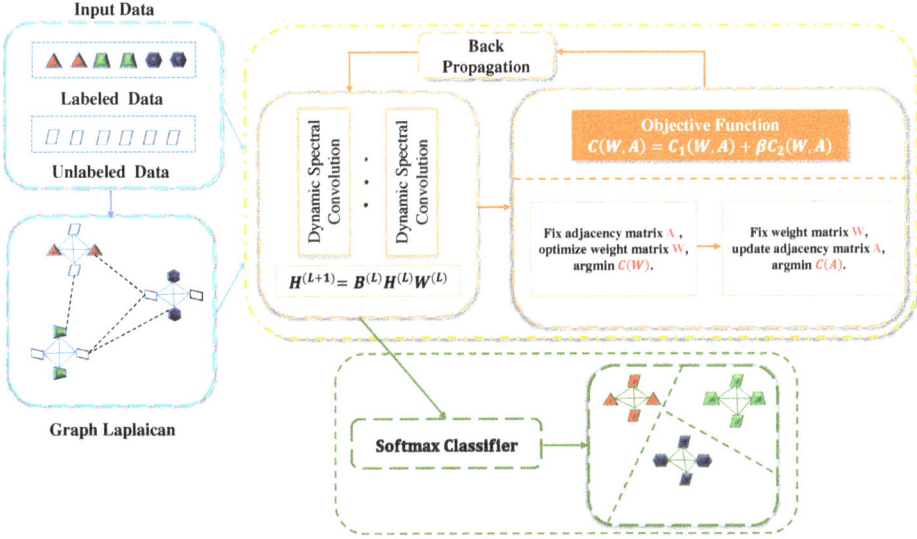

Figure 4: The framework of multi-layer dynamic graph convolution networks based on manifold regularization.

According to Eq. (30), Fu *et al.* [44] proposed a novel dynamical convolution layer rule $f(X, A, W)$. In addition, it designed a two-layer manifold regularized dynamic GCN to demonstrate the advantages of dynamic graph structure learning. Figure 4 describes the framework of multi-layer manifold regularized dynamic GCN. (Note, $B^{(L)} = \frac{2}{\lambda_{max}} \tilde{D}^{-\frac{1}{2}} \left(A^{(L)} + I_N \right) \tilde{D}^{-\frac{1}{2}}$.)

$$H^{(L+1)} = \frac{2}{\lambda_{max}} \tilde{D}^{-\frac{1}{2}} \left(A^{(L)} + I_N \right) \tilde{D}^{-\frac{1}{2}} H^{(L)} W^{(L)}, \tag{30}$$

$$H^{(2)} = \frac{2}{\lambda_{max}} \tilde{D}^{-\frac{1}{2}} \left(A^{(1)} + I_N \right) \tilde{D}^{-\frac{1}{2}}$$
$$\left[\sigma \left(\frac{2}{\lambda_{max}} \tilde{D}^{-\frac{1}{2}} \left(A^{(0)} + I_N \right) \tilde{D}^{-\frac{1}{2}} X W^{(0)} \right) \right] W^{(1)}. \tag{31}$$

3.4 *Experiment*

SAT-6 [65] dataset consists of totally 405,000 remote sensing images collected from six land cover categories, such as trees, barren land, grassland, roads, buildings and water bodies. Each remote sensing image is 28×28 pixels. To analyze the performance of the proposed two-order GCN,

Hypergraph p-Laplacian GCN and manifold regularized dynamic GCN, we conduct extensive comparative experiments ourselves. In this chapter, we extract edge features of 500 images per class for our experiment. 1,000 images were chosen for testing, 500 images as the validation set and the remaining 500 images were chosen as the training dataset. For the training samples of the semi-supervised classification, we randomly selected 10%, 20%, 30%, 40% and 50% samples as labeled data and the rest samples for unlabeled data. Figure 5 represents some images of SAT-6 dataset.

In Table 1, to avoid any bias introduced by the random partition of samples, we report the mean recognition accuracy by experimenting five times, independently with different methods. For two-order GCN, Hypergraph p-Laplacian GCN and manifold regularized dynamic GCN, we utilize Adam's method with a learning rate to train all models with up to 200 iterations. If the loss values of the cross-entropy loss function in the validation set remain unchanged even after ten consecutive times, two-order GCN, Hypergraph p-Laplacian GCN and manifold regularized dynamic GCN will stop training automatically.

Figure 5: Some images of SAT-6 dataset.

Table 1: Mean recognition accuracy of different methods on SAT-6 dataset.

Method	10%	20%	30%	40%	50%
GCN	33.5 ± 4.3	38.02 ± 3.6	38.88 ± 3.1	39.64 ± 3.4	40.3 ± 2.1
Two-order GCN	42.9 ± 4.7	46.42 ± 4.8	49.52 ± 3.4	51.42 ± 0.3	53.06 ± 0.4
Hypergraph p-Laplacian GCN	52.66 ± 1	53.74 ± 2	54.16 ± 0.1	55.26 ± 1	55.7 ± 0.8
Manifold regularized dynamic GCN	38 ± 2.7	39.6 ± 1.9	39.9 ± 2.8	40.84 ± 4.75	44.1 ± 2.3

Other parameters are summarized as follows: Two-order GCN: 0.01 (learning rate), 16 (hidden units), 0.5 (dropout rate), 5×10^{-4} (L2 regularization); Hypergraph p-Laplacian GCN: 0.1 (learning rate), 8 (hidden units), 0.4 (dropout rate), 5×10^{-5} (L2 regularization); Manifold regularized dynamic GCN: 0.01 (learning rate), 16 (hidden units), 0.5 (dropout rate), 5×10^{-4} (L2 regularization), 0.001 (β). In Table 1, we report the mean recognition accuracy with multiple experiments using different methods.

From these results, we can know that two-order GCN, Hypergraph p-Laplacian GCN and manifold regularized dynamic GCN all achieve a higher recognition accuracy in comparison to GCN. In other words, they all show the superiority of local structure preserving compared to GCN.

Compared with GCN, we can know that two-order GCN, Hypergraph p-Laplacian GCN and manifold regularized dynamic GCN can learn richer sample features to improve the classification performance by the effective convolution fusion of two-order approximation of Chebyshev polynomials-based structure information, Hypergraph p-Laplacian-based structure information, dynamically optimal Laplacian-based structure information and the original data information, respectively. In addition, it also indicates the richer null space of two-order GCN, Hypergraph p-Laplacian GCN and manifold regularized dynamic GCN, i.e., two-order approximation of Chebyshev polynomials, Hypergraph p-Laplacian, and dynamically optimal Laplacian can better describe the local geometry of data compared with graph Laplacian.

4. Conclusions

In the past few years, GRL, especially the GCNs, has been demonstrated to be a powerful tool to extract effective features of the graph structured data. However, it is significant for the existing GCN variants on how to reasonably and effectively capture and utilize the manifold structure information of data distribution. In this chapter, we reviewed the spectral convolution on simple graph domains in detail and showed many extended GCN models based on spectral graph convolution with K-order Chebyshev polynomial. Two-order GCN that simultaneously considers the direct and indirect space local geometric relationships of different samples is developed to improve the classification performance of semi-supervised classification. Considering the superiority of graph p-Laplacian and Hypergraph learning in local geometry preserving in comparison to graph Laplacian, Hypergraph p-Laplacian GCN is proposed to learn more representative sample features on graph structured data. In the most existing GCN-related applications, it is difficult to generate the optimal sample structure information during the learning process because these models pay little attention to optimizing the local structure of data distribution. Thus, a dynamic GRL model named manifold regularized dynamic GCN is proposed to optimize its captured local geometric distribution until model fitting.

5. Future Works

The definition of spectral graph convolution based on K-order Chebyshev polynomial uses the example of graph Laplacian to preserve the local geometry of row vectors on high-dimensional data, while neglecting the manifold structure information of column vectors. In the future works, we will give the definition of spectral example-feature graph convolution by simultaneously introducing feature graph Laplacian and example graph Laplacian matrix to describe local geometry distribution of row vectors and column vectors. And then an effective one-order approximation of spectral dual graph convolution and multi-layers dual GCNs will be presented. In addition, we will explore the superiority of graph Hessian, graph p-Laplacian for GCN in capturing sample structure relationships.

Acknowledgment

This work was supported in part by the National Natural Science Foundation of China under Grant 61671480, in part by the Fundamental Research Funds for the Central Universities, China University of Petroleum (East China) under Grants 18CX07011A and YCX2019080.

References

1. R. Sheikhpour, M. A. Sarram, S. Gharaghani, and M. A. Z. Chahooki, "A survey on semi-supervised feature selection methods," *Pattern Recogn.*, vol. 64, pp. 141–158, 2017.
2. N. F. F. D. Silva, L. F., and E. R. Hruschka, "A survey and comparative study of tweet sentiment analysis via semi-supervised learning," *ACM Comput. Surv. (CSUR)*, vol. 49, no. 1, p. 15, 2016.
3. J. Gu, Z. Wang, J. Kuen, L. Ma, A. Shahroudy, B. Shuai, and T. Chen, "Recent advances in convolutional neural networks," *Pattern Recogn.*, vol. 77, pp. 354–377, 2018.
4. T. H. Chan, K. Jia, S. Gao, J. Lu, Z. Zeng, and Y. Ma, "PCANet: A simple deep learning baseline for image classification?" *IEEE Trans. Image Process.*, vol. 24, no. 12, pp. 5017–5032, 2015.

5. W. Liu, X. Yang, D. Tao, J. Cheng, and Y. Tang, "Multiview dimension reduction via Hessian multiset canonical correlations," *Inform. Fusion*, vol. 41, pp. 119–128, 2018.

6. W. Liu, T. Ma, Q. Xie, D. Tao, and J. Cheng, "LMAE: A large margin auto-encoders for classification," *Signal Process.*, vol. 141, pp. 137–143, 2017.

7. M. Qian, J. Qi, L. Zhang, M. Feng, and H. Lu, "Language-aware weak supervision for salient object detection," *Pattern Recogn.*, vol. 96, p. 106955, 2019.

8. M. Meng, M. Lan, J. Yu, J. Wu, and D. Tao, "Constrained discriminative projection learning for image classification," *IEEE Trans. Image Process.*, vol. 29, pp. 186–198, 2019.

9. Z. Lin, J. Lu, and X. Qiu, "An effective hybrid low delay packet loss concealment algorithm for MDCT-based audio codec," *Appl. Acoust.*, vol. 154, pp. 170–175, 2019.

10. C. Song, Y. Huang, Y.-Z. Huang, N. Jia, and L. Wang, "GaitNet: An end-to-end network for gait based human identification," *Pattern Recogn.*, vol. 96, p. 106988, 2019.

11. F. Li, Z. Zhu, X. Zhang, J. Cheng, and Y. Zhao, "Diffusion induced graph representation learning," *Neurocomputing*, vol. 360, pp. 220–229, 2019.

12. Q. Wang, Z. Mao, B. Wang, and L. Guo, "Knowledge graph embedding: A survey of approaches and applications,": *IEEE Trans. Knowledge Data Eng.*, vol. 29, no. 12, pp. 2724–2743, 2017.

13. B. Perozzi, R. Al-Rfou, and S. Skiena, "Deepwalk: Online learning of social representations," in *Proc. 20th ACM SIGKDD Int. Conf. Knowledge Discovery and Data Mining*, ACM, pp. 701–710, 2014.

14. T. Mikolov, K. Chen, G. Corrado, and J. Dean, "Efficient estimation of word representations in vector space," in *Proc. 1st Int. Conf. Learning Representations*, IEEE, 2013.

15. J. Tang, M. Qu, M. Wang, M. Zhang, J. Yan, and Q. Mei, "Line: Large-scale information network embedding," in *Proc. 24th Int. Conf. World Wide Web*, International World Wide Web Conferences Steering Committee, pp. 1067–1077, 2015.

16. A. Grover and J. Leskovec, "node2vec: Scalable feature learning for networks," in *Proc. 22nd ACM SIGKDD Int. Conf. Knowledge Discovery and Data Mining*, ACM, pp. 855–864, 2016.

17. M. Bernstein, V. De Silva, J. C. Langford, and J. B. Tenenbaum, "Graph approximations to geodesics on embedded manifolds", Technical report, Department of Psychology, Stanford University, pp. 961–968, 2000.

18. J. D. Carroll and P. Arabie, "Multidimensional scaling," *Measurement, Judgment and Decision Making*, Academic Press, pp. 179–250, 1998.

19. S. T. Roweis and L. K. Saul, "Nonlinear dimensionality reduction by locally linear embedding," *Science*, vol. 290, no. 5500, pp. 2323–2326.

20. M. Belkin and P. Niyogi, "Laplacian eigenmaps and spectral techniques for embedding and clustering,". in *Advances in Neural Information Processing Systems*, pp. 585–591, 2002.

21. D. L. Donoho and C. Grimes, "Hessian eigenmaps: Locally linear embedding techniques for high-dimensional data," *Proc. Natl. Acad. Sci.*, vol. 100, no. 10, pp. 5591–5596, 2003.

22. W. Liu and D. Tao, "Multiview hessian regularization for image annotation," *IEEE Trans. Image Process.*, vol. 22, no. 7, pp. 2676–2687, 2013.

23. D. Tao, L. Jin, W. Liu, and X. Li, "Hessian regularized support vector machines for mobile image annotation on the cloud," *IEEE Trans. Multimedia*, vol. 15, no. 4, pp. 833–844, 2013.

24. W. Liu, X. Ma, Y. Zhou, D. Tao, and J. Cheng, "p-Laplacian regularization for scene recognition," *IEEE Trans. Cybernet.*, vol. 49, no. 8, pp. 2927–2940, 2018.

25. W. Allegretto and H. Y. Xi, "A Picone's identity for the p-Laplacian and applications," *Nonlin. Anal.: Theory, Meth. Appl.*, vol. 32, no. 7, pp. 819–830, 1998.

26. W. Liu, Z. J. Zha, Y. Wang, K. Lu, and D. Tao, "p-Laplacian regularized sparse coding for human activity recognition," *IEEE Trans. Industr. Electr.*, vol. 63, no. 8, pp. 5120–5129, 2016.

27. I. Goodfellow, J. Pouget-Abadie, M. Mirza, B. Xu, D. Warde-Farley, S. Ozair, and Y. Bengio, "Generative adversarial nets," in *Advances in Neural Information Processing Systems*, pp. 2672–2680, 2014.

28. M. Guo, E. hou, D. A. Huang, S. Song, S. Yeung, and L. Fei-Fei, "Neural graph matching networks for fewshot 3D action recognition," in *Proc. Eur. Conf. Computer Vision (ECCV)*, pp. 653–669, 2018.

29. L. Yao, C. Mao, and Y. Luo, "Graph convolutional networks for text classification," in *Proc. AAAI Conf. Artificial Intelligence*, vol. 33, pp. 7370–7377, 2019.

30. J. Qiu, J. Tang, H. Ma, Y. Dong, K. Wang, and J. Tang, "DeepInf: Social influence prediction with deep learning," in *Proc. 24th ACM SIGKDD Int. Conf. Knowledge Discovery & Data Mining*, ACM, pp. 2110–2119, 2018.

31. F. Monti, M. Bronstein, and X. Bresson, "Geometric matrix completion with recurrent multi-graph neural networks," in *Advances in Neural Information Processing Systems*, pp. 3697–3707, 2017.

32. J. Shang, C. Xiao, T. Ma, H. Li, and J. Sun, "GAMENet: Graph augmented memory networks for recommending medication combination," in *Proc. AAAI Conf. Artificial Intelligence*, vol. 33, pp. 1126–1133, 2019.

33. S. Kearnes, K. McCloskey, M. Berndl, V. Pande, and P. Riley, "Molecular graph convolutions: Moving beyond fingerprints," *J. Comput.-aided Molecular Design*, vol. 30, no. 8, pp. 595–608, 2016.

34. Y. Hoshen, "Vain: Attentional multi-agent predictive modeling," in *Advances in Neural Information Processing Systems*, pp. 2701–2711, 2017.

35. M. Simonovsky and N. Komodakis, "Dynamic edge-conditioned filters in convolutional neural networks on graphs," in *Proc. IEEE Conf. Computer Vision and Pattern Recognition*, pp. 3693–3702, 2017.

36. J. Masci, D. Boscaini, M. Bronstein, and P. Vandergheynst, "Geodesic convolutional neural networks on riemannian manifolds," in *Proc. IEEE Int. Conf. Computer Vision Workshops*, pp. 37–45, 2015.

37. D. Boscaini, J. Masci, E. Rodolà, and M. Bronstein, "Learning shape correspondence with anisotropic convolutional neural networks," in *Advances in Neural Information Processing Systems*, pp. 3189–3197, 2016.

38. J. Atwood and D. Towsley, "Diffusion-convolutional neural networks," in *Advances in Neural Information Processing Systems*, pp. 1993–2001, 2016.

39. T. N. Kipf and M. Welling, "Semi-supervised classification with graph convolutional networks," in *Proc. IEEE Int. Conf. Learning Representations*, 2017.

40. S. Fu, W. Liu, S. Li, and Y. Zhou, "Two-order graph convolutional networks for semi-supervised classification," *IET Image Process.*, vol. 13, no. 14, pp. 2763–2771, 2019.

41. S. Fu, W. Liu, Y. Zhou, and L. Nie, "HpLapGCN: Hypergraph p-Laplacian graph convolutional networks," *Neurocomputing*, vol. 362, pp. 166–174, 2019.

42. N. Yadati, M. Nimishakavi, P. Yadav, A. Louis, and P. Talukdar, "Hypergcn: Hypergraph convolutional networks for semi-supervised classification," in *Proc. IEEE Int. Conf. Multimedia and Expo*, 2018.

43. Y. Feng, H. You, Z. Zhang, R. Ji, and Y. Gao, "Hypergraph neural networks," in *Proc. AAAI Conf. Artificial Intelligence*, vol. 33, pp. 3558–3565, 2019.

44. S. Fu, W. Liu, Y. Zhou, Z. J. Zha, and L. Nie, "Dynamic graph convolutional networks by manifold regularization," in *Proc. 28th Int. Joint Conf. Artificial Intelligence Workshops*, 2019.

45. Y. Huang, Q. Liu, S. Zhang, and D. N. Metaxas, "Image retrieval via probabilistic hypergraph ranking," in *2010 IEEE Computer Society Conference on Computer Vision and Pattern Recognition*, IEEE, pp. 3376–3383, 2010.

46. S. Agarwal, J. Lim, L. Zelnik-Manor, P. Perona, D. Kriegman, and S. Belongie, "Beyond pairwise clustering," in *2005 IEEE Computer Society Conference on Computer Vision and Pattern Recognition (CVPR'05)*, vol. 2, IEEE, pp. 838–845, 2005.

47. J. A. Rodriguez, "On the Laplacian spectrum and walk-regular hypergraphs," *Lin. Multilin. Algebra*, vol. 51, no. 3, pp. 285–297, 2003.

48. J. Y. Zien, M. D. Schlag, and P. K. Chan, "Multilevel spectral hypergraph partitioning with arbitrary vertex sizes," *IEEE Trans. Comput.-aided Design Integra. Cir. Syst.*, vol. 18, no. 9, pp. 1389–1399, 1999.

49. T. C. Hu and K. Moerder, "Multiterminal flows in a hypergraph," *VLSI Circuit Layout: Theory and Design*, IEEE Press, pp. 87–93, 1985.

50. M. Bolla, "Spectra, Euclidean representations and clusterings of hypergraphs," *Discrete Math.*, vol. 117. nos. 1–3, pp. 19–39, 1993.

51. D. Zhou, J. Huang, and B. Schölkopf, "Learning with hypergraphs: Clustering, classification, and embedding," in *Advances in Neural Information Processing Systems*, pp. 1601–1608, 2007.

52. O. Duchenne, F. Bach, I. S. Kweon, and J. Ponce, "A tensor-based algorithm for high-order graph matching," *IEEE Trans. Pattern Anal. Mach. Intell.*, vol. 33, no. 12, pp. 2383–2395, 2011.

53. A. Banerjee, I. Dhillon, J. Ghosh, S. Merugu, and D. S. Modha, "A generalized maximum entropy approach to bregman co-clustering and matrix approximation," *J. Mach. Learn. Res.*, vol. 8, pp. 1919–1986, 2007.

54. S. Y. Tan and H. S. Tan, "A microcellular communications propagation model based on the uniform theory of diffraction and multiple image theory," *IEEE Trans. Antennas Propag.*, vol. 44, no. 10, pp. 1317–1326, 1996.

55. W. Allegretto and H. Y. Xi, "A Picone's identity for the p-Laplacian and applications," *Nonlin. Anal.: Theory, Meth. Appl.*, vol. 32, no. 7, pp. 819–830, 1998.

56. D. Luo, H. Huang, C. Ding, and F. Nie, "On the eigenvectors of p-Laplacian," *Mach. Learn.*, vol. 81, no. 1, pp. 37–51, 2010.

57. H. Takeuchi, "The spectrum of the p-Laplacian and p-harmonic morphisms on graphs," *Illinois J. Math.*, vol. 47, no. 3, pp. 939–955, 2003.

58. T. Bühler and M. Hein, "Spectral clustering based on the graph p-Laplacian," in *Proc. 26th Ann. Int. Conf. Machine Learning*, ACM, pp. 81–88, 2009.

59. X. Ma, W. Liu, D. Tao, and Y. Zhou, "Ensemble p-Laplacian regularization for remote sensing image recognition," *Cognitive Comput.* 2018, 10.1007/s11063-019-10000-4.

60. X. Ma, W. Liu, S. Li, D. Tao, and Y. Zhou, "Hypergraph p-Laplacian regularization for remotely sensed image recognition," *IEEE Trans. Geosci. Remote Sens.*, vol. 57, no. 3, pp. 1585–1595, 2018.

61. D. I. Shuman, S. K. Narang, P. Frossard, A. Ortega, and P. Vandergheynst, "The emerging field of signal processing on graphs: Extending high-dimensional data analysis to networks and other irregular domains," *IEEE Sig. Process. Magaz.*, vol. 30, no. 3, pp. 83–98, 2013.

62. J. Bruna, W. Zaremba, A. Szlam, and Y. LeCun, "Spectral networks and locally connected networks on graphs," in *Proc. IEEE Int. Conf. Learning Representations*, 2013.

63. D. K. Hammond, P. Vandergheynst, and R. Gribonval, "Wavelets on graphs via spectral graph theory," *Appl. Comput. Harmonic Anal.*, vol. 30, no. 2, pp. 129–150, 2011.

64. M. Defferrard, X. Bresson, and P. Vandergheynst, "Convolutional neural networks on graphs with fast localized spectral filtering," in *Advances in Neural Information Processing Systems*, pp. 3844–3852, 2016.

65. M. Papadomanolaki, M. Vakalopoulou, S. Zagoruyko, and K. Karantzalos, "Benchmarking deep learning frameworks for the classification of very high resolution satellite multispectral data," *ISPRS Ann. Photogram., Remote Sens. Spatial Inform. Sci.*, vol. 3, no. 7, 2016.

Part IV

Deep Learning for Medical Sensing

https://doi.org/10.1142/9789811218842_0010

Chapter 10

Deep Retinal Image Non-Uniform Illumination Removal

Chongyi Li[*,||], Huazhu Fu[†,**], Miao Yang[‡,††], Runmin Cong[§,¶,‡‡]
and Chunle Guo[*,§§]

*School of Electrical and Information Engineering,
Tianjin University, Tianjin 300072, China
†Inception Institute of Artificial Intelligence (IIAI), Abu Dhabi, UAE
‡School of Electronic Engineering, Jiangsu Ocean University,
Lianyungang 222005, China
§Institute of Information Science, Beijing Jiaotong University,
Beijing 100044, China
¶Beijing Key Laboratory of Advanced Information
Science and Network Technology, Beijing Jiaotong University,
Beijing 100044, China
||lichongyi@tju.edu.cn
**hzf@ieee.org
††lemonmiao@gmail.com
‡‡rmcong@bjtu.edu.cn
§§guochunle@tju.edu.cn

Abstract

Retinal images have been widely used by clinicians for early diagnosis of ocular diseases. However, the quality of retinal images is often clinically unsatisfactory due to eye lesions and imperfect imaging processes.

235

The non-uniform or poor illumination on retinal images hinders the patho-logical information and further impairs the diagnosis of ophthalmologists. To solve these issues, we propose a deep learning-based retinal image non-uniform illumination removal called NuI-Go, which combines the powerful capabilities of convolutional neural networks (CNNs) with the characteristics of retinal images with non-uniform illumination. Concretely, the proposed NuI-Go consists of three Recursive non-local encoder–decoder residual blocks (NEDRBs) for progressively enhancing the degraded retinal images. Each NEDRB contains a feature encoder module that captures the hierarchi-cal feature representations, a non-local context module that models the context information, and a feature decoder module that recovers the details and spatial dimension. Extensive experimental results demonstrate that the proposed method performs favorably against the state-of-the-art methods on both synthetic dataset and real retinal images. Besides, we further demon-strate the advantages of the proposed method for improving the perfor-mance of retinal vessel segmentation.

Keywords: retinal image, image enhancement, quality degradation, deep learning, vessel segmentation, non-uniform illumination

1. Introduction

Retinal images have been widely used by clinicians for early diagnosis of ocu-lar diseases, including glaucoma, diabetic retinopathy, pathological myopia, etc. [1]. However, retinal images often suffer from non-uniform or poor illumination issue due to imperfect imaging process [2, 3]. Such quality degraded retinal images significantly impair the diagnosis of ophthalmologists and also affect the performance of computer-aided analysis of retinal diseases [4], such as retinal vessel segmentation, optical disc segmentation and vascu-lar structure recognition. However, traditional image enhancement methods, such as HE, CLAHE [5], Gamma correction, Multi-scale Retinex, etc., can-not be directly extended to retinal images. Therefore, it is meaningful to design non-uniform or poor illumination recovery methods for retinal images.

Some retinal image enhancement methods have been proposed in recent years. However, the performance of existing methods, ranging from tradi-tional methods to deep learning-based methods, is unsatisfactory. Different from existing retinal image enhancement methods, our method combines the capacity of deep networks with image physical formation model to study the issue of retinal image non-uniform illumination, which results in more

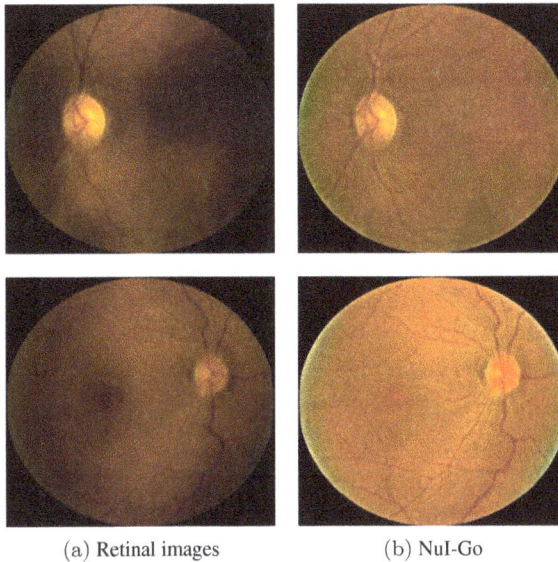

(a) Retinal images   (b) NuI-Go

Figure 1: Sample results on the real retinal images.

reasonable and reliable enhancement performance. In Fig. 1, we present two sample results of our NuI-Go. As shown, the non-uniform illumination seriously affects the visual quality of retinal images and covers some key information, such as vessels. Our NuI-Go can effectively correct the illumination of the retinal image while well preserving its details and color.

The contributions of this work are as follows:

— To the best of our knowledge, it is the first method to introduce deep learning to the problem of retinal image non-uniform illumination removal, which effectively corrects the illumination of a retinal image while well preserving its details and color.

— We propose a simple yet effective non-local encoder–decoder network to progressively enhance degraded retinal images with non-uniform illumination. The proposed network can effectively integrate local and non-local information, which is important for modeling the context.

— We propose a non-uniform illuminated retinal images synthesis approach, which provides the paired training data for supervised learning.

— Benefiting from the novel network architecture and the proposed physical model-based non-uniform illumination synthesis approach, our method achieves impressive performance on retinal image non-uniform illumination removal, and also has significant advantages in retinal image vessel segmentation.

2. Related Work

Numerous methods have been proposed to improve the quality of degraded images in recent years. In this chapter, we mainly focus on image light enhancement methods which can be categorized into traditional image enhancement approaches and retinal image enhancement approaches.

2.1 Traditional image enhancement approaches

HE-based methods [6–11] adjust the histogram distribution of an image to improve its contrast. However, HE- based methods tend to over-enhance the images due to impertinently changing the dynamic range of the image and some local HE methods are time-consuming. Unlike the HE-based methods, the Retinex theory [12] that assumes that the image can be decomposed into reflectance and illumination has been widely used in low-light image enhancement thanks to its physical explanation. The reflectance component is commonly treated as the enhanced image [13–15]. In Ref. [15], a Retinex-based non-uniform illumination enhancement method that preserves naturalness while enhancing details was proposed. Fu *et al.* [16] proposed a weighted variation model to simultaneously estimate the reflectance and illumination of an input image. Guo *et al.* [17] estimated the coarse illumination map by searching the maximum intensity of each pixel in RGB channels, then refined the coarse illumination map by a structure prior. According to a new Retinex model, Li et *al.* [18] estimated the illumination map by solving an optimization problem. However, there are some limitations of Retinex-based image enhancement due to the gaps between real-world high-quality images and the reflectance component. Thus, the results of Retinex-based methods look unnatural and unrealistic.

The last decade has witnessed an increasing interest in image light enhancement, especially with the emergence of deep learning. Most deep learning-based methods heavily rely on paired data for supervised training [19–27]. Usually, these paired data are generated by synthesis. In Ref. [28], the random Gamma correction was used to simulate low-light images. With synthetic image patches, an autoencoder network was proposed to build the relations between the low-light images and the corresponding normal images [28]. Chen *et al.* [29] obtained a paired low/normal light image dataset by changing the settings of exposure time and ISO. Then, a deep RetinexNet which combines the Retinex theory and deep network was trained on this synthetic dataset. Li *et al.* [30] proposed a deep model for weakly illuminated image enhancement, called LightenNet. The LightenNet takes a weakly illuminated image as input and outputs its illumination map that is subsequently used to obtain the enhanced image based on a Retinex model. To achieve the training data, the authors degraded clear images according to a Retinex model. Ren *et al.* [31] proposed a low-light image enhancement method via a deep hybrid network, which consists of two subnetworks to learn the global content and salient structures of the clear image. Besides, the authors used the training data from MIT-Adobe FiveK dataset [32]. Wang *et al.* [33] proposed an underexposed photo enhancement network by estimating the illumination map, where a new underexposure image dataset including 3,000 underexposed images and the corresponding retouched reference images are collected. The paired training data exist with some limitations, such as limited lighting conditions and scenarios, ideal assumptions and inaccurate formation models, requiring complex expertise in photo editing, to name a few. When coming to real-world conditions, the deep models trained on such synthetic data tend to introduce annoying artifacts and color casts [34].

2.2 *Retinal image enhancement approaches*

Recently, some methods [3, 4, 35–37] specially designed for retinal image enhancement have been proposed. Feng *et al.* [35] proposed a retinal image contrast enhancement method based on Coutourlet transform, which modifies the Contourlet coefficients in corresponding sub-bands by a nonlinear function. The potential reason for the success is that the Contourlet transform has better performance in representing edges than wavelets for its anisotropy and directionality, and is therefore well-suited for multi-scale edge enhancement. Saha *et al.* [36] proposed a two-step method to process the non-uniform or poor illumination of retinal images, including an illumination correction algorithm and a color restoration algorithm. Subjective and objective experiments were carried out to demonstrate that this method does not create false color or artifacts and can improve the performance of automated pathology detection and classification. Zhou *et al.* [3] proposed a retinal image enhancement method based on luminosity and contrast adjustment. Specifically, a luminance gain matrix, which is obtained by Gamma correction of the value channel in the HSV (hue, saturation and value) color space, is used to enhance the R, G and B (red, green and blue) channels, respectively. Contrast is then enhanced in the luminosity channel of CIE Lab color space by CLA-HE [5]. Cheng *et al.* [4] proposed a structure-preserving guided retinal image filter, called SGRIF. The proposed SGRIF consists of a step of global structure transferring and a step of global edge-preserving smoothing.

In addition, such a method reconstructs degraded retinal images based on the attenuation and scattering model. In Ref. [37], a non-uniform illuminated fundus image enhancement method was proposed, which reduces the blurriness of fundus images based on the cataract physical model and enhances the images with an objective of contrast perfection with no preamble of artifacts.

Despite the prolific work, the performance of retinal image enhancement remains largely unsatisfactory. Recently, much attention has been devoted to deep learning strategies for image enhancement and impressive performance is achieved [38–41]; however, deep learning-based retinal image enhancement has not been fully explored. To the best of our knowledge, the proposed NuI-Go is the first deep learning-based non-uniform illumination removal method for retinal image. Besides, the proposed retinal image non-uniform illumination synthesis approach is the first physical model and domain knowledge-based approach, which can be used for deep models' training and also for the evaluations of retinal image enhancement methods.

3. Proposed Method

As illustrated in Fig. 2, our NuI-Go network consists of three recursive non-local encoder–decoder residual blocks (NEDRBs) that share the same structure. Please note that these three NEDRBs have different weights. This method takes a degraded retinal image as input and produces a set of enhanced results in a coarse-to-fine scheme. In what follows, we explain the proposed NuI-Go network and the losses used for network optimization in detail.

3.1 *Our NuI-Go network*

We suggest tackling the retinal image non-uniform illumination problem in multiple states, where a simple yet effective NEDRB is used in each stage. Note that, the performance gains by adding more states ($T>3$) will become marginal, but at the cost of more training and testing time. Thus, we choose the third stage's result as our final result. The step-wise results are provided in Fig. 3. In each state, the input of the NEDRB is the enhanced result of previous NEDRB. In this way, our NuI-Go can progressively enhanced the input and diversify the training data, which reduces the difficulty of training a large model and improves the generalization capability of the network, thus achieving better performance of removing non-uniform illumination. Besides, such a recursive manner has several advantages than previous networks: (1) it is computationally efficient, which is desired for practical applications; (2) it

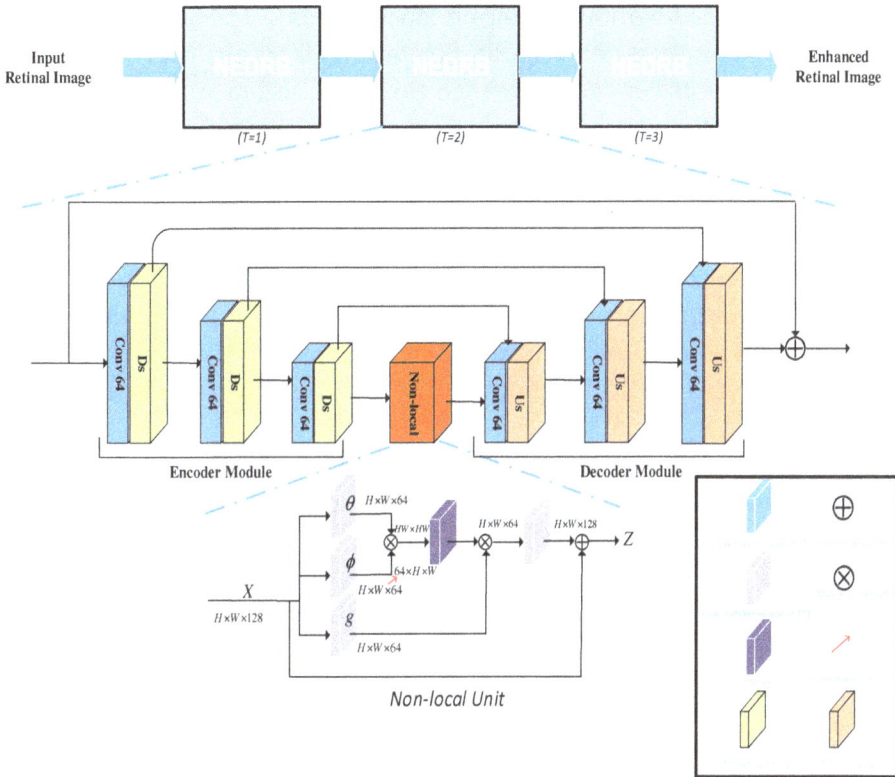

Figure 2: The architecture of the NuI-Go, including three NEDRBs. Each NEDRB contains a feature encoder module, a non-local context module, and a feature decoder module. The detailed structure of non-local unit is presented at the bottom. An input retinal image is fed to the NuI-Go. The enhanced retinal image is achieved after processing by three successive NEDRBs.

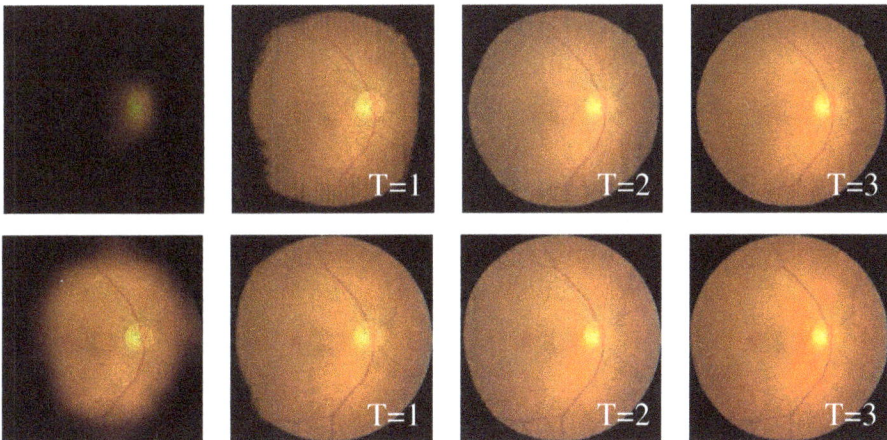

Figure 3: Stage-wise results of our NuI-Go network. From left to right are the input retinal images, the results of stage $T = 1$, 2 and 3, respectively. It is clear that the results become clearer with the increasing of stages.

is easy to be trained without requiring fancy training strategies; (3) it allows to stop the inference at any stage in practical applications (i.e., our method provides a trade-off between inference time and accuracy). We explain the proposed NEDRB and the key non-local unit as follows.

3.1.1 *NEDRB*

NEDRB is an encoder–decoder structure with an embedded non-local unit. It contains a feature encoder module that captures the hierarchical feature representations, a non-local context module that models the context information, and a feature decoder module that recovers the details and spatial dimension.

　　The motivations of the design of our NEDRB are summarized as follows: (1) The encoder–decoder structure can fully exploit hierarchical feature representations and reconstruct the enhanced results. (2) The non-local unit captures long-range dependencies for spatial context modeling. (3) The symmetric skipconnections provide long-range information compensation and reuse. At last, we enforce the NEDRB to learn the difference between the degraded retinal image and its ground truth, which facilitates gradient backpropagation and pixel-wise correction.

3.1.2 *Non-local unit*

While non-local self-similarity has demonstrated good performance in traditional image restoration tasks, the effectiveness of non-local features is little studied in deep learning-based image enhancement. In addition, the retinal image shares structured texture and similar color, which provides a strong prior for correcting the non-uniform illumination regions by the context.

　　Therefore, instead of employing the multi-scale convolution to extract local context information used in traditional encoder–decoder networks, we incorporate the non-local unit [42] into our encoder–decoder structure, which enlarges the receptive field to the entire image. Our intuition is to incorporate non-local features into local encoder–decoder features for building richer feature representations that combine both non-local and local information. As shown in Fig. 2(right), the non-local unit computes the response at a position as weighted sum of the features at all spatial positions, where "H × W × 64" denotes height × width × 64 channels of feature maps X (reshaping is performed when noted). Specifically, it can be implemented by simple convolution, softmax, etc. (for embedded Gaussian version's non-local unit), where the unknown parameters can be adaptively learned. Embracing the benefit of non-local context information that captures the long-range feature correlation, our network can effectively focus on the nonuniform illumination regions and predict the potential color and details by the long-range dependencies (context modeling).

In our NuI-Go, the down-sampling layer is implemented by $2 \times$ max pooling while the up-sampling layer is implemented by transposed convolution with the kernel size of 3×3 and stride 2. All convolutional layers have the kernel size of 3×3 and output 64 feature maps, except for the convolutional layers before the results, where each convolutional layer has the kernel size of 1×1 and outputs 3 residual maps. We pad zeros before each convolutional layer to ensure the results have the same size as the input image. For the embedded Gaussian version's non-local unit, we follow the implementation of [42]. For the computational efficiency, each convolutional layer in the non-local unit has the kernel size of 1×1.

3.2 *Loss function*

The perceptual loss has demonstrated much better visual performance than the per-pixel losses. However, it usually fails to achieve good quantitative performance [43]. To remit this issue, we incorporate the l_1 loss into our network optimization, which ensures the results have good quantitative scores. We only add the l_1 loss to stage $T = 3$ while the stages $T = 1$ and $T = 2$ are optimized by only using the perceptual loss. Such an optimization scheme saves the training time.

We employ the perceptual loss based on the VGG-19 network ϕ [44] pretrained on the ImageNet dataset [45]. Let $\phi_j(x)$ be the jth convolutional layer. We measure the distance between the feature representations of the enhanced retinal image $\hat{R}$ and the ground truth image R as: $L^3_{per} = \Sigma_{i=1}^{M} |\phi_j(\hat{R}_i) - \phi_j(R_i)|$, where M is the number of each batch in the training procedure. Similar to the L^3_{per} for $T = 3$, the losses of $T = 1$ and $T = 2$ are denoted as L^1_{per} and L^2_{per}. l_1 loss measures the difference between $\hat{R}$ and R as: $L^3_{l_1} = \Sigma_{i=1}^{M} |\hat{R} - R|$.

The total loss is the linear aggregation of the multi-term loss:

$$L_{total} = L^1_{per} + L^2_{per} + L^3_{per} + \lambda L^3_{l_1},\tag{1}$$

where λ is the weight of $L^3_{l_1}$. We pick up the weight according to our training data.

4. Experiments

We first introduce the synthesis of non-uniform illuminated retinal images and the details of training and implementation of our method. Then, extensive experiments on the synthetic and real-world retinal images are conducted. At last, the practical application of our method is performed.

4.1 *Non-uniform illumination retinal image synthesis*

In the real world, it is almost impossible to obtain sufficient pairs of retinal images with non-uniform illumination and the corresponding clear retinal image. To train the proposed NuI-Go, we approximate the non-uniform illumination degradation model of retinal images by combining the domain knowledge of retinal images with the human-lens attenuation and scattering model. Inspired by the human-lens attenuation and scattering model [46], the retinal image degradation can be reduced to a Retinex model [47] when only the effects of illumination are considered:

$$I_c(i, j) = R_c(I, j)L(i, j), c \in \{r, g, b\}, \tag{2}$$

where (i, j) denotes the pixel coordinates; I is the observed retinal image; R is the reflectance of the retinal image (i.e., the desired result); and L is the flash illumination of the fundus camera. This simplified model can be described as one where the observed retinal image I is decomposed into the product of the reflectance R and the illumination mask L. Thus, in this chapter, we expect to estimate the reflectance of retinal image R from the observed non-uniform illumination retinal image I.

According to Eq. (2), given a clear retinal image R, a non-uniform illumination mask L is needed to generate the non-uniform illumination retinal image I. We first transform the retinal image to the Lab color space and search the well-lit regions in the lightness channel by different thresholds. When the pixel values are larger than the current threshold, they are set to 1. Or these pixels serve as the non-uniform illumination pixels by Gamma correction with random γ values ranging from 0.1 to 0.5. The purpose is to preserve the original lightness of the well-lit regions and meanwhile make the low-light regions darker as the non-uniform illumination mask. The potential reason for such operations is that some regions in the retinal images are seldom affected by non-uniform or poor illumination, such as optic cup region. In this chapter, we use five thresholds [0.1:0.1:0.5] empirically.

After that, a coarse non-uniform illumination mask is generated. Considering the retinal images usually have smoothed and region-wise non-uniform illumination, we smooth the coarse non-uniform illumination mask by simple 8 × down-sampling and up-sampling operations. Here, other smooth approaches also can be used.

At last, a retinal image with non-uniform illumination can be synthesized as $I_c = R_c(i,j)L(i,j)$. Figure 4 presents a visual example of the synthetic image, where the synthetic retinal image by our proposed approach has realistic nonuniform illumination.

4.2 *Training and implementation details*

We collect 2500 well-lit retinal images from the publicly available training set of Kaggle's Diabetic Retinopathy Detection Challenge (KDRDC)[1] that

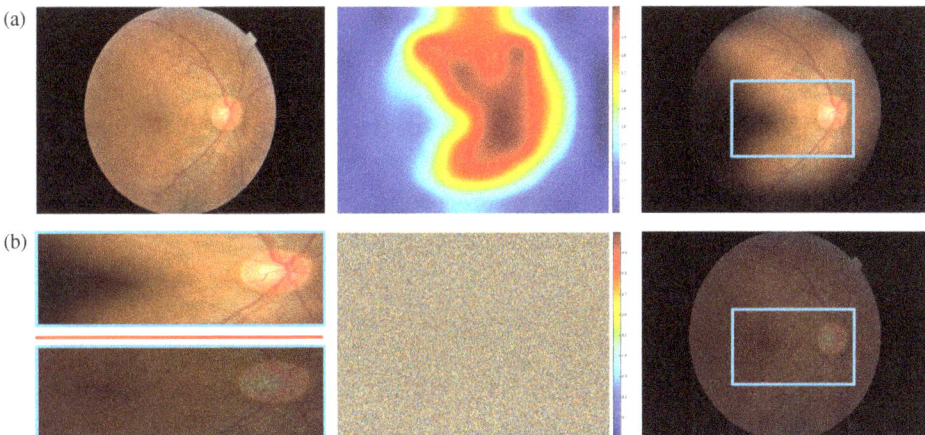

Figure 4: A visual example of the non-uniform illumination synthesis via different approaches. (a) Our proposed physical model and domain knowledge-based nonuniform synthesis approach. From left to right are the well-lit retinal images, the smoothed illumination mask and the synthetic retinal image. Different colors represent different values in the illumination masks. (b) The traditional non-uniform illumination synthesis approach based on random noise. From left to right are the amplified details, our details (up) and the details of traditional approach (down), the random noise map for simulating non-uniform illumination, and the synthetic retinal image by random noise. It is clear that the traditional synthesis approach neglects the characteristics of retinal images, and this results in unperfect synthetic retinal image. In contrast, our approach generates realistic retinal image with non-uniform illumination, especially the optic cup regions.

---

[1] https://www.kaggle.com/c/diabetic-retinopathy-//detection/.

provides retinal images taken under a variety of imaging conditions. Following the approach mentioned before, we generate 12,500 non-uniform illumination retinal images and resize them to 256×256 due to our limited memory. We randomly split these synthetic images into two parts: 10,000 images for training and the rest as a testing dataset (denoted as Test A).

We implement our network with TensorFlow on a PC with an Intel(R) i7 6700 CPU, 32GB RAM and an NVIDIA GeForce GTX 1080Ti GPU. During training, a batch size of 8 is applied. The filter weights are initialized by Xavier. We use ADAM for network optimization, and fix the learning rate to $1e^{-4}$. We compute perceptual loss at layer relu5_4 of the VGG-19 network. The weight λ is set to 100 empirically. The runtime of the NEDRB and the NuI-Go is 0.088 s and 0.34 s for an image with a size of 256×256, which is fast for practical applications.

4.3 Evaluations

In this section, we conduct experiments on retinal image enhancement and retinal vessel segmentation. The compared methods include recent retinal image enhancement method (SGF [4]), traditional image enhancement method (CLAHE [5]) and state-of-the-art low-light image enhancement method (LIME [17]). The results are produced by using publicly available source codes with recommended parameter settings.

4.3.1 Image enhancement

We compare the image enhancement performance of different methods on synthetic retinal images (**Test A**) and real-world retinal images (i.e., the testing set of Kaggle's Diabetic Retinopathy Detection Challenge (KDRDC), including 53,576 images taken under a variety of imaging conditions, denoted as **Test B**. The visual comparisons on synthetic and real-world retinal images are presented in Fig. 5.

As shown in Fig. 5, the methods of CLAHE [5] and SGF [4] have little effect on the non-uniform illumination regions. LIME [17] improves the brightness of retinal images; however, there still remain some poor illumination regions in the results due to the limited generalization capability. Such results may lead to a clinical misdiagnosis. In contrast, our NuI-Go can effectively correct the non-uniform illumination and preserve the details and natural look of retinal images. More visual results on synthetic and real retinal images are provided in Figs. 6 and 7, respectively.

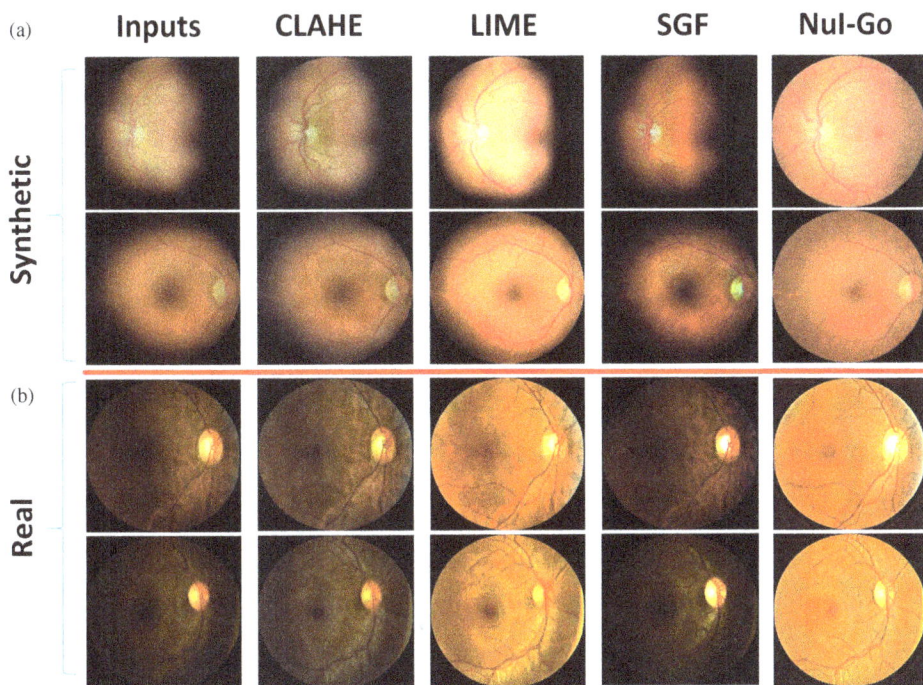

Figure 5: Visual comparisons. (a) The results of synthetic images from **Test A**. (b) The results of real-world images from **Test B**. From left to right are the inputs, the results of CLAHE [5], LIME [17], SGF [4] and our NuI-Go.

We further use the full-reference image quality assessment metrics PSNR and SSIM [48] for quantitative evaluations on **Test A** since the corresponding ground truth images are available. A higher PSNR value indicates the similarity in terms of pixel-wise values. A higher SSIM value indicates a result that is closer to the ground truth in terms of structural properties. The average quantitative scores on **Test** A are reported in Table 1.

As presented in Table 1, our NuI-Go achieves the best performance across all metrics. In addition, our method has overwhelming advantages than the compared methods (CLAHE [5], LIME [17], SGF [4]) in terms of the average quantitative scores on the synthetic dataset. For example, the average PSNR value of our method is 32.2668 while the second best value (the result of LIME [17]) is only 16.3671. For the average SSIM value, it has a similar tendency to that of PSNR. Such a result indicates the effectiveness of the proposed non-uniform illumination removal method.

For **Test B**, we employ a commonly used non-reference image quality assessment algorithm NIQE [49] to assess the perceptual quality. A smaller

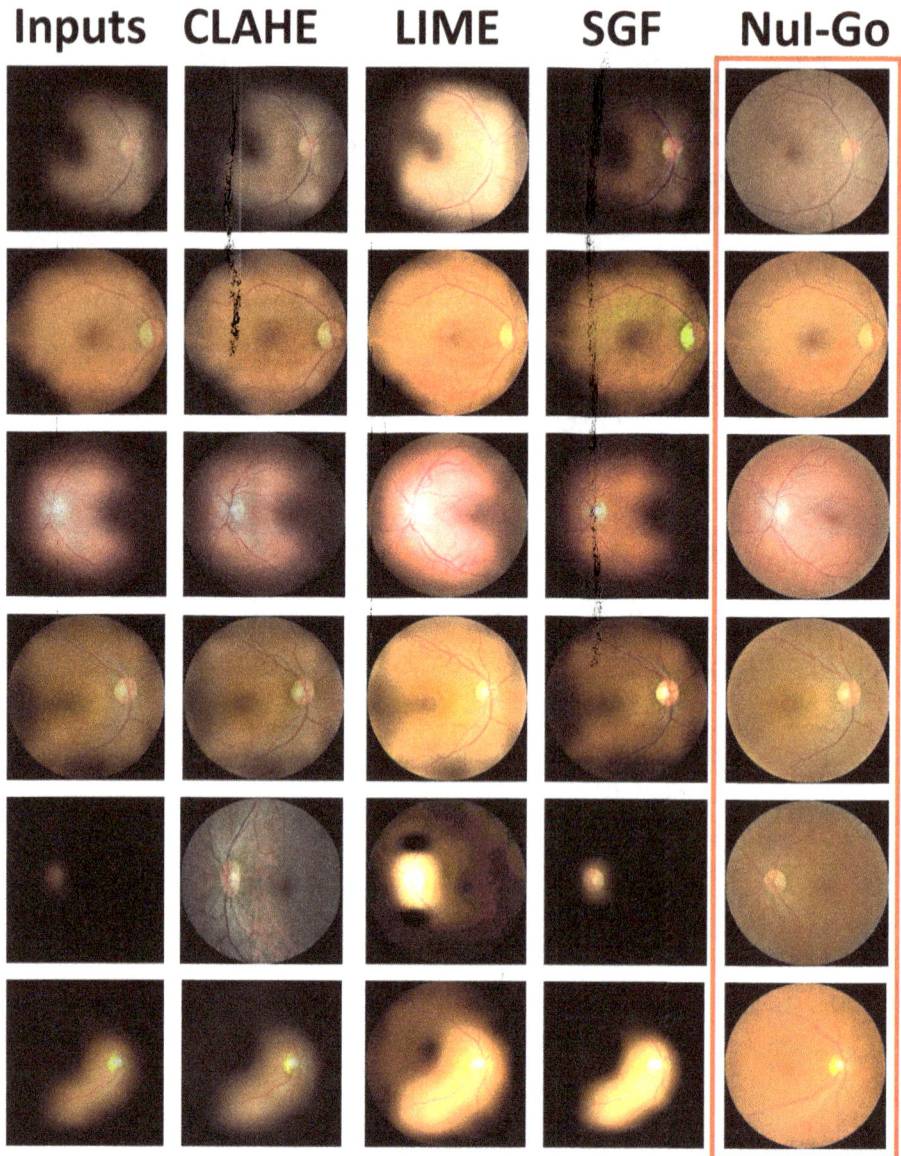

Figure 6: Visual comparisons on synthetic retinal images. From left to right are the inputs from **Test A**, the results of CLAHE [5], LIME [17], SGF [4] and our NuI-Go. Red box indicates our results.

NIQE score indicates better perceptual quality. Besides, we also conduct a user study to provide realistic feedback about subjective quality. We first randomly select 50 images from **Test B**, and then invite 10 participants with image enhancement expertise to rank the results of different methods from 1 to 5 where 1 is the worst quality and 5 is the best quality. This user study is

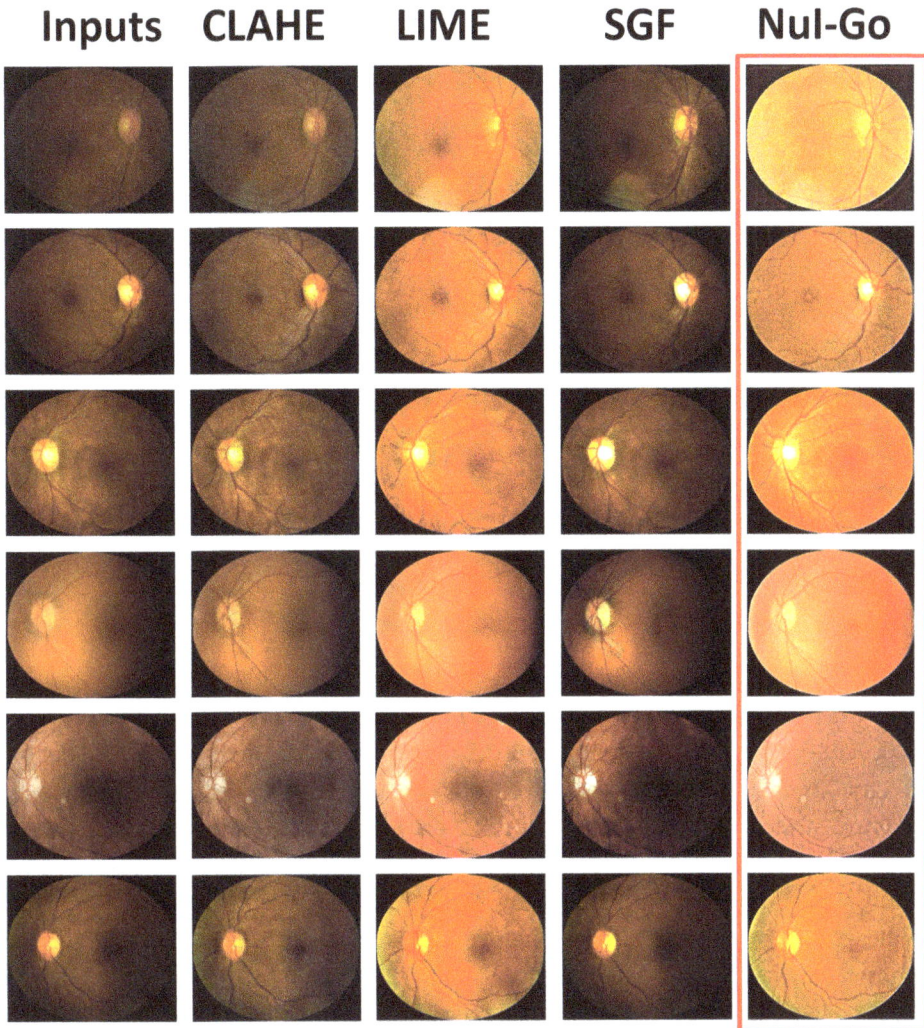

Figure 7: Visual comparisons on real retinal images. From left to right are the inputs from **Test B**, the results of CLAHE [5], LIME [17], SGF [4] and our NuI-Go. Red box indicates our results.

Table 1: Quantitative evaluations on **Test A** in terms of PSNR (dB) and SSIM.

Metrics	Inputs	CLAHE [5]	LIME [17]	SGF [4]	NuI-Go
PSNR (dB)	14.6994	16.3671	14.3072	12.2262	**32.2668**
SSIM	0.6450	0.5065	0.6902	0.5558	**0.8281**

Note: For each case, the best result is in bold.

Table 2. The average non-reference NIQE scores and user study scores on Test B.

Metrics	Inputs	CLAHE [5]	LIME [17]	SGF [4]	NuI-Go
NIQE	7.8087	6.9539	7.5331	8.8697	**6.9110**
User Study	2.1452	2.5708	3.2416	2.4122	**3.8558**

Note: For each case, the best result is in bold.

repeated five times. The average values in terms of NIQE [49] and user study are shown in Table 2.

In Table 2, our method achieves the best performance in terms of non-reference NIQE [49] and subjective user study. Such results demonstrate our method can generalize to real retinal images and effectively improve the visual quality of degraded retinal images. Besides, it also indicates the reasonability of our non-uniform illumination synthesis approach. Though our model is trained on synthetic training data, it is also useful for real retinal images.

4.3.2 *Retinal vessel segmentation*

To evaluate the performance of our method in practical applications, we use recent CE-Net [50] that is trained on DRIVE dataset [51] (20 images for training and 20 images for testing) as a basic retinal vessel segmentation model. We found that the existing vessel segmentation datasets do not contain the non-uniform or poor illumination retinal images. However, the non-uniform illumination retinal images are ubiquitous in the real world. Therefore, we degrade the testing set of DRIVE dataset [51] and the STARE dataset [52] by the approach mentioned before. The testing data did not appear in the training set. The first manual annotation in these two datasets is used as the ground truth for performance evaluation. We enhance these testing images by different methods and feed the results to the CE-NET [50]. At last, we measure the average accuracy (Acc) and the area under receiver operation characteristic curve (AUC) by the formula mentioned in [50]. First, we present several sample results before and after processing by our method and the corresponding segmentation results in Fig. 8. Secondly, we report the Acc and AUC scores in Table 3.

As shown in Fig. 8, our method can effectively unveil the vessel under dark regions. As a result, the vessel of retinal images processed by our method

Table 3. Quantitative evaluations on DRIVE dataset/STARE dataset in terms of Acc and AUC.

Metrics	Inputs	CLAHE [5]	LIME [17]	SGF [4]	NuI-Go
Acc	0.9508/0.9369	0.9497/0.9331	0.9518/0.9384	0.9382/0.9322	**0.9543/0.9417**
AUC	0.9329/0.7976	0.9570/0.8509	0.9746/0.9076	0.9074/0.8014	**0.9756/0.9213**

Note: For each case, the best result is in bold.

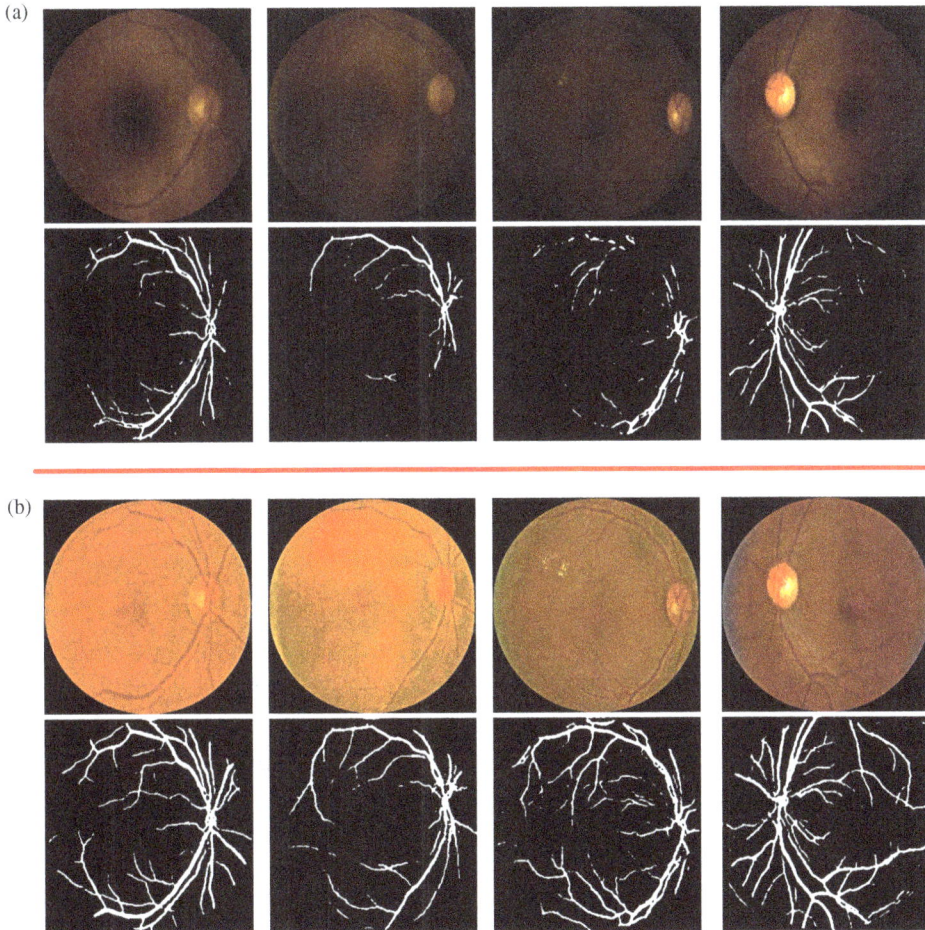

Figure 8: Visual retinal vessel segmentation results. (a) The input images from the degraded DRIVE dataset [51] and STARE dataset [52] and the corresponding segmentation results. (b) Our enhanced results and the corresponding segmentation results. The segmentation results are produced by the CE-Net [50].

can be easily segmented by the CE-Net [50] that is trained by using well-lit retinal images. Thus, despite the retinal image segmentation algorithms not considering the non-uniform illumination issue in their designs and training procedures, our method could be used as an effective pre-processing step for further improving their performance.

As presented in Table 3, the results enhanced by our NuI-Go achieve better segmentation performance on the DRIVE [51] and STARE [52] datasets. Compared with the degraded images (inputs), the Acc increases from 0.9508/0.9369 to 0.9543/0.9417 and the AUC increases from 0.9329/0.7976 to 0.9756/0.9213, respectively. In addition, it is interesting that some enhancement algorithms decrease the performance of retinal vessel segmentation, which is undesired in practical applications. To demonstrate the robustness of our method on real retinal images, we provide several sample segmentation results for real retinal images in Fig. 9. It is obvious that after enhancing by our method, some retinal vessels hidden in the dark are recovered and thus can be accurately segmented by the CE-Net [50]. The results in Fig. 9 further demonstrate the effectiveness and robustness of our method.

5. Conclusion

In this chapter, we mainly introduce the enhancement of retinal images and propose a deep learning-based retinal image non-uniform illumination removal method. In the proposed method, the non-uniform illumination is progressively removed via recursive residual learning, while the details and color of retinal images are recovered by the combination of local and non-local features. As a key component, we introduce non-local unit into the encoder–decoder network, which models the context effectively. Besides, we propose a non-uniform illumination of retinal image synthesis approach, which is useful for the supervised learning of deep networks. Experiments on synthetic and real-world retinal images show that the proposed method outperforms existing image enhancement methods. Additionally, experimental results suggest that our method can effectively improve retinal vessel segmentation. With the proposed non-unifom illumination removal method, the diagnosis of ophthalmologists and the performance of computer-aided analysis of retinal diseases can be improved. Morveover, the proposed rertinal image synthesis approach can call for the development of deep learning-based computer-aided diagnosis.

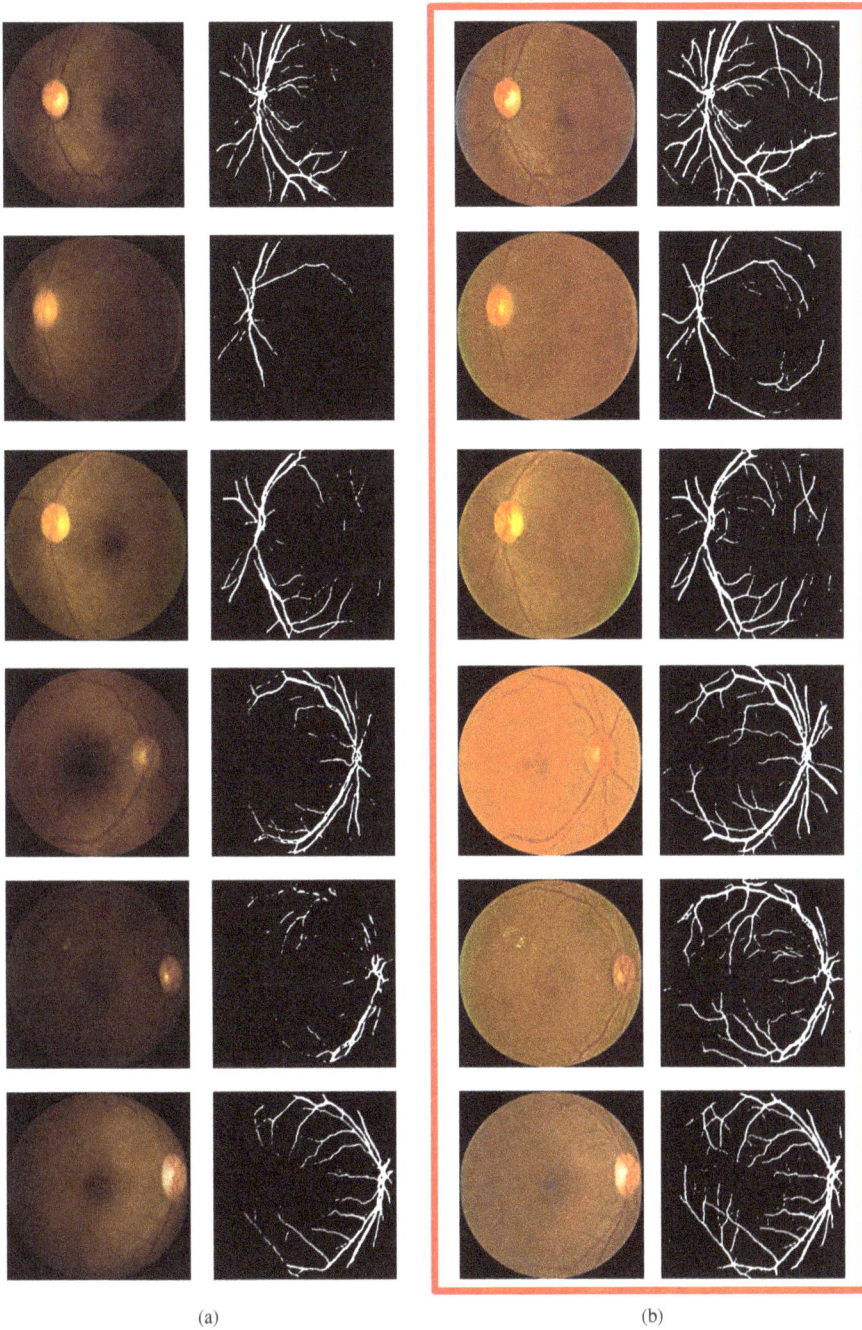

Figure 9: Visual retinal vessel segmentation results of real retinal images. (a) The input images from **Test B** and the corresponding segmentation results. (b) Our enhanced results and the corresponding segmentation results. The segmentation results are produced by the CE-Net [50]. Red box indicates our results.

References

1. M. D. Abramoff, M. K. Garvin, and M. Sonka, "Retinal imaging and image analysis," *IEEE Rev. Biomed. Eng.*, vol. 3, pp. 169–208, 2010.
2. A. Hani and H. A. Nugroho, "Retinal vasculature enhancement using independent component analysis," *J. Biomed. Sci. Eng.*, vol. 2, no. 7, pp. 543–549, 2009.
3. M. Zhou, K. Jin, S. Wang, J. Ye, and D. Qiao, "Color retinal image enhancement based on luminosity and contrast adjustment," *IEEE Trans. Biomed. Eng.*, vol. 65, no. 3, pp. 521–527, 2018.
4. J. Cheng, Z. Li, Z. Gu, H. Fu, D. Wong, and J. Liu, "Structure-preserving guided retinal image filtering and its application for optic disc analysis," *IEEE Trans. Med. Imag.*, vol. 37, no. 11, pp. 2536–2546, 2018.
5. K. Zuiderveld, *Contrast Limited Adaptive Histogram Equalization*, Academic Press, San Diego, 1995.
6. D. Coltuc, P. Bolon, and J. Chassery, "Exact histogram specification," *IEEE Trans. Image Process.*, vol. 15, no. 5, pp. 1143–1152, 2006.
7. H. Ibrahim and N. Kong, "Brightness preserving dynamic histogram equalization for image contrast enhancement," *IEEE Trans. Consum. Electron.*, vol. 53, no. 4, pp. 1752–1758, 2007.
8. J. Stark, "Adaptive image contrast enhancement using generalizations of histogram equalization," *IEEE Trans. Image Process.*, vol. 9, no. 5, pp. 889–896, 2000.
9. T. Celik and T. Tjahjadi, "Contextual and variational contrast enhancememt," *IEEE Trans. Image Process.*, vol. 20, no. 12, pp. 3431–3441, 2001.
10. C. Lee, C. Lee, and C. Kim, "Contrast enhancement based on layered difference representation of 2d histograms," *IEEE Trans. Image Process.*, vol. 22, no. 12, pp. 5372–5384, 2013.
11. C. Li, J. Guo, R. Cong, Y. Pang, and B. Wang, "Underwater image enhancement by dehazing with minimum information loss and histogram distribution prior," *IEEE Trans. Image Process.*, vol. 25, no. 12, pp. 5664–5677, 2016.GF
12. E. Lee, "The retinex theory of color vision," *Scientific American*, vol. 237, no. 6, pp. 108–128, 1977.
13. Y. Li and M. Brown, "Single image layer separation using relative smoothness," in *Proc. IEEE Int. Conf. Comput. Vis. Pattern Rec. (CVPR)*, 2014, pp. 2752–2759.
14. C. Rother, M. Kiefei, L. Zhang, B. Scholkopf, and P. Gehler, "Recovering intrinsic images with a global sparsity prior on reflectance," in *Advances in Neural Inf. Process. Sys.*, 2011, pp. 765–773,.
15. S. Wang, J. Zheng, H. Hu, and B. Li, "Naturalness preserved enhancement algorithm for non-uniform illumination images," *IEEE Trans. Image Process.*, vol. 22, no. 9, pp. 3538–3548, 2013.
16. X. Fu, D. Zeng, Y. Huang, X. Zhang, and X. Ding, "A weighted variational model for simultaneous reflectance and illumination estimation," in *Proc. IEEE Int. Conf. Comput. Vis. Pattern Rec. (CVPR)*, 2016, pp. 2782–2790.

17. X. Guo, Y. Li, and H. Ling, "LIME: Low-light image enhancement via illumination map estimation," *IEEE Trans. Image Process.*, vol. 26, no. 2, pp. 982–993, 2017.

18. M. Li, J. Liu, W. Yang, X. Sun, and Z. Guo, "Structure-revealing low-light image enhancement via robust retinex model," *IEEE Trans. Image Process.*, vol. 27, no. 6, pp. 2828–2841, 2018.

19. C. Li, C. Guo, J. Guo, P. Han, H. Fu, and R. Cong, "PDR-Net: Perception-inspired single image dehazing network with refinement," *IEEE Trans. Image Multimedia*, vol. 22, no. 3, pp. 704–716, 2019.

20. H. Fu, J. Cheng, Y. Xu, D. Wong, J. Liu, and X. Cao, "Joint optica disc and cup segmentation based on multi-lable deep network and polar transformation," *IEEE Trans. Med. Imag.*, vol. 37, no. 7, pp. 1597–1605, 2018.

21. H. Fu, J. Cheng, Y. Xu, C. Zhang, D. Wong, J. Liu, and X. Cao, "Disc-aware ensemble network for glaucoma screen from fundus image," *IEEE Trans. Med. Imag.*, vol. 37, no. 11, pp. 2493–2501, 2018.

22. C. Li, R. Cong, J. Hou, S. Zhang, Y. Qian, and S. Kwong, "Nested network with two-stream pyramid for salient object detection in optical remote sensing images," *IEEE Trans. Geosci. Remote Sens.*, vol. 57, no. 11, pp. 9156–9166, 2019.

23. H. Fu, Y. Xu, S. Lin, D. Wong, B. Mani, M. Mahesh, T. Aung, and J. Liu, "Angle-closure detection in anterior segment OCT based on multi-level deep network," *IEEE Trans. Cybernet.*, 2019.

24. C. Li, R. Cong, S. Kwong, J. Hou, H. Fu, G. Zhu, D. Zhang, and Q. Huang, "ASIF-Net: Attention steered interweave fusion network for RGB-D salient object detection," *IEEE Trans. Cybernetics*, 2020.

25. C. Guo, C. Li, J. Guo, R. Cong, H. Fu, and P. Han, "Hierarchical features driven residual learning for depth map super-resolution," *IEEE Trans. Image Process.*, vol. 28, no. 5, pp. 2425–2557, 2019.

26. C. Li, S. Anwar, and F. Porikli, "Underwater scene prior inspired deep underwater image and video enhancement," *Pattern Rec.*, vol. 98, pp. 1–11, 2020.

27. C. Li, J. Guo, F. Porikli, H. Fu, and Y. Pang, "A cascaded convolutional neural network for single image dehazing," *IEEE Access*, vol. 6, pp. 24877–24887, 2018.

28. K. Lore, A. Akintayo, and S. Sarkar, "LLNet: A deep autoencoder approach to natural low-light image enhancement," *Pattern Rec.*, vol. 61, pp. 650–662, 2017.

29. W. Chen, W. Wang, W. Yang, and J. Liu, "Deep retinex decomposition for low-light enhancement," in *Proc. British Mach. Vis. Conf. (CVPR)*, 2018, pp. 1–12.

30. C. Li, J. Guo, F. Porikli, and Y. Pang, "LightenNet: A convolutional neural network for weakly illuminated image enhancement," *Pattern Rec. Lett.*, vol. 104, pp. 15–22, 2018.

31. W. Ren, S. Liu, L. Ma, Q. Xu, X. Xu, X. Cao, J. Du, and M. Yang, "Low-light image enhancement via a deep hybrid network," *IEEE Trans. Image Process.*, vol. 28, no. 9, pp. 4364–4375, 2019.

32. V. Bychkovsky, S. Pairs, E. Chan, and F. Durand, "Learning photographic global tonal adjustment with a database of input/output image pairs," in *Proc. IEEE Int. Conf. Comput. Vis. Pattern Rec. (CVPR)*, 2011, pp. 97–104.

33. R. Wang, Q. Zhang, C. Fu, X. Shen, W. Zheng, and J. Jia, "Underexposed photo enhancement using deep illumination estimation," in *Proc. IEEE Int. Conf. Comput. Vis. Pattern Rec. (CVPR)*, 2019, pp. 6849–6857,.

34. C. Guo, C. Li, J. Guo, C. Loy, J. Hou, S. Kwong, and R. Cong, "Zero-reference deep curve estimation for low-light image enhancement," in *Proc. IEEE Int. Conf. Comput. Vis. Pattern Rec. (CVPR)*, 2020.

35. P. Feng, Y. Pan, B. Wei, W. Jin, and D. Mi, "Enhacing retinal image by the Contourlet transform," *Pattern Rec. Lett.*, vol. 28, no. 4, pp. 516–522, 2007.

36. S. Saha, A. Fletcher, D. Xiao, and Y. Kanagasingam, "A novel method for authomated correction of non-uniform/poor illumination of retinal images without creating false artifacts," *J. Vis. Commun. Image Rep.*, vol. 51, pp. 95–103, 2018.

37. A. Mitra, S. Roy, S. Soy, and S. Setua, "Enhancement and restoration of non-uniform illuminated Fundus image of retina obtained through thin layer of cataract," *Comput. Methods Programs in Biomedicine*, vol. 156, pp. 169–178, 2018.

38. W. Wu, K. Cao, C. Li, C. Qiao, and C. Loy, "Transgaga: Geometry-aware unsupervised image-to-image translation," in *Proc. IEEE Int. Conf. Comput. Vis. Pattern Rec. (CVPR)*, 2019, pp. 8012–8021.

39. Y. Deng, C. Loy, and X. Tang, "Aesthetic-driven image enhancement by adversarial learning," in *Proc. Int. Conf. Multimedia (ACMMM)*, 2018, pp. 870–878.

40. S. Gu, W. Zuo, S. Guo, Y. Chen, C. Chen, and L. Zhang, "Learning dynamic guidance for depth image enhancement," in *Proc. IEEE Int. Conf. Comput. Vis. Pattern Rec. (CVPR)*, 2017, pp. 3769–3778.

41. C. Dong, C. Loy, K. He, and X. Tang, "Image super-resolution using deep convolutional networks," *IEEE Trans. Pattern Anal. Mach. Intell.*, vol. 38, no. 2, pp. 295–307, 2015.

42. X. Wang, R. Girshick, A. Gupta, and K. He, "Non-local neural networks," in *Proc. IEEE Int. Conf. Comput. Vis. Pattern Rec. (CVPR)*, 2018, pp. 7794–7803.

43. J. Johnson, A. Alahi, and L. Fei-Fei, "Perceptual losses for real-time style transfer and super-resolution," in *Proc. Eur. Conf. Comput. Vis. (ECCV)*, 2016, pp. 694–711.

44. K. Simonyan and A. Zisserman, "Very deep convolutional networks for large-scale image recognition," in *Proc. Int. Conf. Learn. Representations (ICLR)*, 2015.

45. J. Deng, W. Dong, R. Socher, L. Li, and L. Fei-Fei, "ImageNet: A large-scale hierarchical image database," in *Proc. IEEE Int. Conf. Comput. Vis. Pattern Rec. (CVPR)*, 2009, pp. 248–255.

46. E. Peli and T. Peli, "Restoration of retinal images obtained through cataracts," *IEEE Trans. Med. Imag.*, vol. 8, no. 4, pp. 401–406, 1989.

47. E. Land, "An alternative technique for the computation of the designator in the Retinex theory of color vision," *Nati. Acad. Sci.*, vol. 83, no. 10, pp. 3078–3080, 1986.

48. Z. Wang, A. Bovik, H. Sherikh, and E. Simoncelli, "Image quality assessment: From error visibility to structural similarity," *IEEE Trans. Image Process.*, vol. 13, no. 8, pp. 600–612, 2004.

49. A. Mittal and R. Soundararajan, "Making a completely blind image quality analyzer," *IEEE Signal Process. Lett.*, vol. 20, no. 3, pp. 209–212, 2013.

50. Z. Gu, J. Cheng, H. Fu, K. Zhou, H. Hao, Y. Zhao, T. Zhang, S. Gao, and J. Liu, "CE-Net: Contex encoder network for 2D medical image segmentation," *IEEE Trans. Med. Imag.*, vol. 38, no. 10, pp. 2281–2292, 2019.

51. J. Staal, M. Abramoff, M. Niemeijer, M. Viergever, and B. Ginneken, "Ridge-based vessel segmentation in color images of the retina," *IEEE Trans. Med. Imag.*, vol. 23, no. 4, pp. 501–509, 2004.

52. A. Hoover, V. Kouznetsova, and M. Goldbaum, "Locating blood vessels in retinal images by piece-wise threshold probing of a matched filter response," in *Proc. American Med. Inform. Asso. Symp.*, 1998, pp. 931–935.

Chapter 11

A Comparative Analysis of Efficient CNN-based Brain Tumor Classification Models

Tanveer Hussain*,‡, Amin Ullah*,§, Umair Haroon*,¶, Khan Muhammad†,‖ and Sung Wook Baik*,**

*Intelligent Media Laboratory, Department of Software,
Sejong University, Seoul 143-747, Republic of Korea
†Department of Software, Sejong University, Seoul 143–747,
Republic of Korea
‡tanveerkhattak3797@gmail.com
§aminullah@ieee.org
¶umair3797@gmail.com
‖khan.muhammad@ieee.org
**sbaik@sejong.ac.kr

Abstract

There have been a lot of research works from the past decades and currently going on in the field of medical imaging. Brain tumor classification into various types is an interesting sub-domain of medical imaging, where many researchers are enthusiastic to serve humanity through computer- aided diagnosis (CAD). With the development of deep neural networks (DNNs), there has been a revolutionary change in the accuracy of classification and regression problems, particularly in the medical domain. Inspired by this fact, the proposed technique is based on convolutional neural network (CNN) classification strategy, which efficiently and accurately is able to classify a brain medical resonance imaging (MRI) into three classes (i.e., meningioma, glioma and pituitary tumor). In this chapter, we fine-tune two CNN models (i.e., SqueezeNet and GoogLeNet) for our specific problem of tumor classification into various grades. SqueezeNet model poses AlexNet-level accuracy with 50 times fewer parameters and fast computing capability, while GoogLeNet is one of the first architectures which introduced the inception module that helped significantly in dropping off the number of trainable weights in a network. The main target of the proposed method is to achieve higher level of accuracy alongside efficiency. The proposed method has been trained on a dataset with 3064 slices from 233 patients, having 708, 1426, 930 cases of meningioma, glioma and pituitary tumors, respectively.

Keywords: brain tumor, classification, tumor diagnosis, deep learning, convolutional neural networks

1. Introduction

In the present era of technology, automatic analysis of medical images for different diagnosis is increasing day by day. Tumor is a deadly disease and identification of tumors in the very early stages has much better chance of being cured properly as compared to late identification. Various handcrafted features-based tumor segmentation techniques are proposed to identify and categorize the tumor according to its different stages. Magnetic resonance imaging (MRI) data has been widely used for making different algorithms for the classification or segmentation of tumors. Thus, MRI data plays a vital role in diagnosing human brain tumors. The conventional methods for detection and classification observe MRI data manually through human vision, perception and inspection by a specialist. Thus, it depends upon the experience of the radiologist who observes the MRI data. It is not obvious or guaranteed 100% that the decision made by a specialist is correct. Further, specialist-based identification or classification proves to be impractical when dealing with huge amounts of data, which is also non-reproducible. Thus, this is the motivation for using computer-assisted tools to address such problems.

Brain tumor classification is broadly divided into two types, i.e., normal, abnormal and further classification of abnormal classes. In the proposed technique, brain MRI is classified on the basis of pathological types, which is very hard and challenging as compared to simple binary classification of normal or abnormal classes. Researchers suggested several traditional machine learning and convolutional nueral network (CNN)-based methods to segment and recognize brain tumor. Classification of deep learning-based brain MRI includes two main steps: features extraction and tumor classification. The accuracy of the classifier depends upon the features extraction step as if the extracted features from MRI are representative and capture the tumor patterns, then classification step outputs the accurate results. For instance, in mainstream existing literature, several overlapping feature extraction methods are used to describe the brain tumor including texture features [1, 2], intensity values, local binary patterns, first-order statistics [3], gray level co-occurrence matrix (GLCM) [4], wavelet transform [5] and Gabor filters [6]. In a research work [6], researchers suggested novel 3D voxel classification-based brain tumor segmentation technique by making use of Gabor features and AdaBoost as a classifier. Selvaraj *et al.* [7] proposed the support vector machine (SVM)-based abnormal and normal slices of MRI classifier which contains first- and second-order statistics computation. Javed *et al.* [1] treated the problem of multi-class classification with fuzzy weights, low-level texture features and used SVM for classification. John *et al.* [5] used GLCM and discrete wavelet transform for the detection and classification of tumors and also claimed that it has better accuracy compared to spatial domain. Low-level features and intensity values are able to represent texture effectively, but the recent studies prove that bag of words representation is discriminative and much more robust while dealing with medical image retrieval and classification of MRI data [8–11]. Bag of words technique is basically for the text retrieval domain, but it can be accurately adapted to pictorial data analysis techniques. Important concept of Patch-based bag of words demonstrations is actually a generic form of intensity histograms. Intensity values or pixels are just swapped with image patches, which is the salient difference.

The low-level features and bag of words-based detection and classification techniques are not sophisticated enough to give satisfying results while dealing with such delicate type of data. Thus, there is a need for satisfying results based upon which diagnosis or treatment should be suggested by the doctors. The problem of classification achieves higher and satisfying level of accuracy while dealing with CNNs. Thus, motivated by the effective usage of CNNs, our proposed technique makes use of deep neural network (DNNs) in the field of medical imaging and classifies brain tumor with precise accuracy. Thus, making the field of classification fully automatic whereby the specialist can easily decide whether the case is in danger or in the safe zone. Besides the high level of accuracy, the deep neural architecture followed by the proposed Softmax classifier is highly compressed, thus resulting in a very low format and size model. This output model is then usable in even smartphones and other resource-constrained devices like Raspberry Pi, etc.

2. Related Work

An approach mainly used by radiologists to analyze the internal composition of the human body is based on MRI data. The details about human organs, their structure, and information of tiny tissues can be studied deeply through MRI technology. It is very helpful for the treatment and identification of various dangerous diseases. It is being widely used in brain tumor segmentation and classification into multi-grades or normal and abnormal classes. Many researchers have contributed to this area of research and there are abundant ongoing researches to assist radiologists. A considerable amount of techniques is offered by the researchers for brain tumor segmentation and classification, as well as developed tools and software to assist radiologists. Some of the researchers proposed individual segmentation or classification technique or even the integration of both the strategies. In recent years, several types of computer-aided tumor detection and classification methods have been proposed by researchers. For instance, Arimura et al. [12] presented a computer-aided diagnosis (CAD) system to assist the neuroradiologists in detecting brain diseases. This method is decomposed into several steps, which include extracting features based on image processing technique and classification using machine learning classifiers. This method is purely based on low-level features and conventional machine learning techniques that achieve very low accuracy. Similarly, some other conventional features and machine learning-based approaches are presented in Refs. [13–15]. Recently, a classifier based on SVM [16] classified brain tumor into different grades using first- or second-order statistical measurements. The accuracy achieved by this system is 80% on 21 patient's data and combining different grades of tumor to test the algorithm. Their training step includes pre-segmented tumor portions and the machine is trained on these portions to classify the grade of tumor.

Recently, with the success of DNNs and CNN, relatively higher rate of testing accuracy as compared to typical classification and features selection methods is seen. Furthermore, the features used in deep learning models for classification are learned spontaneously and are based on the input given to the model. Different weights are assigned to them such that they can produce effective classifiers. Patil et al. [17] presented an approach based on deep learning, CNN, and compared gigs on Back Propagation Neural Network. They downloaded training data from BRATS 2014, which comprises MRI data of 213 patients. The key point of this proposed system is adding more layers to the convolutional structure, which does not necessarily improve the performance of grading brain tumor. Selvaraj et al. [7] fine-tuned CNN features of MRI data for brain tumor classification. In this approach, they selected deep features from a pre-trained Fast (CNN-F) structure which has total of eight layers containing five convolutional layers and three fully connected layers. The accuracy achieved by their system without using any dropout is 81.81% while tuning with dropout has 77.27% accuracy.

Our proposed technique focuses on Zulpe *et al.*'s [4] technique and it achieves much higher accuracy as compared to theirs. The proposed technique fine-tuned SqueezeNet and GoogLeNet models for the brain tumor MRI classification.

3. Proposed Methodology

This section explores the working of the suggested technique and describes the main concepts of brain tumor classification in an efficient way. The proposed system is a two-fold process comprising the training phase, which includes substeps of loading the deep convolutional model, fine-tuning the existing model followed by the final testing phase. We fine tune two different deep learning models (GoogLeNet and SqueezeNet) for better results. In the first phase, a lightweight SqueezeNet [18] model is loaded to easily handle brain tumor classification in real-time situations. Most of the recent researchers try to focus on increasing accuracy over different computer vision datasets and ignore the need of efficient algorithms. There are multiple architectures that can achieve much better accuracy but with too many parameters. SqueezeNet can achieve the AlexNet-level accuracy with 50 times fewer parameters and much less model size of about 0.5 MB. In the second phase, GoogLeNet [19] model is loaded, which is also the winner of the ILSVRC (ImageNet Large Scale Visual Recognition Competition) 2014. SqueezeNet and GoogLeNet are trained on large ImageNet dataset, and in this system we used these pre-trained weights and added up our data to train the model, thus the model is more effective and efficient. After the training process models are downloaded and are ready to run through proposed algorithm to predict real-time brain tumors. Proposed framework is shown in Fig. 1 and each step is explained in what follows.

3.1 *Fine-tuning SqueezeNet model*

In order to modify the SqueezeNet architecture, the final layer is updated as per requirements of our framework, that is classification of tumor into three classes. The overall architecture of SqueezeNet model is given in Fig. 2, where the macro-architecture is actually system-level organization of multiple modules

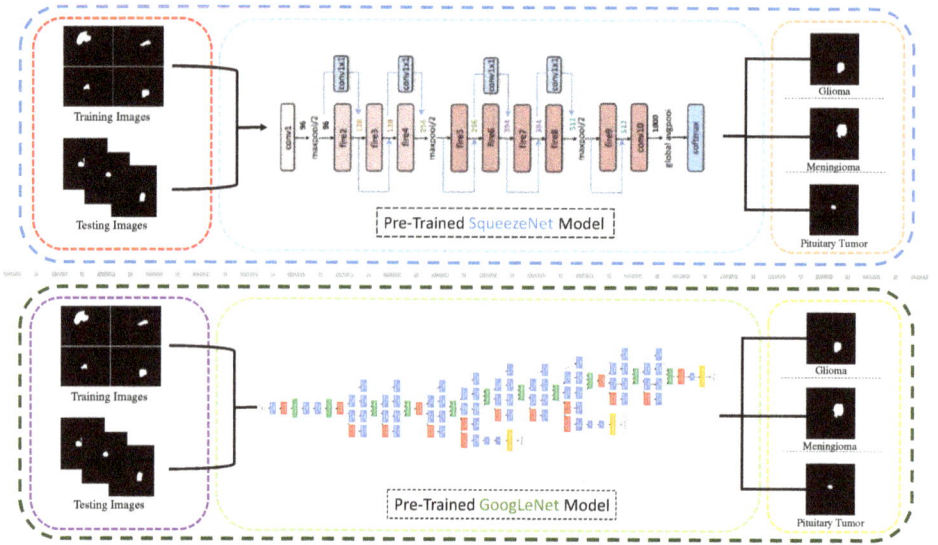

Figure 1: Overall proposed framework.

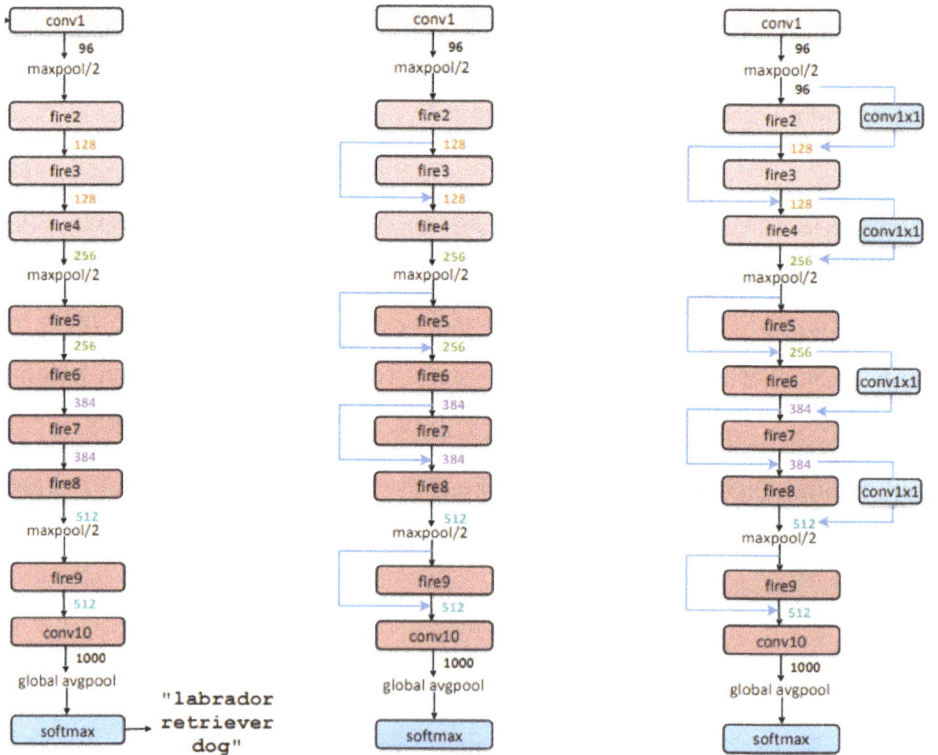

Figure 2: SqueezeNet [18] architecture.

and integrating them into an end-to-end CNN architecture. The main strategies while designing this architecture include: (1) to replace 3×3 filters with 1×1 filters that contain $9 \times$ less parameters than 3×3 filters, (2) to reduce the input channels to 3×3 filters that perform down-sampling in the architecture such that all the convolutional layers have large activation maps. After the conv-10 layer, the Softmax layer is also modified, which contains only 3-number of classes according to requirements of our work which actually was trained on 1,000 classes of ImageNet dataset. We modified the Softmax layer of CNN architecture of SqueezeNet to three brain tumor classes, as it is responsible for the final. This layer is modified because output of Softmax layer is used to represent categorical distribution, like probability distribution over three different classes in our system. In conv-10 layer, all the global and local features of the input are learned completely and these features are then input to Softmax classifier.

3.2 *SqueezeNet architecture*

SqueezeNet architecture has first convolutional layer (conv-1), and next 8-fire modules (fire 2–9) with a final convolutional layer (conv-10). Max pooling is also applied with 2 stride after conv-1, fire4, fire8 and last conv-10.

Testing phase involves checking the accuracy of the trained model over a number of test images distributed at the beginning into training, validation and testing sets. Algorithm 1 shows the mechanism of overall proposed system.

Algorithm 1: Computing the class of brain tumor

Input: Brain MRI image

Functions:

P = Train (I)

Train (I):

1. Load the proposed CNN network architecture

2. Resize the image into 227

3. Pass the image into the loaded CNN network

4. At the last Softmax layer, make a prediction

5. **Return** P

label = search for the prediction in label txt file

Output: label

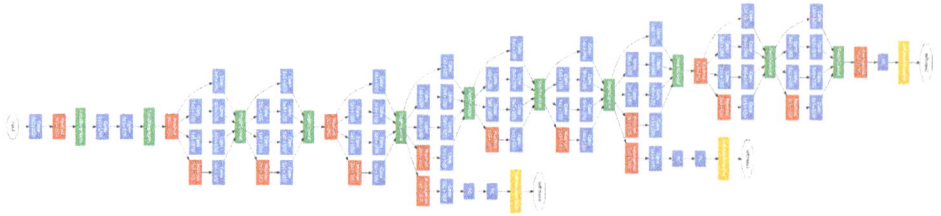

Figure 3: GoogLeNet [19] architecture (left–right).

3.3. *Fine tuning GoogLeNet model*

For the training process of GoogLeNet model, the last three layers of the GoogLeNet architecture have been modified according to our proposed dataset. A detailed architecture of GoogLeNet model is shown in Fig. 3. GoogLeNet architecture uses a series of nine interconnected inception modules, after the last inception module we add a new average pooling layer to our model which also connects our trained layers of the model to our new modified layers. After average pooling, we modify the architecture by adding a fully connected layer to our fine-tuned model. At the end, we modify our Softmax layer to 3 classes to output probabilities in accord to our desired dataset.

3.4 *GoogLeNet architecture*

GoogLeNet is one of the first architectures which introduced the inception module that helped significantly in dropping off the number of trainable weights in a network. It contains a parallel combination of convolution layers, i.e., 1×1, 3×3 and 5×5 convolutions alongside a single max-pooling layer. Furthermore, a convolutional Layer of 1×1 is also included in the network before the other two convolutions, i.e., 3×3 and 5×5, due to dimensionality reduction. GoogLeNet architecture consists of nine inception modules, holding 22 layers, four max-pooling and an average-pooling layer. In all convolutional layers and inception modules, the ReLU activation function is used.

4. Experimental Results

In this section, the performance of the proposed system is evaluated using different accuracy assessment metrics. The experimental settings with the benchmark dataset are described in subsequent subsections.

4.1 *Brain tumor dataset*

The proposed system is evaluated on brain tumor [20] benchmark dataset. This dataset consists of 2D slices of brain T1-weighted MRI. These T1-weighted MR images are gathered from Nanfang Hospital, Guangzhou, China and General Hospital, Tianjing Medical University, China, recorded in the period 2005–2010. They collected a total of 3,064 slices of MRI from 233 different patients and divided them into three different categories.

i.   Meningioma, found in 708 MRI slices.
ii. Glioma, found in 1,426 slices of MRI.
iii. Pituitary tumor, found in 930 MRI slices.

All these MRI slices have a plane resolution of 512 × 512. The gap between two slices is 1 mm, while the thickness of each slice is 6 mm. The ground truth of the dataset is manually segmented and delineated by three different experienced radiologists of those hospitals. The samples of all classes with the highlighted tumor are presented in Fig. 4.

4.1.1. *Related research review*

This section contains the contribution of researchers on Brain Tumor Dataset. First of all, in 2018 Afshar and Mohammadi *et al.* [21] proposed a technique

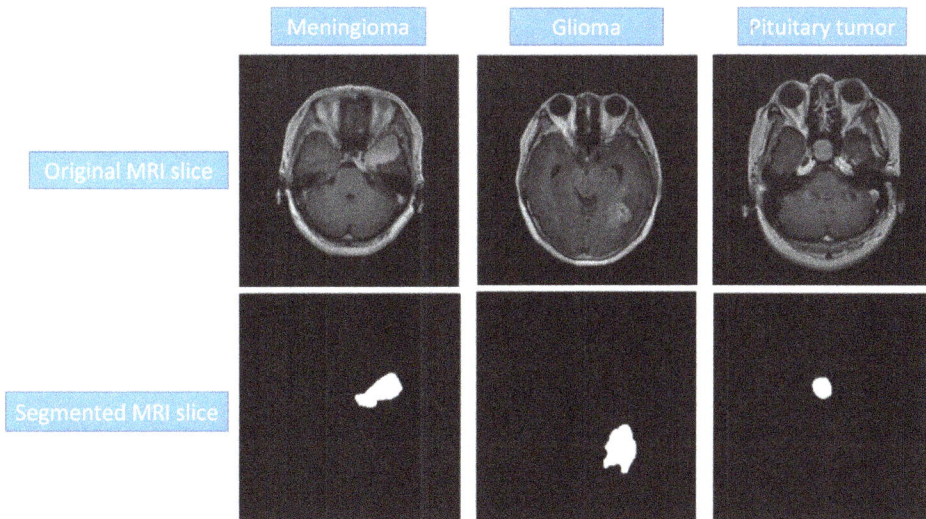

Figure 4:   Brain tumor dataset MRI sample of original and segmented slices of each class.

named capsule network architecture for brain tumor classification using standard brain tumor dataset. After that, Abiwinanda *et al.* [22] proposed a CNN model for classification with an accuracy of 84.19%. In 2019, Hussain *et al.* [23] improved that accuracy up to 90% using Genetic algorithm for classification.

Table 1 highlights the different techniques and approaches for brain tumor classification using different datasets such as brain tumor dataset, immune miRNA disease dataset, Radiopaedia, (DS-66, DS-160, DS-255), etc. Immune miRNA disease dataset is retrieved from two curated databases named as Human miRNA associated disease database (HMDD) and miR2Disease [24] . Radiopaedia [25] dataset is divided into four different classes containing 121 MR images, i.e., Menigiomas, Gliomas (I), Gliomas (II) and Glioblastomas. Three MRI benchmark datasets (DS-66, DS160 and DS-255) containing, respectively, 66, 160, and 255 images are openly available on the "Harvard Medical School" website [26, 27]. There are some other datasets also mentioned in Table 1 including MNIST, CIFAR-10 and SVHN [28].

Table 1:   Detailed related research work.

Paper	Method/classifier	Datasets	Accuracy
[21]	Capsule Network Architecture	Brain Tumor Dataset	86.56%
[28]	Dense Capsule Networks (DCNet) and Diverse Capsule Networks (DCNet++)	MNIST CIFAR-10 SVHN	99.75% 0.31% with seven-fold decrease in number of parameters 96.90%
[22]	CNN Model	Brain Tumor Dataset	training accuracy 98.51% validation accuracy 84.19%
[29]	SVM PCA	Immune miRNA disease dataset	90.2%
[23]	Genetic algorithm	Brain Tumor Dataset	90.9
[30]	CNN based multiple grades classification system for brain tumor	Radiopaedia Brain Tumor	87.38 90.67
[31]	DWT and BoW	Three dataset DS-66 DS-160 DS-255	100%, 100%, and 99.61%
[32]	Multi-level Dense Capsule Networks	MNIST dataset SVHN dataset	99.75% 96.90%

The MNIST dataset is a digit dataset which has 60,000 + training and 10,000 + testing images with size of 28 × 28 each. The CIFAR-10 dataset is used for object recognition. CIFAR-10 dataset is known to be a subset of the 80M minute images dataset, it also contains 60,000 color images of size 32 × 32. It consists of 10 classes with 6,000 images each. SVHN (Street view House Numbers) is a real-world image dataset which consists of 73,257 training and 26,032 testing digits.

4.2 *Experimental setup*

The experiments of the proposed technique are conducted by 75%, 25% and 25% of training, testing and validation sets, respectively. The platform used for the simulation of the proposed methodology is MATLAB 2017b running over a GeForce-Titan-X GPU with Ubuntu 16.04 64 bit-version operating system. The detailed explanation of the above datasets used in the experiments are described in the coming section.

4.3 *Evaluation of the proposed system*

We conducted two types of experiments: first testing is performed on SqueezeNet model using the pre-trained weights of ImageNet dataset. We used two different approaches for experiments, first one is accuracy of the proposed system using 10 epochs, while in the second type of experiment the epochs are increased to 30. The increase in number of epochs to 30 shows higher accuracy than 10 epochs. The second experiments are conducted on GoogLeNet model, which also uses pre-trained weights of ImageNet dataset and we tested the same approach of experiments as mentioned before.

4.3.1 *Performance using SqueezeNet CNN model*

In this section, we discussed the results achieved using SqueezeNet model for 10 and 30 epochs. The confusion matrix of this system is described in Fig. 5. The detailed accuracy metrics of the fine-tuned SqueezeNet model for 10 epochs is shown in Fig. 5 of the system after 10 epochs is given in Fig. 5. The learning process of our fine-tuned model for 10 epochs is given in Fig. 6 our model with the 10 epochs is explained in Fig. 6.

We also trained our deep neural network for 30 epochs. The detailed confusion matrix of these results is described in Fig. 7, which demonstrates the accuracy of every class and the overall accuracy of the proposed method.

Confusion Matrix

Figure 5:   Confusion matrix of SqueezeNet model on 10 epochs.

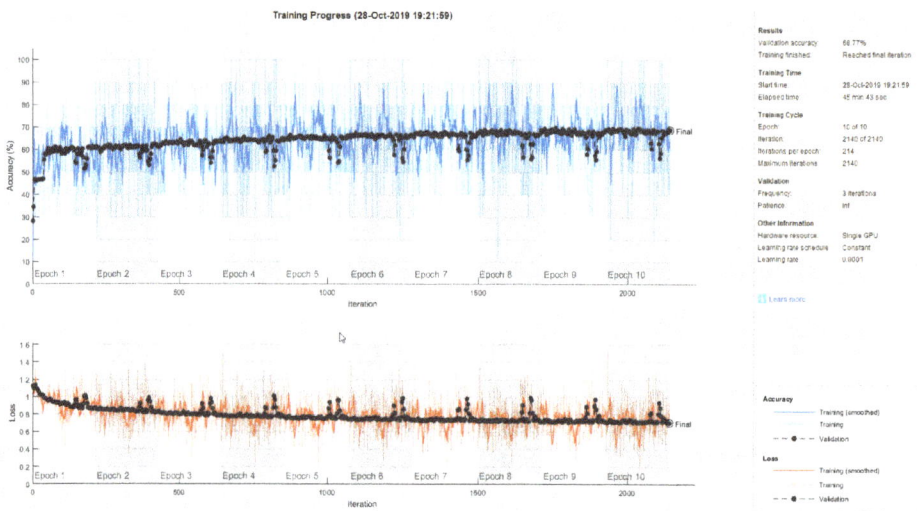

Figure 6:   Accuracy of SqueezeNet model after 10 epochs.

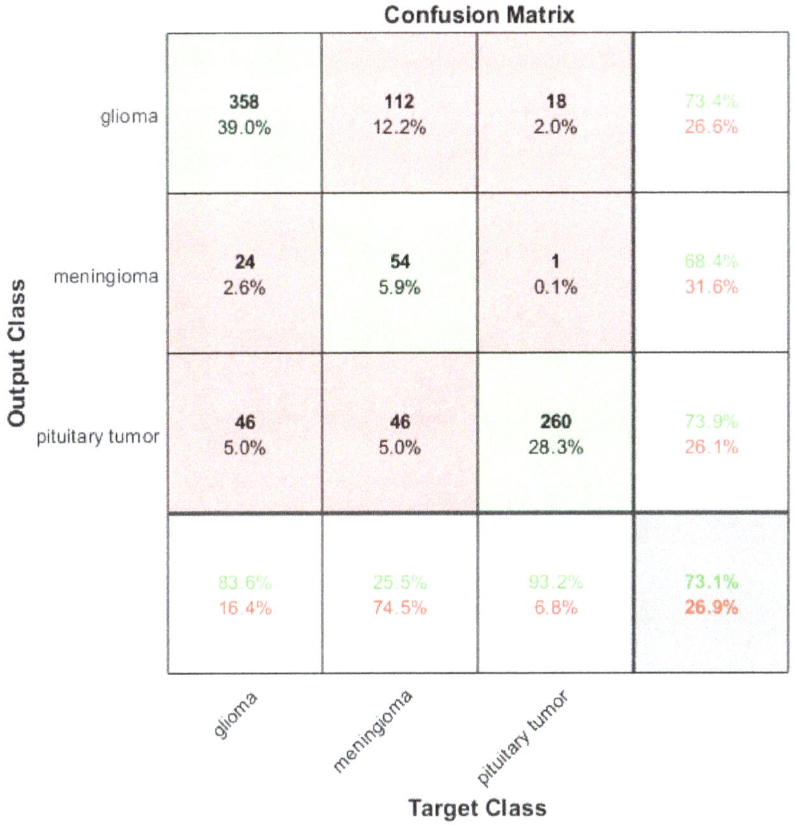

Figure 7:   Confusion matrix of SqueezeNet model trained for 30 epochs.

In this type of experiment, we can see that the accuracy is increased from 68.8% to 73.1% as compared to the 10 epochs. Figure 8 shows the accuracy and training loss of our model in detail.

4.3.2 *GoogLeNet results using 10 epochs*

In this section, the performance of GoogLeNet model is discussed. The confusion matrix of this system is described in Fig. 9. The obtained overall accuracy of the system after 10 epochs is 66.4%. The detailed explanation of our model accuracy for 10 epochs is shown in Fig. 10.

The comprehensive confusion matrix of the model for 30 epochs is shown in Fig. 11. The accuracy study shows that our model obtains much higher accuracy when we increase the number of epochs. Accuracy on 10 epochs was

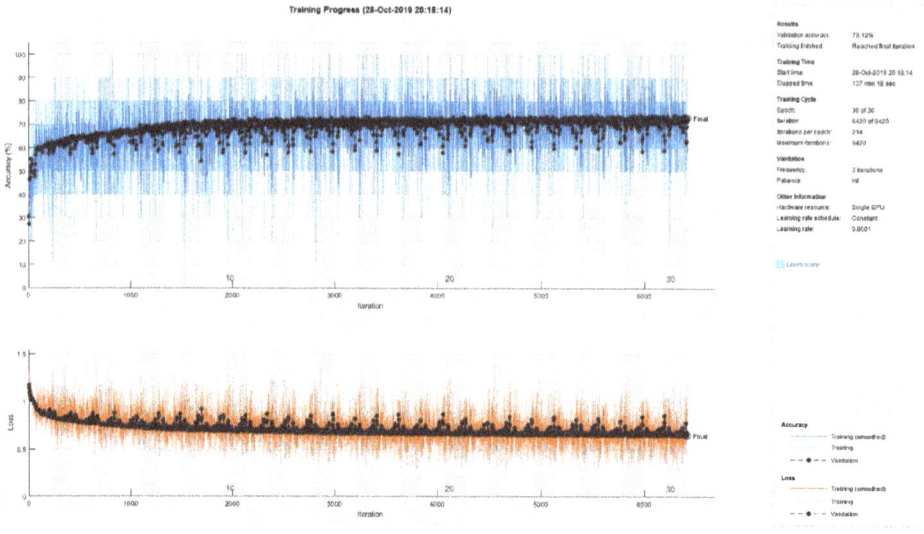

Figure 8: Accuracy of SqueezeNet model after 30 epochs.

Figure 9: Confusion matrix of GoogLeNet model on 10 epochs.

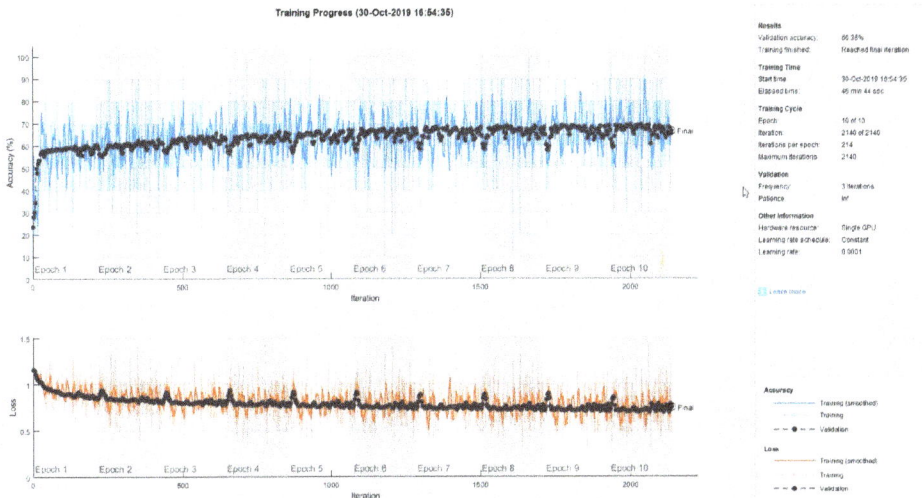

Figure 10: Accuracy of GoogLeNet model after 10 epochs.

Figure 11: Confusion matrix of GoogLeNet model on 30 epochs.

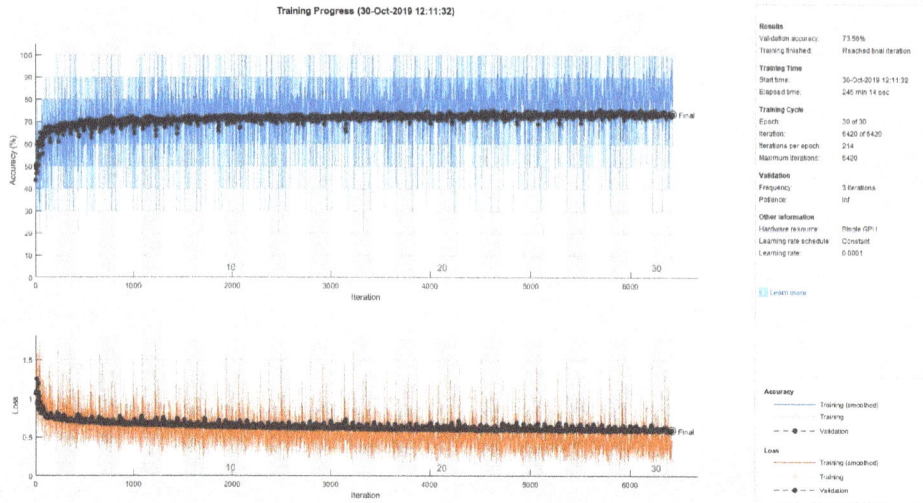

Figure 12:   Accuracy of GoogLeNet model after 30 epochs.

66.4% and it increased up to 73.6% on 30 epochs. In Fig. 12, the detailed summary of accuracy and loss of the model is shown.

4.4 *Comparison with other state-of-the-art methods*

For evaluation of the proposed system, we considered top state-of-the-art techniques including GK fuzzy clustering method and a hybrid intelligent approach for brain tumor classification. GK fuzzy technique is purely based on clustering, which is not a suitable option while dealing with medical image data. Medical image data requires highly accurate and trustable features on the basis of which decision is completely trustworthy because medical data is very delicate in nature. The second compared method uses discrete wavelet transform domain and proposes a hybrid technique comprising of several stages. In this method, they used low-level features and traditional principal component analysis. These methods have used small size datasets which are not generic to test data and thus are not implementable in real-time. The next method is enhancing performance of classification through brain tumor region partition and augmentation, but the accuracy of this method is not enough. The statistical results compared with state-of-the-art are given in Table 2. The proposed technique considers the delicate nature of medical image data and uses deep level convolutional features and guarantees very high level of accuracy. Based on the experimental results and comparison with state-of-the-art methods, we are positive that our implemented method is a complimentary alternative to the manual brain tumor classification approaches.

Table 2: Comparison of proposed technique with state-of-the-art techniques.

Method	Accuracy (%)
GK fuzzy clustering [3]	95.13
Hybrid intelligent technique [33]	97
Enhanced performance of brain tumor [20]	91.28
Proposed methods	98.43

5. Conclusion

The proposed technique concluded in this chapter is an aid for radiologists to classify brain tumor MRI images into three classes, i.e., meningioma, glioma and pituitary tumor. The proposed technique is based on Deep Learning which efficiently and accurately classifies the MRI images into their respective classes. For better and efficient results, we used transfer learning approach for two different CNN models including SqueezeNet and GoogLeNet. We used these pre-trained models to produce more accurate results to assist the radiologists to diagnose brain tumor. Our fine-tuned SqueezeNet model classified the MRI images with an accuracy of 73.1% and GoogLeNet achieved approximately 73.6% accuracy. In future, we will explore the detection and localization of brain tumor using deep CNN models. Further, we want to explore some efficient CNN models [34] and resource restricted devices [35] to make our system available in smart healthcare centers [36].

Acknowledgment

This work was supported by the National Research Foundation of Korea (NRF) grant funded by the Korea Government (MSIP) (No. 2016R1A2 B4011712).

References

1. U. Javed, M. M. Riaz, A. Ghafoor, and T. A. Cheema, "MRI brain classification using texture features, fuzzy weighting and support vector machine," *Prog. Electromag. Res.*, vol. 53, pp. 73–88, 2013.
2. S. Patil and V. Udupi, "A computer aided diagnostic system for classification of brain tumors using texture features and probabilistic neural network," *Int. J. Comput. Sci. Eng. Inf. Technol. Res.*, vol. 3, pp. 61–66, 2013.

3. A. Shankar, A. Asokan, and D. Sivakumar, "Brain tumor classification using Gustafson-Kessel (GK) fuzzy clustering algorithm," *Brain,* vol. 1, pp. 68–72, 2016.

4. N. Zulpe and V. Pawar, "GLCM textural features for brain tumor classification," *Int. J. Comput. Sci. Iss., (IJCSI)* vol. 9, p. 354, 2012.

5. P. John, "Brain tumor classification using wavelet and texture based neural network," *Int. J. Sci. Eng. Res.,* vol. 3, pp. 1–7, 2012.

6. J. Jiang, Y. Wu, M. Huang, W. Yang, W. Chen, and Q. Feng, "3D brain tumor segmentation in multimodal MR images based on learning population-and patient-specific feature sets," *Comput. Med. Imag. Graph.,* vol. 37, pp. 512–521, 2013.

7. H. Selvaraj, S. T. Selvi, D. Selvathi, and L. Gewali, "Brain MRI slices classification using least squares support vector machine," *Int. J. Intell. Comput. Med. Sci. Image Process.,* vol. 1, pp. 21–33, 2007.

8. U. Avni, H. Greenspan, E. Konen, M. Sharon, and J. Goldberger, "X-ray categorization and retrieval on the organ and pathology level, using patch-based visual words," *IEEE Trans. Med. Imag.,* vol. 30, pp. 733–746, 2010.

9. A. Bosch, X. Munoz, A. Oliver, and J. Marti, "Modeling and classifying breast tissue density in mammograms," in *2006 IEEE Computer Society Conference on Computer Vision and Pattern Recognition (CVPR'06),* 2006, pp. 1552–1558.

10. L. Chen, G. E. Remondetto, and M. Subirade, "Food protein-based materials as nutraceutical delivery systems," *Trends Food Sci. Technol.,* vol. 17, pp. 272–283, 2006.

11. M. Huang, W. Yang, M. Yu, Z. Lu, Q. Feng, and W. Chen, "Retrieval of brain tumors with region-specific bag-of-visual-words representations in contrast-enhanced MRI images," *Comput. Math. Meth., Med.,* vol. 2012, 2012.

12. H. Arimura, T. Magome, Y. Yamashita, and D. Yamamoto, "Computer-aided diagnosis systems for brain diseases in magnetic resonance images," *Algorithms,* vol. 2, pp. 925–952, 2009.

13. J. Lu, G. Getz, E. A. Miska, E. Alvarez-Saavedra, J. Lamb, D. Peck, *et al.,* "MicroRNA expression profiles classify human cancers," *Nature,* vol. 435, p. 834, 2005.

14. S. Ramaswamy, P. Tamayo, R. Rifkin, S. Mukherjee, C.-H. Yeang, M. Angelo, *et al.,* "Multiclass cancer diagnosis using tumor gene expression signatures," *Proc. Natl. Acad. Sci.,* vol. 98, pp. 15149–15154, 2001.

15. D. N. Louis, H. Ohgaki, O. D. Wiestler, W. K. Cavenee, P. C. Burger, A. Jouvet, *et al.,* "The 2007 WHO classification of tumours of the central nervous system," *Acta Neuropathologica,* vol. 114, pp. 97–109, 2007.

16. Soltaninejad, Mohammadreza, Xujiong Ye, Guang Yang, Nigel Allinson, and Tryphon Lambrou. "Brain tumour grading in different MRI protocols using SVM on statistical features," *Med. Image Underst. Anal.,* British Machine Vision Association Pages, pp. 259–264, 2014.

17. S. Patil and V. Udupi, "A computer aided diagnostic system for classification of brain tumors using texture features and probabilistic neural network," *Int. J. Comput. Sc. Eng. Inf. Technol. Research,* vol. 3, 2013.

18. F. N. Iandola, S. Han, M. W. Moskewicz, K. Ashraf, W. J. Dally, and K. Keutzer, "SqueezeNet: AlexNet-level accuracy with 50x fewer parameters and <0.5 MB model size," 2016, arXiv preprint arXiv:1602.07360.

19. C. Szegedy, W. Liu, Y. Jia, P. Sermanet, S. Reed, D. Anguelov, *et al.,* "Going deeper with convolutions," in *Proc. IEEE Conf Computer Vision and Pattern Recognition,* 2015, pp. 1–9.

20. J. Cheng, W. Huang, S. Cao, R. Yang, W. Yang, Z. Yun *et al.,* "Enhanced performance of brain tumor classification via tumor region augmentation and partition," *PloS One,* vol. 10, p. e0140381, 2015.

21. P. Afshar, A. Mohammadi, and K. N. Plataniotis, "Brain tumor type classification via capsule networks," in *2018 25th IEEE Int. Conf. Image Processing (ICIP),* 2018, pp. 3129–3133.

22. N. Abiwinanda, M. Hanif, S. T. Hesaputra, A. Handayani, and T. R. Mengko, "Brain tumor classification using convolutional neural network," in *World Congress on Medical Physics and Biomedical Engineering 2018,* 2019, pp. 183–189.

23. L. Hussain, S. Saeed, I. A. Awan, A. Idris, M. S. A. Nadeem, and Q.-u.-A. Chaudhry, "Detecting brain tumor using machines learning techniques based on different features extracting strategies," *Curr. Med. Imag.,* vol. 15, pp. 595–606, 2019.

24. Q. Jiang, Y. Wang, Y. Hao, L. Juan, M. Teng, X. Zhang *et al.,* "miR2Disease: a manually curated database for microRNA deregulation in human disease," *Nucleic Acids Res.,* vol. 37, pp. D98–D104, 2008.

25. A. Shetty and E. Ranschaert, "Radiopaedia," ed: Retrieved from Modified CT severity index: https://radiopaedia.org/articles, 2018.

26. S. Wang, Y. Zhang, T. Zhan, P. Phillips, Y.-D. Zhang, G. Liu *et al.,* "Pathological brain detection by artificial intelligence in magnetic resonance imaging scanning (invited review)," *Prog. Electromag. Res.,* vol. 156, pp. 105–133, 2016.

27. S. Wang, S. Lu, Z. Dong, J. Yang, M. Yang, and Y. Zhang, "Dual-tree complex wavelet transform and twin support vector machine for pathological brain detection," *Appl. Sci.,* vol. 6, p. 169, 2016.

28. S. S. R. Phaye, A. Sikka, A. Dhall, and D. Bathula, "Dense and diverse capsule networks: Making the capsules learn better," 2018, arXiv preprint arXiv:1805.04001.

29. A. Prabahar and J. Natarajan, "Prediction of microRNAs involved in immune system diseases through network based features," *J. Biomed. Inform.,* vol. 65, pp. 34–45, 2017.

30. M. Sajjad, S. Khan, K. Muhammad, W. Wu, A. Ullah, and S. W. Baik, "Multi-grade brain tumor classification using deep CNN with extensive data augmentation," *J. Comput. Sci.,* vol. 30, pp. 174–182, 2019.

31. W. Ayadi, W. Elhamzi, I. Charfi, and M. Atri, "A hybrid feature extraction approach for brain MRI classification based on Bag-of-words," *Biomed. Signal Process. Control,* vol. 48, pp. 144–152, 2019.

32. S. S. R. Phaye, A. Sikka, A. Dhall, and D. R. Bathula, "Multi-level dense capsule networks," in *Asian Conference on Computer Vision,* 2018, pp. 577–592.

33. E.-S. A. El-Dahshan, T. Hosny, and A.-B. M. Salem, "Hybrid intelligent techniques for MRI brain images classification," *Digital Signal Process.,* vol. 20, pp. 433–441, 2010.

34. Muhammad, Khan, Tanveer Hussain, and Sung Wook Baik, "Efficient CNN based summarization of surveillance videos for resource-constrained devices," *Pattern Recogn. Lett.,* vol. 130, pp. 370–375, 2020.

35. Hussain, Tanveer, Khan Muhammad, Javier Del Ser, Sung Wook Baik, and Victor Hugo C. de Albuquerque, "Intelligent Embedded Vision for Summarization of Multiview Videos in IIoT." *IEEE Trans. Industr. Inform.,* vol. 16, no. 4, pp. 2592–2602, 2019.

36. Hussain, Tanveer, Khan Muhammad, Salman Khan, Amin Ullah, Mi Young Lee, and Sung Wook Baik. "Intelligent baby behavior monitoring using embedded vision in IoT for smart healthcare centers." *J. Artif. Intell. Syst.,* vol. 1, no. 1, pp. 110–124, 2019.

Chapter 12

Classification of Travel Patterns Including Wandering Based on Bi-directional Long Short-Term Memory Networks

Nhu Khue Vuong[*,‡], Yong Liu[†,§], Syin Chan[†,¶], Chiew Tong Lau[†,‖],
Zhenghua Chen[*,**], Min Wu[*,††] and Xiaoli Li[*,‡‡]

[*]*Institute for Infocomm Research, Singapore*
[†]*School of Computer Engineering,*
Nanyang Technological University, Singapore
[‡]*vuong_nhu_khue@i2r.a-star.edu.sg*
[§]*stephenliu@ntu.edu.sg*
[¶]*asschan@ntu.edu.sg*
[‖]*asctlau@ntu.edu.sg*
[**]*Chen_Zhenghua@i2r.a-star.edu.sg*
[††]*wumin@i2r.a-star.edu.sg*
[‡‡]*xlli@i2r.a-star.edu.sg*

Abstract

Classification of travel patterns including wandering is important for early recognition of cognitive deterioration and other health conditions in people with dementia (PWD). In this chapter, we develop machine learning (ML) and deep learning (DL) models to recognize dementia-related wandering patterns based on the orientation data available in mobile devices. In particular, we use DL with long short-term memory networks (LSTM) and bi-directional LSTM to detect direct, pacing, lapping and random travel patterns. Experimental results on a real dataset collected from 14 subjects show that deep LSTM classifiers improve the classification accuracy by 2% compared to traditional ML classifiers. The results and proposed methodology can be further improved so as to develop useful healthcare applications for dementia-related wandering monitoring and management.

Keywords: classification, dementia, dementia related wandering, detection, wandering behavior, wandering patterns, wandering management

1. Introduction

Dementia is a condition which affects the brain and usually leads to memory loss, mental decline and deterioration of intellectual functions, and its global prevalence is expanding rapidly [1]. Wandering is a common but dangerous behavior among people with dementia (PWD). A five-year study of clinical records in patients with preclinical dementia [2] found that wandering is the earliest symptom followed by cognitive complaints. Evidence also shows that changes in locomotion patterns begin many years prior to the onset of dementia [3] and gait-related motor disturbances present in all subtypes of dementia, even in the early and preclinical stages [4], [5]. Therefore, monitoring wandering patterns is very important to provide information on the cognitive deterioration and health conditions of the wanderers [6].

Some features of wandering patterns are further demonstrated to be associated with cognitive declination, health status, and other physiological effects in PWD. Patients with mild cognitive impairment were found to have greater variability in walking speeds and activity levels than those without the diagnosis [7]. Increased stride time variability is also correlated negatively with cognitive performance and increased the risk of falls in community-dwelling elders [8]. Algase *et al.* [9] found that pacing patterns among PWD who wander are signals of agitation and discomfort. Hence, it is essential to measure the wandering patterns in PWD in order to identify and quantify their declines in cognitive functions.

In Martino-Saltzman *et al.* [10], four types of travel patterns, namely direct travel, pacing, lapping and random travel, were defined to characterize wandering behavior of PWD. The findings in [9, 10] indicate that direct travel is not wandering, while the other three types of patterns show different severity of dementia. To classify these four types of travel patterns, manual approach by visually inspecting the recorded movements of PWD from videotapes is common in the gerontology research community. However, it is labor-intensive and susceptible to human error. In addition, informed consent is required as visual observation is considered invasive of a subject's privacy.

Many other methods have been used for automated classification of travel patterns, including wandering. Localization modalities such as RFID [11] and WiFi [12] had been used to track and analyze the sequences of locations traversed to recognize travel patterns, including wandering. Lin *et al.* [13] and Delaunay and Guerin [14] used GPS trajectories and developed algorithms based on GPS data to detect loop-like patterns. Gochoo *et al.* [15] had employed binary passive infrared sensor and deep learning (DL) to classify travel patterns, including wandering. Recently, inertial sensors from mobile and wearable devices have been used to track the orientation information while a person travels in order to determine travel patterns [16]. Clearly, inertial sensors are cheaper and more readily available compared to other methods using video cameras, RFID [11], infrared sensors [15] and ultrawide band localization systems, hence can be widely used for monitoring dementia patients [17].

In this chapter, we classify wandering patterns using orientation data collected from inertial sensors in mobile devices. We propose a framework using both traditional ML classifiers and DL classifiers (e.g., LSTM) for automated detection of travel patterns from the orientation data. We conduct experiments on human subjects' locomotion data and the results demonstrate the superior performance of the proposed DL approach for detection of dementia-related wandering patterns. The chapter is organized as follows. Section 2 provides the background of dementia-related wandering and formulates the problem. Section 3 introduces the methodology followed by the details of ML and DL approaches. Section 4 presents and discusses the experimental results and Section 5 concludes this chapter.

2. Problem Formulation

2.1 *Background*

Martino-Saltzman [10] modeled travel patterns for PWD as direct, pacing, lapping and random patterns as shown in Fig. 1. Direct is single straightforward movement, which is efficient travel and not regarded as wandering. Pacing is moving back and forth between two physical locations. Lapping is circumvolution around a physical position. Random is walking without rules and in no particular order. Pacing, lapping and random are inefficient and they constitute different types of wandering.

Figure 1 gives one example each for the four travel patterns: direct, pacing, lapping and random. For each pattern, the left plot depicts the subject's

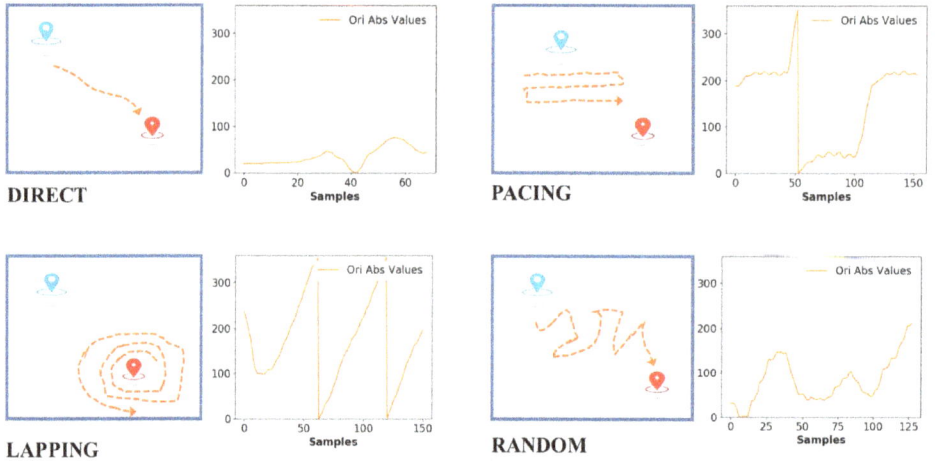

Figure 1: Illustration of wandering patterns. The arrows with dash-line depict the path taken by the subject. The graph on the right of each pattern shows the orientation (absolute) value for each sampled data point.

walking trail when the subject exhibits the corresponding travel pattern. The right plot shows the corresponding azimuth values (or orientation signals) as the person travels (x-axis: the data samples, y-axis: orientation signal values). The values of the orientation signals range from 0° to 360°, and in Fig. 1, the orientation signals are quite distinct for each type of travel pattern.

2.2 *Problem formulation*

Given a time series collected from inertial sensors, we formulate the wandering pattern detection as a four-class classification problem which aims to discriminate between direct, pacing, lapping and random patterns. For simplicity, we only consider time series of single patterns in this study. That means each time series, which can be different in length, is associated with one of the four classes only. We have a collection $X = (x_i)_{i=1}^{n}, y = (y_i)_{i=1}^{n}$ of n labeled samples where each sample x_i is a time series vector and the associated label y_i is either one of the four classes (direct, pacing, lapping and random). Our problem is to develop models to classify these time series.

3. Methodology

Figure 2 presents the methodology of our study, which consists of dataset preparation, the flow of training and evaluation of traditional machine learning (ML) approach and DL approach.

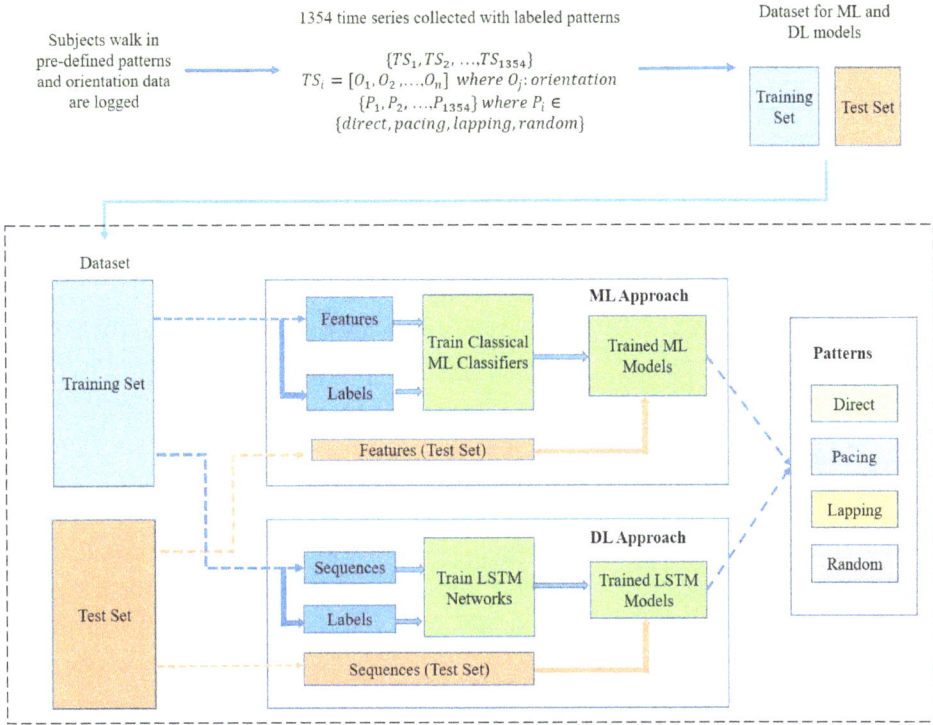

Figure 2: The framework of our study: data preparation, approaches and evaluation.

The methodology depicted in Fig. 2 comprises two major blocks: the data preparation and the approach. In general, the data preparation is to prepare and collect the relevant data for our study. Subjects are recruited to walk in pre-defined patterns so that orientation data as well as the patterns walked are logged. These orientation time series data and patterns are the dataset prepared for our study. The dataset is split into training and test sets, which are directly input to the approach block in Fig. 2 and are employed to examine the performance of both ML and DL approaches.

For ML approach, features are first extracted from training time series and then the selected features (usually after features selection step) and the labels of the training time series are used to train the ML models. The trained ML models will be used for examining the performance on the test set by extracting the selected features from test set's time series and feeding these features to the trained ML models which will subsequently predict the corresponding travel patterns, which is either direct, pacing, lapping or random. The predicted patterns provided by the models will be compared with the ground truth (the labels in the test set) to evaluate the performance of the trained ML models.

For DL approach, training time series are first made into sequences of the same length in order to feed into DL models. The sequences and labels of training dataset are subsequently employed to train DL networks (i.e., LSTM networks in our study). The trained LSTM networks will then be used to evaluate the performance on the test set. To perform testing, time series from the test set are transformed into sequences and then fed into the trained networks to predict the corresponding travel patterns. Similar to the ML approach, the predicted patterns are then compared with the labels in the test set to gauge the performance of the DL models.

We will describe the details of each block in the proposed methodology in the following sections.

3.1 *Dataset preparation*

The orientation data were collected using a smart phone. An application was developed and installed on the phone to log the following inertial data: timestamp (ms), acceleration force along the x-, y- and z-axis, including gravity (m/s^2), linear acceleration of the device (m/s^2), orientation of the device or the angle between magnetic north and the device's y-axis (degree). The data sampling rate was 20 Hz.

3.1.1 *Subjects*

Fourteen non-dementia subjects were recruited from the university campus and laboratory. The criteria for inclusion were the subject is mentally healthy and capable of independent ambulation. The data collection exercise was approved by the Institutional Review Board (IRB) of Nanyang Technological University (NTU), Singapore. The research objectives were explained to each subject and their consent to participate in the research was obtained before collecting the data.

3.1.2 *Data collection procedure*

Before the collection exercise, subjects were given explainations and examples of each type of travel pattern. Each subject was asked to replicate each pattern once to affirm their understanding before the actual collection. All subjects held a mobile phone in their hand (palm facing up). This was more convenient for the subject to interact with the user interface on the phone screen to start/stop logging the data. To ensure data quality, the subjects were instructed not to change the direction of holding the phone during data logging. So, the phone was held in the horizontal position, i.e., parallel to the ground with the screen facing upwards.

During this collection exercise, the data collector personnel first informed the subjects of the travel patterns that were deemed to be collected. The subjects would then be expected to walk in the corresponding the subjects started walking, they clicked on a toggle button on the phone screen to trigger data logging, and the inertial data of that travel episode would then start to be logged. When the subjects stopped walking, they toggled the button to stop data logging for that travel episode. The inertial data would be then automatically stored as a text file in the phone. To ensure data quality and correctness, the subjects' movements during their walk were closely monitored and observed. We discarded the data if the travel patterns exhibited by the subjects did not agree with our data collection plan.

3.1.3 *Dataset description*

Table 1 summarizes the statistics of patterns collected from each subject.

The subjects are labeled from 1 to 14 and their demographics information including sex and age (as of the year 2020) is also reported. Sixty-four

Table 1: Statistics of the dataset collected.

S	X	Age	D	P	L	R	Total
1	M	32	28	25	34	27	114
2	M	61	32	28	25	27	112
3	F	30	25	29	31	19	104
4	F	60	7	28	52	37	124
5	F	30	4	11	8	6	29
6	F	30	25	25	26	25	101
7	M	28	25	25	25	25	100
8	M	65	25	26	26	25	102
9	F	59	27	25	26	27	105
10	M	30	25	25	26	25	101
11	M	61	7	4	4	3	18
12	M	36	50	50	50	50	200
13	M	41	39	18	20	23	100
14	M	60	7	11	14	12	44

Note: S: Subjects, X: male/female, D: direct, P: pacing, L: lapping, R: random.

percent of the subjects are male and 36% are female. The mean age of the subjects is 44 years old. Approximately 43% of the subjects are senior (above 55 years old).

In Table 1, the number of direct (D), pacing (P), lapping (L) and random (R) patterns collected from each subject are recorded. We aim to collect 100 patterns (i.e., 100 time series of orientation and patterns) per subject; however, that is not compulsory and, more importantly, it is subject to the subject's availability and health conditions to participate and perform as many travel patterns as per their convenience.

In total, 1354 time series with labeled patterns are collected from 14 subjects. To give an idea of the orientation time series collected, Fig. 3 shows the plots of all-time series categorized according to their labeled patterns. For each plot in Fig. 3, the x-axis represents the data samples of each time series collected and the y-axis shows the orientation values captured for each time series. We can see that the length of each time series varies within the same type of pattern and across different types of patterns. In general, the length of direct pattern time series is shorter than the one of pacing, lapping and random patterns. For pacing and lapping, the time series' length sometimes goes up to over 1000 samples. Since the lengths of all-time series collected

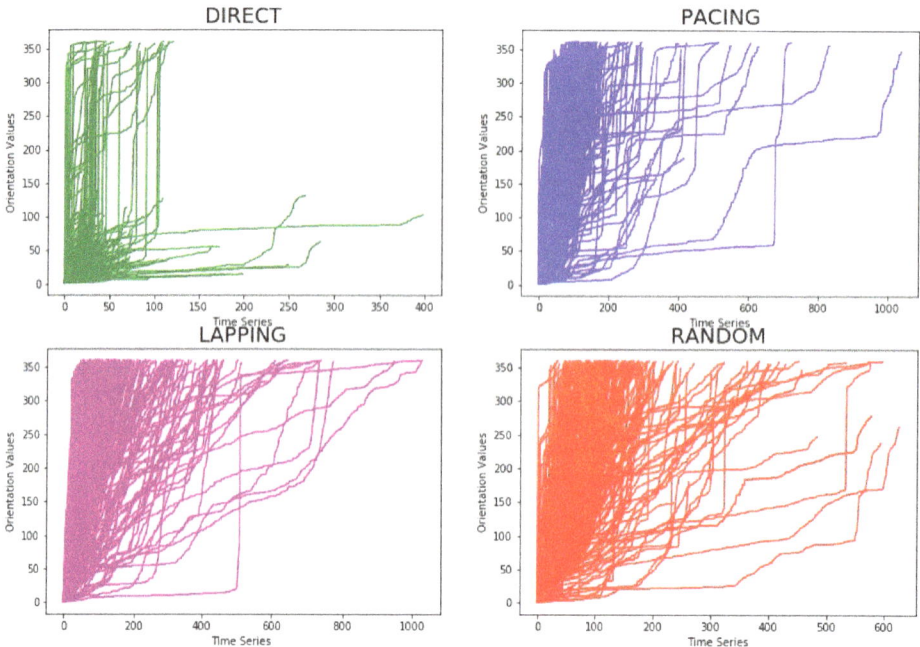

Figure 3:   Line plots of 1354 time series collected which are categorized per class.

are not identical, it is critical to ensure that they should be pre-processed to have the same length before feeding into DL networks.

To generalize the performance evaluation, we randomly generate 10 training and testing sets from the total 1354 collected time series. Each subject's dataset is split into training and testing parts of equal size so as to avoid overtraining and long training time. That means the training and testing sets of each of the 10 randomly generated sets have 677 time series and patterns each. To generalize the performance, the evaluation of all the models and approaches is averaged across the 10 randomly generated sets.

3.2 *Traditional machine learning approach*

We employ traditional ML classifiers for travel pattern classification. Figure 2 illustrates the training process of ML models with features extracted from original time series as inputs and their corresponding labels as ground truth outputs. For evaluation, the extracted features from time series of the test set are fed into the trained ML classifiers which then yield the predicted labels for the travel patterns. Table 2 details the list of 15 features used in our ML models and experiments. All features in Table 2 are extracted using the tsfresh library package [18]. In particular, the 12th feature (linear_trend feature) has four attributes extracted, including intercept, p-value, slope and standard error [18]. This feature is used to specifically represent the linear trend of direct patterns. All four attributes of this feature are used for model training and evaluation. The last three features (13–15th) in Table 2 are included in order to represent the repetitive attributes of pacing and lapping patterns. The rest of the features are commonly used statistical features for classification tasks.

To train and evaluate classification models, five ML classifiers from Sklearn and Xgboost libraries [19] are used, namely Artificial Neural Networks (ANN), DecisionTree (DT), Support Vector Machine (SVM), Random Forest (RF) and Xgboost (XGB). We briefly describe these classifiers and the parameter settings used in our experiments to train the five ML classifiers as follows.

- ANN is a multi-layer perceptron classifier. Back propagation is used to train the network with the objective of minimizing the squared error between the network output and target values. We use LBFGS (Limited-memory Broyden-Fletcher-Goldfarb-Shanno) for log-loss function optimization because it can converge faster and perform better for small datasets.

Table 2: Details of time series features extracted.

S/N	Feature	Description
1	Mean (x)	Returns the mean of x
2	Length (x)	Returns the length of x
3	Median (x)	Returns the median of x
4	Kurtosis (x)	Returns the kurtosis of x (calculated with the adjusted Fisher–Pearson standardized moment coefficient G2).
5	Skewness (x)	Returns the sample skewness of x (calculated with the adjusted Fisher–Pearson standardized moment coefficient G1).
6	Standard_deviation (x)	Returns the standard deviation of x
7	Maximum (x)	Calculates the highest value of the time series x.
8	Minimum (x)	Calculates the lowest value of the time series x.
9	Quantile $(x, 0.25)$	Calculates the first quantile of x.
10	Quantile $(x, 0.75)$	Calculates the third quantile of x.
11	Number_peaks (x, n)	Calculates the number of peaks of at least support n in the time series x. $n = 4$ in our experiments
12	Linear_trend $(x, param)$	Calculate a linear least-squares regression for the values of the time series versus the sequence from 0 to length of the time series minus one. The parameters (param) used in our experiments include intercept, p-value, slope and standard error.
13	Ratio_value_number_ to_time_series_length (x)	Returns a factor which is 1 if all values in the time series occur only once, and below one if this is not the case.
14	Percentage_of_ reoccurring_values_ to_all_values (x)	Returns the ratio of unique values that are present in the time series more than once.
15	Percentage_of_ reoccurring_ datapoints_to_all_ datapoints (x)	Returns the percentage of unique values that are present in the time series more than once.

- DT uses information gain to evaluate how well a given feature separates the training examples into their target class. The set of classification rules are represented in the form of a tree. The maximum depth of the tree is set to be 10 in our experiments.
- RF is an ensemble classifier that consists of many DTs and outputs the most popular class. A tree is grown from independent random vectors

using a training set, resulting in a classifier. After a large number of trees are generated, RF outputs the class that is the mode of the class's output by individual trees. The maximum depth and number of estimators are set to 20 and 50, respectively.

- SVM maximizes the margin between the training examples and the class boundary. SVM generates a hyperplane which provides a class label for each data point described by a set of feature values. We use linear kernel and set the penalty parameter to 0.025 for the SVM classifier.
- XGB library implements the gradient boosting decision tree algorithm which uses gradient boosting and decision trees to enhance both speed and performance. Gradient boosting is to make new and stronger models by capitalizing and reducing the errors of prior models and then adding them together to make the final classification or prediction. Gradient descent algorithm is used to minimize the loss when adding new models.

3.3 *Deep learning approach*

We apply deep LSTM networks to detect travel patterns. LSTM is well-known for the capability to encode temporal information by using memory cells with a small number of gates so as to preserve useful information with long-term dependencies. LSTM has been used for a number of human activity recognition with time series data and achieved remarkable performance [20]. Orientation signals of dementia-related travel patterns are typical time series with temporal and spatial dependency. Hence, LSTM is a good candidate for this problem.

3.3.1 *Bi-directional long short-term memory*

The general LSTM network structure [21] is shown in Fig. 4(a). LSTM tackles the problem of vanishing and exploding of the gradient through memory cells and several gates (input, forget, output, input modulation gate) which have the ability to retain useful information with temporal dependency. These mechanisms enable LSTM networks to capture complex temporal dynamics presenting in orientation signals of human walking activity and travel patterns.

Meanwhile, we also use bi-directional LSTM, an improved version of LSTM, which consists of backward and forward layers (Fig. 4(b)) so as to consider both the past and future information. In particular, the forward layer considers the past information of an orientation sequence and encodes past orientation information into the current state. The backward layer considers the future information of an orientation sequence and encodes it into the

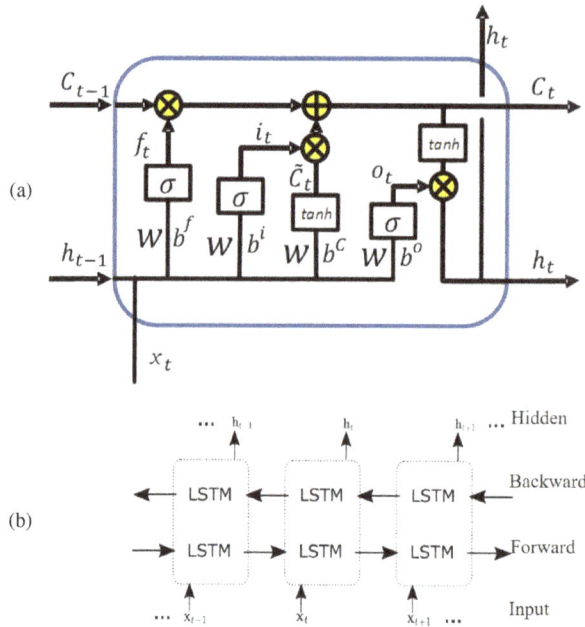

Figure 4: (a) LSTM network, (b) bi-directional LSTM network [21].

current state. This mechanism is apparently relevant for the identification of travel patterns because the future orientation signals or walking steps are also meaningful in determining the type of the travel pattern possibly exhibited by the human subject.

Similar to ML approaches, the trained LSTM models are then used to predict the travel patterns based on input sequences from time series in the test set. Note that the time series in both training and testing sets are pre-processed to have the same sequence length, then directly fed into LSTM models (Fig. 2).

3.3.2 Network architecture

In this work, we construct the LSTM/Bi-LSTM network for classification of travel patterns by using an array of layers containing a sequence input layer, an LSTM/Bi-LSTM layer, a dense fully connected layer, a Softmax layer and a classification output layer (Fig. 5).

- The sequence input layer inputs sequence data into the LSTM/Bi-LSTM. We use masking and padding (pre-padding) to ensure that all the sequences in each batch have the same length.

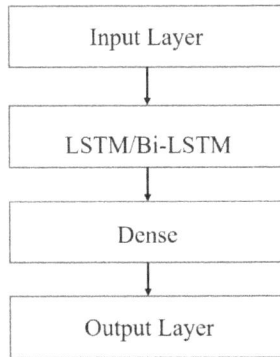

Figure 5: LSTM/Bi-LSTM network architecture.

- The single LSTM/Bi-LSTM hidden layer is used to define the DL model for our problem. More layers are harder and longer to train, but they do not necessarily improve the performance.
- The dense layer is used to interpret the features extracted by the LSTM/Bi-LSTM hidden layer.
- The Softmax and classification output layers are combined in our model and used to make predictions of the travel patterns. Softmax activation is chosen because it allows the model to interpret the outputs as probabilities. The output for the model is a four-element vector depicting the probability of a given input belonging to each of the four travel patterns. Cross entropy function and Adam optimizer are chosen for our multi-class classification problem because they have been shown to work well in most practical applications.

4. Results and Discussion

4.1 *Experimental settings*

4.1.1 *Evaluation metrics*

Experimental results are obtained in terms of accuracy, precision, recall, training and testing time.

- Accuracy is the fraction of correctly classified patterns and the total number of patterns.
- Recall is the true positive rate or sensitivity. It is the ratio of the number of true positives and the sum of the number of true positives and false negatives. It represents the classifier's ability to find all the positive samples.

- Precision is the ratio of the number of true positives and the sum of the number of true positives and the number of false positives. It represents the classifier's ability to not label a sample that is negative as positive.
- Training and testing time are the time in seconds that each classifier takes to train and test the given dataset.

4.1.2 *Hardware specifications*

The experiments are run on a desktop computer with a 64-bit Intel Core i7-6700 CPU, 3.4 GHz speed, 64 GB RAM and an AMD Radeon R5 340X graphics card with GPU.

4.2 *LSTM/Bi-LSTM hyper-parameters settings*

The settings of several hyper-parameters related to network structure and algorithm training and testing of the LSTM/Bi-LSTM models are described in what follows. For convenience, the selection of these hyper-parameters is based on the accuracy (acc) and/or training time (tr_t) metrics only.

4.2.1 *Selection of sequence length*

The lengths of all 1354 orientation time series collected are different (see Fig. 6). In order to feed them into LSTM networks, we pre-pad sequences with zeros and use a masking layer to ensure all training and testing sequences are of the same length. Based on the distribution of sequences' lengths, we

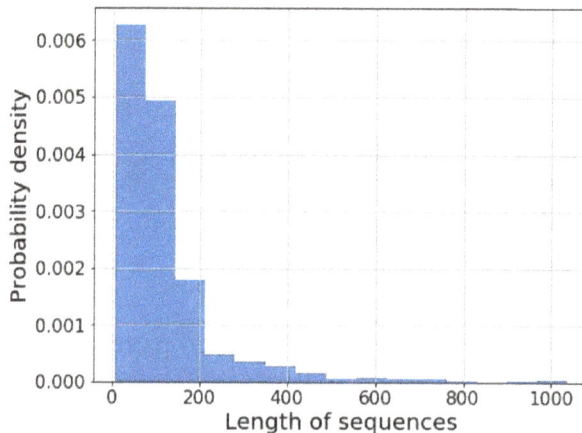

Figure 6:   Histogram of the length of 1354 time series collected.

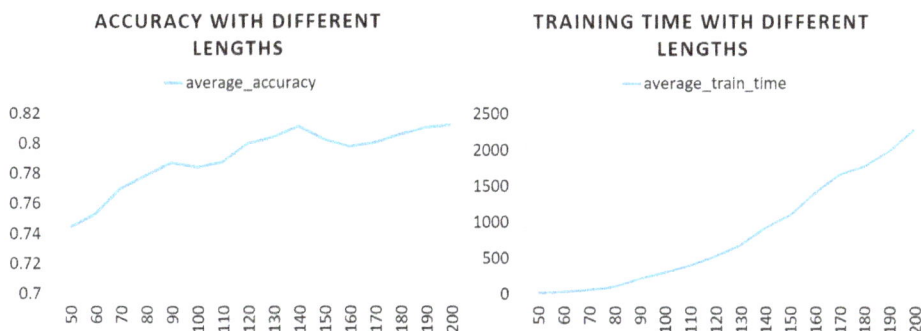

Figure 7:   Accuracy and training time with different values of sequence lengths.

fix the length of each sequence to be 450 timestamps because that is able to cover almost 97% of data and can save computational cost compared to larger lengths.

To understand the effect of various sequence lengths, we conducted a small study to explore the performance and training time trends with the sequence length varying from 50 to 200 with an increment of 10 in each trial. In Fig. 7, we report the accuracy and training time performed by LSTM models with different values of sequence lengths. The x-axis in Fig. 7 is the sequence length and the y-axis represents the accuracy (percentage) and the training time (seconds) both of which are averaged over 10 randomly generated datasets mentioned in Section 3.1.3. We can observe that the training time increases exponentially as the length increases whereas the accuracy fluctuates when the values of sequence length vary from 50 to 200. For the length window from 50 to 200, the best performance is achieved when the sequence length is 140.

We also conducted an experiment to check the performance when the sequence length is set to 500. The accuracy is improved by 1.67% (81.13–82.80%), but the training time increases more than nine times (8569s/917s) as compared with the performance achieved when the sequence length is set to 140. It is not efficient to report the performance of all possible values of sequence lengths. Hence, setting the sequence length to 450 is reasonable so as to retain most information of all sequences in the dataset and reduce the amount of truncated information as well as training time.

4.2.2 *Dropout probability*

A dropout layer is usually included to reduce model overfitting in LSTM/ Bi-LSTM networks. To experiment with the effect of adding a dropout layer,

Table 3: Model performance with different dropout probabilities.

| | Dropout probability | | | | | |
| | 0.2 | | 0.4 | | 0.6 | |
Dataset	acc	tr_t	acc	tr_t	acc	tr_t
train_test_set_1	0.738	333	0.711	343	0.676	375
train_test_set_2	0.723	356	0.769	377	0.722	387
train_test_set_3	0.779	370	0.758	367	0.733	406
train_test_set_4	0.767	365	0.720	346	0.719	392
train_test_set_5	0.757	364	0.748	378	0.682	446
train_test_set_6	0.703	335	0.711	365	0.686	398
train_test_set_7	0.733	355	0.775	402	0.754	436
train_test_set_8	0.754	359	0.738	371	0.728	371
train_test_set_9	0.773	368	0.708	431	0.695	439
train_test_set_10	0.728	356	0.708	363	0.686	395
AVERAGE	0.746	356	0.735	374	0.708	405

we have included a dropout layer to the LSTM model using three different values of dropout probability (0.2, 0.4 and 0.6). To limit the training time, we set the sequence length to 140, which yielded the best performance for sequence length in the range from 50 to 200 as reported in Section 4.2.1. The number of epochs is set to be 20.

Table 3 reports the model performance with an additional dropout layer in terms of accuracy and training time. It is shown that adding the dropout layer decreases the model performance compared to the original model without the dropout layer (Fig. 6 where sequence length is 140) and the average performance declines as the dropout probability increases. We therefore exclude the dropout layer in our LSTM/Bi-LSTM models in the experiments in this study.

4.2.3 *Number of hidden nodes*

Table 4 reports the performance of the LSTM model when we vary the number of hidden nodes. Similar to the above experiments, we set the sequence length and the number of epochs to be 140 and 20, respectively. The dropout layer is also excluded in this experiment. The number of hidden nodes is set to be 100, 128, 200, 256 and 300. On average, the model achieves highest accuracy of 81% when the number of hidden nodes is equal to 256. Therefore,

Table 4: Model performance with different numbers of hidden nodes.

Dataset	Number of hidden nodes									
	100		128		200		256		300	
	acc	tr_t	acc	tr_t	acc	tr_t	acc	tr_t	acc	tr_t
train_test_set_1	0.753	444	0.732	1339	0.778	1358	0.798	2353	0.785	3269
train_test_set_2	0.748	409	0.785	1334	0.773	1340	0.803	2278	0.785	3325
train_test_set_3	0.748	420	0.773	1339	0.810	1408	0.798	2352	0.803	3209
train_test_set_4	0.770	408	0.807	1343	0.806	1367	0.829	2327	0.813	3243
train_test_set_5	0.754	459	0.789	1363	0.798	1414	0.811	2368	0.809	3320
train_test_set_6	0.745	475	0.789	1361	0.794	1416	0.806	2414	0.813	3276
train_test_set_7	0.710	435	0.798	1347	0.801	1387	0.813	2555	0.820	3438
train_test_set_8	0.763	418	0.776	1377	0.801	1429	0.809	2532	0.803	3358
train_test_set_9	0.753	419	0.769	1339	0.809	1428	0.820	2480	0.801	3378
train_test_set_10	0.714	429	0.784	1323	0.795	1388	0.791	2319	0.814	3268
AVERAGE	0.749	432	0.780	1349	0.797	1394	0.810	2406	0.803	3313

we set the number of hidden nodes to be 256 in all subsequent experiments with LSTM/Bi-LSTM models.

4.2.4 *Learning rate, batch size, number of epochs*

For learning rate, we use the default value of Adam optimizer, which is 0.001, for all experiments. For the batch size, we set it to be 64 in all experiments.

For the number of epochs, we try to tune this parameter by varying its value from 25 to 40 with increments of 5. The performance results are reported in Table 5. Increasing the number of epochs will apparently increase the training time; hence, we do not report the training time in Table 5. Similar to the above tuning experiments, we set the sequence length and the number of hidden nodes to be 140 and 256, respectively. From Table 5, we can see that the accuracy does not vary a lot when we change the number of epochs from 25 to 40. Combined with the result reported in Table 4 where the number of hidden nodes is set to 256, we observe that the validation accuracy starts decreasing when the number of epochs increases from 20 to 30. The validation accuracy goes up but declines again when the number of epochs is set to 35 and 40, respectively. Consequently, we set the number of epochs to be 30 in our subsequent experiments to prevent overfitting.

Table 5:   Model performance with different numbers of hidden nodes.

	Number of epochs			
	25	30	35	40
Dataset	acc	acc	acc	acc
train_test_set_1	0.7997	0.7732	0.8056	0.7894
train_test_set_2	0.8100	0.7850	0.7982	0.7850
train_test_set_3	0.7747	0.7938	0.8100	0.8056
train_test_set_4	0.8159	0.8189	0.8262	0.7997
train_test_set_5	0.8100	0.8189	0.8012	0.8100
train_test_set_6	0.8056	0.7968	0.7776	0.8100
train_test_set_7	0.8144	0.8056	0.8247	0.8292
train_test_set_8	0.8041	0.8130	0.8100	0.7968
train_test_set_9	0.8115	0.8203	0.8218	0.7982
train_test_set_10	0.7953	0.8071	0.8041	0.8056
AVERAGE	0.8041	0.8032	0.8080	0.8029

In short, the sequence length, the number of hidden nodes and the number of epochs used in our DL models are 450, 256 and 30 respectively. The performance results of LSTM and Bi-LSTM models reported in Section 4.3 are based on these settings.

4.3 Experimental results and discussion

In Fig. 8, we show the boxplot of each classifier's accuracy performance across 10 test sets. The mean values indicating average accuracy of each classifier are marked with triangle shapes in each plot. We can observe that Bi-LSTM achieves the best performance in terms of average accuracy (82.9%). For ML approaches, XGB and RF have comparable average accuracy (80.4%), and perform better than ANN, DT and SVM. For DL approaches, Bi-LSTM (82.9%) achieves slightly higher but more consistent average accuracy than LSTM (82.4%). Overall, DL approaches (i.e., LSTM and Bi-LSTM) outperform traditional ML approaches for wandering pattern detection.

In Table 6, we provide the detailed precision and recall performance of all 10 test sets for both ML and DL classifiers in this study. The average

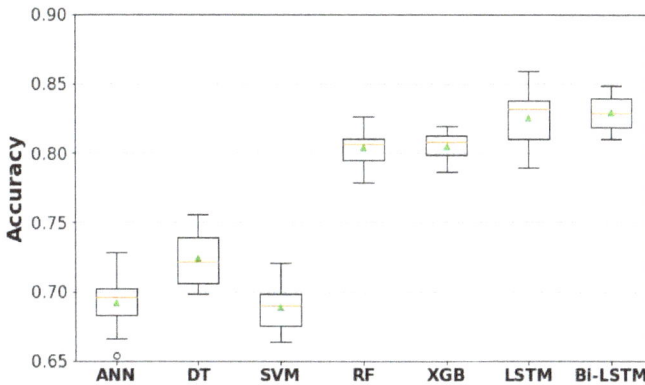

Figure 8: Performance of ML and DL approaches across 10 test sets.

precision and recall over 10 test sets show that XGB performs best in terms of finding all positive samples and discerning positive samples from negative samples for ML classifiers being compared. For DL classifiers, LSTM and Bi-LSTM have the same average precision performance, but Bi-LSTM is superior to LSTM in identifying positive samples (i.e., recall performance). Overall, Bi-LSTM is the best classifier among all ML and DL models being compared in terms of precision and recall.

To further detail the performance of XGB and Bi-LSTM, which are the most outstanding classifiers for ML and DL approaches in this study, we provide the confusion matrices in Fig. 9. It can be found that Bi-LSTM is more capable of detecting wandering patterns, including pacing, lapping and random. XGB is better in recognizing direct travel patterns. This shows the potential of leveraging the advantages of each classifier to further improve the accuracy in detecting wandering and non-wandering patterns.

In Table 7, we summarize the average training and testing time of all classifiers. DT is the most superior in both training and testing time. DL classifiers take much longer to train as compared to ML classifiers. However, ML classifiers require time to learn, extract and select features before feeding to models, which is not easy to quantify. Nonetheless, the testing time performance of both LSTM and Bi-LSTM should still be acceptable for real-time interference (~0.1 s). In reality, training could be conducted offline. Therefore, it is promising to use the trained DL models for real-time applications to detect travel patterns, including wandering.

Table 6: Detailed precision and recall performance of all 10 test sets.

Dataet	Precision							Recall						
	ANN	DT	SVM	RF	XGB	LSTM	Bi-LSTM	ANN	DT	SVM	RF	XGB	LSTM	Bi-LSTM
1	0.69	0.71	0.69	0.78	0.79	0.81	0.83	0.70	0.70	0.67	0.78	0.78	0.79	0.83
2	0.71	0.70	0.74	0.83	0.82	0.82	0.83	0.70	0.70	0.71	0.82	0.81	0.79	0.81
3	0.71	0.75	0.66	0.80	0.82	0.84	0.83	0.70	0.74	0.66	0.80	0.82	0.84	0.83
4	0.72	0.76	0.71	0.79	0.80	0.84	0.82	0.73	0.76	0.69	0.79	0.80	0.84	0.81
5	0.73	0.75	0.71	0.82	0.81	0.84	0.84	0.71	0.75	0.69	0.82	0.81	0.84	0.84
6	0.69	0.70	0.68	0.81	0.80	0.81	0.82	0.66	0.70	0.67	0.81	0.80	0.81	0.82
7	0.69	0.71	0.73	0.81	0.82	0.86	0.85	0.69	0.71	0.70	0.81	0.82	0.86	0.85
8	0.72	0.74	0.73	0.81	0.81	0.83	0.83	0.68	0.74	0.69	0.80	0.81	0.82	0.82
9	0.69	0.73	0.67	0.81	0.81	0.85	0.82	0.70	0.72	0.67	0.81	0.81	0.85	0.82
10	0.67	0.72	0.69	0.79	0.79	0.83	0.85	0.66	0.72	0.68	0.79	0.78	0.82	0.85

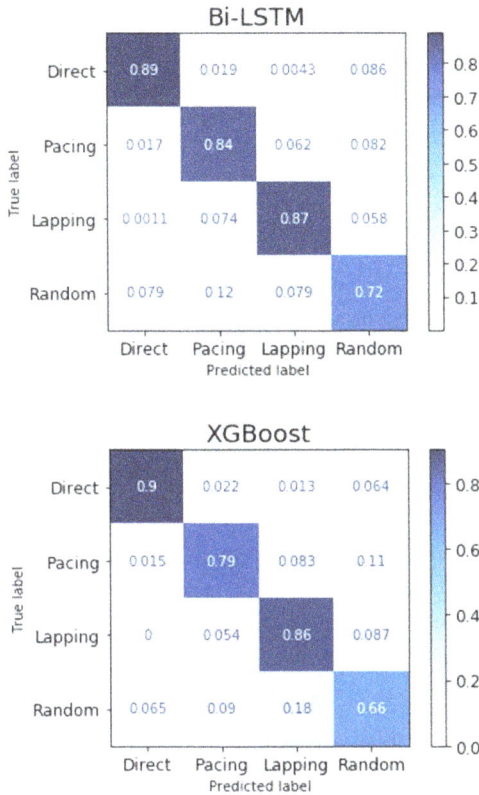

Figure 9: Confusion matrices of XGB and Bi-LSTM.

Table 7: Training and testing time of ML and DL approaches.

Time (s)	ANN	DT	SVM	RF	XGB	LSTM	Bi-LSTM
Training	0.326	0.005	0.298	0.065	0.305	1761.6	6640.7
Testing	0.001	0.001	0.005	0.007	0.008	0.029	0.105

5. Conclusion

In this chapter, we present a framework using ML and DL approaches to detect travel patterns including wandering based on orientation information from inertial sensors in mobile devices. We have also evaluated the framework using a real dataset collected from 14 subjects. The experimental results show that bi-directional LSTM achieves the best classification accuracy in general. Handcrafted features approach using XGB classifier is able to discern direct patterns better. In the future, we plan to combine the advantages of both ML and DL approaches in order to further improve the performance. In addition, we would also like to develop classification models that are able to recognize individual travel patterns from trajectories in which two or more travel patterns occur in the same travel episode.

References

1. M. Prince, R. Bryce, E. Albanese, A. Wimo, W. Ribeiro, and C. P. Ferri, "The global prevalence of dementia: a systematic review and meta-analysis," *Alzheimer's & Dementia*, vol. 9, no. 1, pp. 63–75, 2013.
2. I. H. Ramakers, P. J. Visser, P. Aalten, J. H. Boesten, J. F. Metsemakers, J. Jolles, and F. R. Verhey, "Symptoms of preclinical dementia in general practice up to five years before dementia diagnosis," *Dementia Geriatric Cogn. Disorders*, vol. 24, no. 4, pp. 300–306, 2007.
3. T. Hope, J. Keene, R. H. McShane, C. G. Fairburn, K. Gedling, and R. Jacoby, "Wandering in dementia: a longitudinal study," *Int. Psychogeriatrics*, vol. 13, no. 2, pp. 137–147, 2001.
4. J. Verghese, R. B. Lipton, C. B. Hall, G. Kuslansky, M. J. Katz, and H. Buschke, "Abnormality of gait as a predictor of non-alzheimer's dementia," *N. Engl. J. Med.*, vol. 347, no. 22, pp.1761–1768, 2002.
5. E. Scherder, L. Eggermont, D. Swaab, M. van Heuvelen, Y. Kamsma, M. de Greef, R. van Wijck, and T. Mulder, "Gait in ageing and associated dementias: its relationship with cognition," *Neurosci Biobehav. R*, vol. 31, no. 4, pp. 485–497, 2007.
6. G. Cipriani, C. Lucetti, A. Nuti, and S. Danti, "Wandering and dementia," *Psychogeriatrics*, vol. 14, no. 2, pp. 135–142, 2014.
7. T. Hayes, M. Pavel, and J. Kaye, "An unobtrusive in-home monitoring system for detection of key motor changes preceding cognitive decline," in *26th Conf. Proc. IEEE Eng. Med. Biol. Soc.*, 2004, pp. 2480–2483.
8. J. Verghese, R. Holtzer, R. B. Lipton, and C. Wang, "Quantitative gait markers and incident fall risk in older adults," *The J. Gerontol.: Ser. A*, vol. 64, no. 8, pp. 896–901, 2009.
9. D. L. Algase, C. Beel-Bates, and E. R. Beattie, "Wandering in long-term care," *Ann. Long Term Care*, vol. 11, pp. 33–41, 2003.
10. D. Martino-Saltzman, B. B. Blasch, R. D. Morris, and L. W. McNeal, "Travel behavior of nursing home residents perceived as wanderers and nonwanderers," *The Gerontol.*, vol. 31, no. 5, pp. 666–672, 1991.
11. K. Makimoto, M. Yamakawa, N. Ashida, Y. Kang, and K.-R. Shin, "Japan-Korea joint project on monitoring people with dementia," in *11th World Cong. Internet Med.*, 2006.

12. N. Vuong, S. Goh, S. Chan, and C. Lau, "A mobile-health application to detect wandering patterns of elderly people in home environment," in *35th Conf. Proc. IEEE Eng. Med. Biol. Soc.*, pp. 6748–6751, 2013.

13. Q. Lin, D. Zhang, X. Huang, H. Ni, and X. Zhou, "Detecting wandering behavior based on GPS traces for elders with dementia," in *12th Conf. Proc. IEEE ICARCV*, pp. 672–677, 2012.

14. A. Delaunay and J. Guerin, "Wandering detection within an embedded system for Alzheimer suffering patients," in *AAAI Spring Symp. Series*, 2017.

15. M. Gochoo, T.-H. Tan, V. Velusamy, S.-H. Liu, D. Bayanduuren, and S.-C. Huang, "Device-free non-privacy invasive classification of elderly travel patterns in a smart house using PIR sensors and DCNN," *IEEE Sens. J.*, vol. 18, no. 1, pp. 390–400, 2017.

16. N. Vuong, S. Chan, C. T. Lau, S. Chan, P. L. K. Yap, and A. Chen, "Preliminary results of using inertial sensors to detect dementia-related wandering patterns," in *37th Conf. Proc. IEEE Eng. Med. Biol. Soc.*, 2015, pp. 3703–3706.

17. P. P. Ray, D. Dash, and D. De, "A systematic review and implementation of IOT-based pervasive sensor-enabled tracking system for dementia patients," *J Med Syst*, vol. 43, no. 9, p. 287, 2019.

18. M. Christ, A. W. Kempa-Liehr, and M. Feindt, "Distributed and parallel time series feature extraction for industrial big data applications," arXiv preprint arXiv:1610.07717, 2016.

19. F. Pedregosa *et al.*, "Scikit-learn: Machine learning in Python," *J. Mach. Learn. Res.*, vol. 12, pp. 2825–2830, 2011.

20. S. Yousefi, H. Narui, S. Dayal, S. Ermon, and S. Valaee, "A survey on behavior recognition using WiFi channel state information," *IEEE Commun. Mag.*, vol. 55, no. 10, pp. 98–104, Oct. 2017.

21. Z. Chen, L. Zhang, C. Jiang *et al.*, "WiFi CSI based passive human activity recognition using attention based BLSTM," *IEEE Trans. Mobile Comput.*, pp. 1–12, October 2018.

Index

www.ingramcontent.com/pod-product-compliance
Lightning Source LLC
Chambersburg PA
CBHW081510190326
41458CB00015B/5333